AF598088

# Methods in Molecular Biology

*Series Editor*
John M. Walker
School of Life Sciences
University of Hertfordshire
Hatfield, Hertfordshire, AL10 9AB, UK

For further volumes:
http://www.springer.com/series/7651

# Human Cytomegaloviruses

## Methods and Protocols

Edited by

**Andrew D. Yurochko**

*Department of Microbiology & Immunology, Center for Molecular and Tumor Virology, Feist-Weiller Cancer Center, Louisiana State University Health Sciences Center, Shreveport, LA, USA*

**William E. Miller**

*Department of Molecular Genetics, Biochemistry, and Microbiology, University of Cincinnati College of Medicine, Cincinnati, OH, USA*

*Editors*
Andrew D. Yurochko
Department of Microbiology & Immunology
Center for Molecular and Tumor Virology
Feist-Weiller Cancer Center
Louisiana State University Health Sciences Center
Shreveport, LA, USA

William E. Miller
Department of Molecular Genetics, Biochemistry, and Microbiology
University of Cincinnati College of Medicine
Cincinnati, OH, USA

ISSN 1064-3745 ISSN 1940-6029 (electronic)
ISBN 978-1-62703-787-7 ISBN 978-1-62703-788-4 (eBook)
DOI 10.1007/978-1-62703-788-4
Springer New York Heidelberg Dordrecht London

Library of Congress Control Number: 2014933398



Printed on acid-free paper

Humana Press is a brand of Springer
Springer is part of Springer Science+Business Media (www.springer.com)

# Preface

The book *Human Cytomegaloviruses* in the *Methods of Molecular Biology* Series is designed to be an inclusive document for all of the necessary techniques and approaches to understand and study the pathobiology of human cytomegalovirus. With the significant medical role that human cytomegalovirus plays in human disease (from acute disease following primary infection to chronic disease due to lifelong viral persistence), there is a need to better understand human cytomegalovirus biology and pathogenesis. The book is designed to be a complete reference manual of the modern approaches to the study of human cytomegalovirus. This reference manual should serve as a tool for basic scientists to clinical scientists with interests in the basic fundamental aspects of viral gene expression and replication to those with interests in the specific aspects of viral pathogenesis. The book begins with two chapters designed to serve as a reference for the history of human cytomegalovirus and its associated diseases. Multiple chapters in the book detail the key techniques to culture and grow the virus in model cell types, to the use of primary cells in the study of human cytomegalovirus infection, to the modern molecular techniques required to assess the biological consequences of viral infection. Because of the use of multiple animal models in the study of viral infection, chapters have also been included to understand the basis of and the use of these important systems in understanding viral disease. Finally, the book ends with chapters discussing the relevant development of targeted therapeutics and vaccines as a goal to eventually mitigate human cytomegalovirus disease.

*Shreveport, LA* *Andrew D. Yurochko*
*Cincinnati, OH* *William E. Miller*

## Contents

## Contributors

SURESH B. BOPPANA • *Departments of Pediatrics and Microbiology, The University of Alabama at Birmingham, Birmingham, AL, USA*
WILLIAM J. BRITT • *Departments of Pediatrics and Microbiology, The University of Alabama at Birmingham, Birmingham, AL, USA; Department of Neurobiology, The University of Alabama at Birmingham, Birmingham, AL, USA*
DJURDJICA CEKINOVIC • *Department for Histology and Embryology, Center for Proteomics, School of Medicine, University of Rijeka, Rijeka, Croatia*
GARY C.T. CHAN • *Department of Microbiology & Immunology, SUNY Upstate Medical University, Syracuse, NY, USA*
LUKA ČIČIN-ŠAIN • *Department of Vaccinology, Helmholtz Centre for Infection Research, Braunschweig, Germany*
CHARLES S. COBBS • *California Pacific Medical Center Research Institute, San Francisco, CA, USA*
DONNA COLLINS-MCMILLEN • *Department of Microbiology & Immunology, Center for Molecular and Tumor Virology, Louisiana State University Health Sciences Center, Shreveport, LA, USA*
ANDA CORNEA • *Oregon National Primate Research Center, Oregon Health & Science University, Beaverton, OR, USA*
IRYNA DEKHTIARENKO • *Department of Vaccinology, Helmholtz Centre for Infection Research, Braunschweig, Germany*
PRANAY DOGRA • *Department of Microbiology, University of Tennessee, Knoxville, TN, USA*
ELIZABETH A. FORTUNATO • *Department of Biological Sciences, University of Idaho, Moscow, ID, USA*
GIADA FRASCAROLI • *Institute for Virology, University Medical Center Ulm, Ulm, Germany*
FELICIA GOODRUM • *Department of Immunobiology, BIO5 Institute, The University of Arizona, Tuscon, AZ, USA*
GIORGIO GRIBAUDO • *Department of Life Sciences and Systems Biology, University of Turin, Turin, Italy*
CHRISTIAN HAGEMEIER • *Labor für Pädiatrische Molekularbiologie, Charité-Universitätsmedizin, Berlin, Germany*
MEAGHAN H. HANCOCK • *Vaccine & Gene Therapy Institute, Oregon Health & Sciences University, Beaverton, OR, USA*
LUALHATI E. HARKINS • *Department of Pathology and Laboratory Medicine, Birmingham Veterans Hospital, Birmingham, AL, USA*
LAUREN M. HOOK • *Vaccine & Gene Therapy Institute, Oregon Health & Sciences University, Beaverton, OR, USA*
CARMEN BACA JONES • *Department of Molecular Microbiology & Immunology, Vaccine & Gene Therapy Institute, Oregon Health & Science University, Beaverton, OR, USA*
STIPAN JONJIC • *Department for Histology and Embryology, Center for Proteomics, School of Medicine, University of Rijeka, Rijeka, Croatia*

CRAIG N. KREKLYWICH • *Department of Molecular Microbiology & Immunology, Vaccine & Gene Therapy Institute, Oregon Health & Science University, Beaverton, OR, USA; Department of Surgery, Oregon Health & Science University, Portland, OR, USA*

IGOR LANDAIS • *Vaccine & Gene Therapy Institute, Oregon Health & Sciences University, Beaverton, OR, USA*

VANDA JURANIC LISNIC • *Department for Histology and Embryology, Center for Proteomics, School of Medicine, University of Rijeka, Rijeka, Croatia*

ARIANNA LOREGIAN • *Department of Molecular Medicine, University of Padua, Padua, Italy*

LISA MATLAF • *California Pacific Medical Center Research Institute, San Francisco, CA, USA*

BEATRICE MERCORELLI • *Department of Molecular Medicine, University of Padua, Padua, Italy*

MARTIN MESSERLE • *Institute of Virology, Hannover Medical School, Hannover, Germany*

WILLIAM E. MILLER • *Department of Molecular Genetics, Biochemistry, and Microbiology, University of Cincinnati College of Medicine, Cincinnati, OH, USA*

JAY A. NELSON • *Vaccine & Gene Therapy Institute, Oregon Health & Sciences University, Beaverton, OR, USA*

MACIEJ T. NOGALSKI • *Department of Microbiology & Immunology, Center for Molecular and Tumor Virology, Louisiana State University Health Sciences Center, Shreveport, LA, USA*

CHRISTINE M. O'CONNOR • *Section of Virology, Department of Molecular Genetics, Lerner Research Institute, The Cleveland Clinic, Cleveland, OH, USA*

SUSAN L. ORLOFF • *Department of Molecular Microbiology & Immunology, Vaccine & Gene Therapy Institute, Oregon Health & Science University, Beaverton, OR, USA; Department of Surgery, Oregon Health & Science University, Portland, OR, USA; Portland VA Medical Center, Portland, OR, USA*

GIORGIO PALÙ • *Department of Molecular Medicine, University of Padua, Padua, Italy*

EMMA POOLE • *Department of Medicine, Addenbrooke's Hospital, University of Cambridge, Cambridge, UK*

MATTHEW REEVES • *Department of Medicine, Addenbrooke's Hospital, University of Cambridge, Cambridge, UK*

NINA REUTER • *Institute for Clinical and Molecular Virology, University of Erlangen-Nuremberg, Erlangen, Germany*

JOHN H. SINCLAIR • *Department of Medicine, Addenbrooke's Hospital, University of Cambridge, Cambridge, UK*

CHRISTIAN SINZGER • *Institute for Virology, University Medical Center Ulm, Ulm, Germany*

PATRICIA P. SMITH • *Department of Molecular Microbiology & Immunology, Vaccine & Gene Therapy Institute, Oregon Health & Science University, Beaverton, OR, USA*

TIM E. SPARER • *Department of Microbiology, University of Tennessee, Knoxville, TN, USA*

THOMAS STAMMINGER • *Institute for Clinical and Molecular Virology, University of Erlangen-Nuremberg, Erlangen, Germany*

MARK F. STINSKI • *Department of Microbiology, Carver College of Medicine, University of Iowa, Iowa City, IA, USA*

DANIEL N. STREBLOW • *Department of Molecular Microbiology & Immunology, Vaccine & Gene Therapy Institute, Oregon Health & Science University, Beaverton, OR, USA*

MARCO THOMAS • *Institute for Clinical and Molecular Virology, University of Erlangen-Nuremberg, Erlangen, Germany*
MAHADEVAIAH UMASHANKAR • *Department of Immunobiology, BIO5 Institute, The University of Arizona, Tuscon, AZ, USA*
LÜDER WIEBUSCH • *Labor für Pädiatrische Molekularbiologie, Charité-Universitätsmedizin, Berlin, Germany*
ANDREW D. YUROCHKO • *Department of Microbiology & Immunology, Center for Molecular and Tumor Virology, Feist-Weiller Cancer Center, Louisiana State University Health Sciences Center, Shreveport, LA, USA*
BARBARA ZIELKE • *Institute for Clinical and Molecular Virology, University of Erlangen-Nuremberg, Erlangen, Germany*

# Chapter 1

# History of the Molecular Biology of Cytomegaloviruses

Mark F. Stinski


## Abstract

The history of the molecular biology of cytomegaloviruses from the purification of the virus and the viral DNA to the cloning and expression of the viral genes is reviewed. A key genetic element of cytomegalovirus (the CMV promoter) contributed to our understanding of eukaryotic cell molecular biology and to the development of lifesaving therapeutic proteins. The study of the molecular biology of cytomegaloviruses also contributed to the development of antivirals to control the viral infection.


**Key words** Cytomegaloviruses, Viral DNA, Major immediate early genes, CMV, Promoter

## 1 Introduction

Smith, Weller, and Rowe reported the isolation of the cytomegaloviruses (CMVs) in the mid 1950s [1]. The viruses were species specific, replicated slowly in cell culture, and required a week or more to show cytopathic effect. In the host, symptoms did not appear for weeks after infection. Therefore, the CMVs were placed into a subclassification group of the herpesviruses termed the betaherpesviruses. The biology of CMV infections has been extensively reviewed [1, 2]. The focus of the current chapter will be on the history of the molecular biology of CMVs.

As molecular virology became more common in the late 1960s, several strains of CMVs were investigated. They were referred to as laboratory strains because of their repeated passage in fibroblast cell culture. These are the Towne, AD169, and Davis strains of human CMV and the Smith strain of murine CMV. The early molecular biology of CMVs was determined using these laboratory strains, and the viruses were cultivated in standard cell culture systems. We now know that continued passage of these strains in cell culture caused multiple mutations in the viral genomes. Human CMV genes that affect infection of epithelial and endothelial cells and macrophages were found to be mutated. In addition, a large number of genes were absent from the unique large (UL)

Andrew D. Yurochko and William E. Miller (eds.), *Human Cytomegaloviruses: Methods and Protocols*, Methods in Molecular Biology, vol. 1119, DOI 10.1007/978-1-62703-788-4_1, component of the viral genome, known as the ULb′ region. These genes affect differentially the rate of viral replication in fibroblast and the establishment of latency in CD34+ progenitor cells. Similarly, passage of murine CMV in cell culture caused a loss of virulence in the mouse.

While working with adenoviruses in the early 1970s in the laboratory of Dr. Harry Ginsberg (University of Pennsylvania), I wandered across the street to the Wistar Institute. There I met Dr. Stanley Plotkin, who was working on a CMV strain named Towne. Stanley showed me human fibroblast cells infected with the virus. The infected cells were enlarged and non-refractive. This was in stark contrast to adenovirus-infected fibroblasts, which are condensed and refractive.

## 2 Purification

In the mid 1970s, many laboratories used a popular method for purification of viruses, which was velocity centrifugation in sucrose gradients. However, recovery of infectious human CMV in sucrose gradients was extremely poor. We had observed empirically that storage of the virus in D-sorbitol preserved viral infectivity. Therefore, buffered D-sorbitol in a 20–70 % linear gradient was selected as a medium for velocity centrifugation. Recovery of infectious human CMV was good ($10^9$ PFU/ml). However, there were two partially purified bands of different sedimentation rates. The particles in the bottom band contained viral DNA as determined by $^3$H-thymidine labeling of host cells and had the typical herpesvirus structure as determined by electron microscopy. The particles in the top band lacked DNA and had a dense interior surrounded by a membranous envelope [3]. The particles in the top band were referred to as dense bodies and were considered an aberrant form of viral morphogenesis. Precursor labeling with $^3$H-glucosamine of the partially purified viral particles indicated there were 8–11 glycopolypeptides associated with the membranes of virions and dense bodies, and the viral particles were of a similar glycoprotein composition [3].

## 3 Proteins and Glycoproteins in Infected Cells

In the 1970s, it was well known that the prototypical member of the alphaherpesviruses, herpes simplex virus (HSV), inhibited host cell protein synthesis early after infection. Double-isotopic labeling experiments with human CMV-infected cells indicated that host cell protein synthesis continued while viral proteins were synthesized [4]. Moreover, CMV-infected cells maintained viability for many days even after a high multiplicity of infection (MOI). The viral polypeptides and glycopolypeptides followed a pattern of early and late phases [4]. An inhibitor of viral DNA synthesis or nonpermissive cells (guinea pig embryo fibroblast) allowed for the synthesis of early viral polypeptides, but prevented the appearance of the late viral polypeptides [5]. In this regard, human CMV was following a pattern of temporal viral gene expression similar to the alphaherpesviruses. While HSV reached peak levels of viral DNA synthesis within 12 h after high MOI, human CMV did not reach peak levels until 72 h [5]. This suggested that CMVs differ dramatically from HSV in the rate of viral replication. However, radioactive pulselabeling experiments indicated that CMVs synthesized at least two virus-specific polypeptides (now referred to as IE1-72 and IE2-86) within a very short period after infection (1–2 h) [5]. While these two polypeptides appeared shortly after viral infection, peak viral protein synthesis was not occurring until approximately 3 days after infection, coinciding with peak DNA synthesis. Therefore, it appeared that CMVs were going through a temporal cascade of viral protein synthesis like HSV, but with much slower rates of viral DNA and late protein synthesis. The enlarged cell, the continuation of cellular protein synthesis after infection, the maintenance of cellular viability, and the prolonged phase of viral protein synthesis suggested that CMVs exhibit unique interactions with the host cell.

## 4 Viral DNA

The early reports of human CMV DNA size, assessed by sedimentation in sucrose gradients [6] or by electron microscopy [7], indicated a molecular weight of approximately $100 \times 10^6$, similar to that of HSV. However, others found a class of molecules of approximately $150 \times 10^6$ [8–10]. It was then determined that passage of the virus at high MOI resulted in defective viral DNA molecules of sizes smaller than the $150 \times 10^6$ molecular weight molecules that were packaged into virions. In addition, the particle to PFU ratio increased on passage at high MOI [11]. However, passage of the virus at low MOI resulted in viral DNA molecules of $150 \times 10^6$ molecular weight. Therefore, CMVs had a higher molecular weight DNA than the other herpesviruses, and low MOI passage was necessary to prevent defective viral DNA and particle formation.

To clone the viral genome, it was important to isolate viral DNA from virus passed at low MOI. As DNA cloning became more popular in the late 1970s, several laboratories cloned sections of the viral genome into bacterial plasmids or cosmids [12–14]. We choose to use the bacterial plasmid pACYC184 because it would hold foreign DNA up to approximately 39 kilobases (kb) while replicating in *Escherichia coli* HB101. To cleave the human CMV genome into smaller more manageable fragments, we choose a restriction endonuclease with the least number of cuts known at the time, which was Xba I. The viral genome was then cloned as 21

Xba I fragments [13, 15]. The viral genome of the laboratory strains of human CMV has a unique large and small components flanked by inverted repeat sequences, resulting in the generation of four isomers that occur in equal molar concentrations depending on the orientations of the large and small components. Restriction endonucleases that cleaved viral DNA within the unique and adjacent repeat sequences generated four fragments of 0.5 M and four of 0.25 M concentrations. However, restriction endonucleases that cleaved the viral DNA solely within the unique sequences generated DNA fragments in concentrations of 1.0 M relative to the molarity of the intact DNA [15–17]. The cloning of the viral DNA fragments supported the above interpretation of the viral genome arrangement in the virions.

## 5 Viral RNAs

When the cloned viral DNA fragments became available in the early 1980s, these fragments served as important tools to investigate many additional aspects of CMV molecular biology. Viral RNAs on the polyribosomes of human CMV-infected cells were detected by probing northern blots with $^{32}$P-CMV DNA fragments. The viral RNAs were designated as immediate early, early, or late depending on their time of expression after infection [15, 18]. Immediate early (IE) were those viral RNAs synthesized after infection in the presence of an inhibitor of de novo protein synthesis, such as cycloheximide. Early viral RNAs were synthesized after the immediate early RNAs and proteins but before the onset of viral DNA replication. Late viral RNAs were synthesized after viral DNA replication [18]. In vitro translation of the viral RNAs selected by hybridization to human CMV DNA covalently linked to cellulose demonstrated the temporal expression of viral proteins [18].

LaFemina and Hayward (unpublished data) and DeMarchi [19] had developed physical maps of the Xba I DNA fragments of the Towne and Davis strain DNAs, respectively. In addition, Hind III, Eco RI, and Bam HI physical maps of the Towne and AD169 strains were developed (reviewed in ref. 20). We prepared $^{32}$P-Xba I viral DNA fragments to probe Northern blots of the IE, early and late viral RNAs. The Xba I E DNA fragment, the fifth largest of the 21 Xba I fragments, detected a predominant 1.9 kb viral RNA at IE times after infection, which indicated that the viral RNA was encoded within a region of approximately 20 kb between map units 0.66 and 0.77. There was also a 2.2 kb viral RNA of less abundance coming from the same region [21, 22]. A transcription map of the viral RNAs was developed using Xba I and Bam HI viral DNA fragments as radioactive probes, and the relative abundance and percentage of transcription at the various times after infection were determined. We also developed detailed restriction endonuclease maps of the Xba I E DNA using seven different restriction endonucleases. The precise origin of the predominant 1.9 kb IE viral RNA was determined [23]. The 1.9 kb RNA was selected using a relatively small viral DNA fragment (0.732–0.751 map units) covalently linked to cellulose. In vitro translation showed that the 1.9 kb viral mRNA coded for the abundant IE1-72 viral protein found at early times after infection [23]. We also mapped viral RNAs coming from an adjacent region (map units 0.732–0.739) that were less abundant and designated this region IE2. Thinking that the most abundant IE viral protein was the most important, we mapped the transcription start site for IE1 [24] and determined the structure of the viral mRNA by preparing 5′ or 3′ $^{32}$P-labeled cloned DNA fragments. If viral RNA–DNA hybridization protected the radioactive probe from nuclease digestion (mung bean), then we knew that the probe was in an exon region. In this manner, we determined the size of the protected exon and found that the 1.9 kb viral RNA consisted of four exons and three introns before encountering a polyadenylation signal [24]. However, there was also an alternative splice whereby exons 1, 2, and 3 were spliced to exon 5, and exon 5 was in the IE2 region [25]. We were able to sequence the viral DNAs and identify the donor and acceptor splice sites as well as the exons of IE1 and IE2, which allowed us to predict the amino acid sequence of the IE1-72 and IE2-86 proteins [24–26]. The IE1 and IE2 viral proteins had the first 85 amino acids in common which were encoded by exon 2 (location of the translation start site) and exon 3. Exon 4 contained the carboxy-terminal amino acids and stop codon for IE1-72, while exon 5 contained the carboxy-terminal amino acids and stop codon for IE2-86. We designated IE1 viral gene product as the major immediate early (MIE) gene product because it was the most abundant viral gene product present within the first few hours after infection. However, some of the IE2 gene products were linked to this designation because of the alternate splicing event to exon 5, and consequently, they were also referred to as MIE. In 1984, a rough map of some of the viral mRNAs and proteins specified by human CMV was reviewed [27].

## 6 The Enhancer-Containing MIE Promoter

We reasoned that there must be a very strong promoter upstream of the IE1 1.9 kb transcription start site because of the abundance of the viral gene product. We used HeLa cell extracts and viral DNA templates for in vitro transcription. As we cleaved the viral DNA templates with various restriction endonucleases from large templates to smaller templates, we were able to determine that the transcription was going from right to left between map units 0.739 and 0.751 for the prototype arrangement of the viral genome [28]. We also reasoned there had to be an enhancer/regulatory region upstream of the transcription start site that determined the very strong downstream transcription. We sequenced the upstream DNA region of the Towne strain of human CMV and found repeat sequences containing transcription factor binding sites such as those for CREB/ATF and NF-κB [26, 28]. Similar DNA sequence elements were found for the AD169 strain of human CMV [29]. When the human CMV enhancer-containing promoter (Towne strain) was in a mixture with the late adenovirus promoter (type 5 strain) for in vitro transcription, the human CMV promoter was significantly stronger [28]. To further evaluate the strength of the CMV promoter and the role of the upstream enhancer sequence, we cloned the enhancer-containing promoter upstream of three different heterologous genes designated thymidine kinase (TK), chloramphenicol (CAT), and ovalbumin (OV) and compared the strength of the MIEP to that of other known viral promoters. Again the human CMV MIE enhancer-containing promoter was stronger than other promoters such as that of HSV ICP4 [30], RSV LTR, or SV40 (Unpublished data). Moreover, the extent and type of regulatory sequences upstream of the promoter influenced the level of transcription and the amount of heterologous protein expression [30]. However, the strong CMV promoter did not fit the profile of cytomegaloviruses. HCMV replicated slowly and to low titers relative to herpes simplex virus. Therefore, the acceptance of the CMV promoter as a valuable tool for eukaryotic gene expression came slowly and spread predominantly by "word of mouth." This relatively small region of CMV DNA efficiently competed for cellular RNA polymerase II and the other transcription factors. Between 1985 and 1992, our laboratory received requests for the human CMV MIE enhancer-containing promoter element from around the world. In 1992, the human CMV enhancer-containing promoter became commercially available. In the mid to late 1990s, several pharmaceutical companies were able to efficiently produce eukaryotic therapeutic proteins using the CMV promoter. These therapeutic proteins were used to fight life-threatening diseases such as non-Hodgkin's lymphoma, hemophilia A, and respiratory syncytial virus infections of infants. There was also the development of new therapies for the treatment of rare mammalian diseases involving genetically inherited defective enzymes. The CMV promoter became a major tool for the expression of eukaryotic proteins for molecular biology research, and today, it is used in virtually every laboratory investigating gene expression, signal transduction, cellular physiology, etc. (reviewed in ref. 31).

## 7 Sequence of the Human CMV Genome

The complete sequence of the human CMV genome was reported in 1990, and the CMV enhancer-containing promoter was localized to between 173,731 and 174,280 base pairs of viral DNA [32].

Chee et al. [32] published the protein-coding content of the genome of human CMV AD169. Based on their criteria for an open reading frame (ORF), they predicted over 200 ORFs. They developed a map of the ORFs, predicted viral gene products, and compared the CMV gene products to those other herpesvirus genomes that had already been sequenced. CMV emerged as the most genetically complex of the herpesviruses. Realizing that the wild-type clinical isolates of human CMV had additional ORFs, Murphy et al. [33, 34] compared the ORFs of five human CMV clinical isolates to chimpanzee CMV and prepared a new physical map of the viral genome. Human CMV was found to be remarkably similar to chimpanzee CMV and to have at least 165 genes of which 45 genes were essential for virus replication in human fibroblast cells, and 68 genes were dispensable [35]. Lastly, Gatherer et al. [36] identified previously unrecognized protein-coding and noncoding poly(A) RNAs indicating that CMV transcription is more complex than previously appreciated.

## 8 Viral DNA Replication

In 1990, the Hayward laboratory identified the human CMV origin of lytic DNA replication [37]. This stimulated a series of investigations primary by the Anders and the Pari labs where the transacting factors required for the oriLyt-dependent viral DNA replication were defined [38]. While human CMV used viral proteins homologous to those required for HSV DNA replication, there were some unique features that further defined human CMV as a betaherpesvirus. First, the human CMV oriLyt DNA sequence was very complicated compared to that of HSV [39]. Second, the viral IE2 protein (IE2-86) is contained in a complex with the viral UL84 protein that is necessary to initiate viral DNA synthesis [40, 41], and proteomic analysis showed that the IE2-86-UL84 viral protein complex interacted with other viral and cellular proteins [42]. Finally, an RNA–DNA hybrid structure was found at the origin of replication [43].

## 9 Functions of the IE1 and IE2 Gene Products

Investigating the functions of the human CMV IE1 and IE2 genes was critical for understanding the replication of this important human pathogen. The IE2 gene product (UL122) was found to transactivate expression from early viral promoters [44, 45], and the IE1 gene product (UL123) significantly augmented the activity of IE2. These studies started in the late 1980s and continued for over a decade. Investigators reported the activation of various cellular and viral promoters. They determined the necessary promoter elements for transactivation, the critical domains of the IE2 and IE1 proteins, and the posttranslational modifications of the viral transactivators (reviewed in refs. 46–48). The IE2 gene products are multifunctional, and in addition to their ability to activate early and late viral promoters, they negatively autoregulate the MIE promoter and control cell cycle progression. The IE1 gene product is also multifunctional. It neutralizes cellular intrinsic defense mechanisms associated with the POD or ND10 complexes in the nucleus of the virus-infected cell [49, 50]. It affects an innate immune response of the host cell by inhibiting the activation of type I interferon [51]. Lastly, it also affects cell cycle progression [52]. While the IE2 protein is the master regulator for early viral gene transcription, both the IE1 and IE2 proteins of human CMV were found to be required to efficiently alter cellular repressive chromatin for the activation of viral early promoters [53].

These studies on the MIE viral proteins were important for ongoing investigations into CMV latency and reactivation. In addition, these studies contributed to the development of a drug against the IE2 protein to treat human CMV-induced retinitis (fomivirsen).

## 10 Viral Genetics

Genetic research on CMVs lagged behind the other herpesviruses, such as HSV. It was difficult to isolate recombinant viruses using temperature sensitivity and plaque assay because of the slow replication cycle of CMVs. Greaves et al. [54] reported that mutated human CMVs could be isolated by inserting selective markers (lacZ/gpt cassette) using homologous recombination. This method worked well for mutating viral elements and genes that are nonessential. For example, the IE1 gene of human CMV was nonessential after high MOI, but necessary for efficient viral replication after low MOI [55], and the UL127 gene could be deleted without an effect on viral replication in cell culture [56]. However, viral genes, like IE2, could not be mutated by the above method because this essential viral gene had to be expressed in trans, and the IE2 protein was cytotoxic to the cell. A major advancement in CMV genetics came from the Koszinowski laboratory with the cloning of herpesvirus genomes into bacterial artificial chromosomes (BACs) [57]. Mutations of essential elements or ORFs could be made in the CMV BAC by recombination in bacteria, and the recombinant viral BAC DNA could be purified for transfection into fibroblast cells. In this manner, it was found that the human CMV IE2 and the murine CHV i.e. 3 (the IE2 equivalent gene) genes, are essential for viral replication [58, 59]. In addition, one could delete viral elements or substitute elements. For example, the enhancer could be deleted from murine CMV. The virus could replicate in cell culture, but the mutant virus failed to cause disease in the mouse [60]. The human CMV enhancer could be replaced by the murine CMV enhancer or vice versa. After low MOI, it was found that the recombinant human CMV with the murine CMV enhancer replicated less efficiently [61]. It became clear that the CMV enhancers evolved for the efficient replication of that particular species of CMV. However, if the species of CMVs were closely related, like the human and the chimpanzee, then the viral enhancers could be swapped without an effect on replication in cell culture [61]. In addition, the distal enhancer of human CMV could be deleted, but it was found to be necessary for efficient viral replication at low MOI [62]. The proximal enhancer was found to have essential elements between −39 and −67 relative to the transcription start site, which are necessary for viral replication [63, 64]. When the virion-associated transactivators such as pp71 (UL83) were absent, other sites, such as the CREB site in the presence of the putative SP-1 sites at −55 and −75 of the proximal promoter, were found to be most critical for interaction with the other transcription factor binding sites such as AP-1, NF-1, and NF-κB [65]. Analysis of the enhancers of primate and non-primate CMVs demonstrated that they all have complicated enhancers upstream of the IE1 gene, but they differed in the type and arrangement of the transcription factor binding sites (reviewed in refs. 66–68). This suggested that each species of CMV has a unique approach for the initial expression of the MIE genes.

## 11 Dysregulation of the Host Cell

In the mid 1990s, several laboratories started to investigate how infection with CMV affects the host cell cycle [69–72]. Terminally differentiated cells that are in the $G_0/G_1$ phases of the cell cycle readily support replication of the virus. Human CMV can infect cells during all phases of the cell cycle, but expression of the MIE promoter does not occur during the S phase [73]. Moreover, cells with high levels of cyclin A (A2) cdk1/cdk2 exhibit diminished transcription from the MIE promoter [74]. Human CMV is different from many cellular mitogens in that it inhibits cyclin A and induces cyclin E and B, possibly leading to increased expression of the MIE promoter. However, as the infection progresses, it must also control cyclin B/cdk1 or an abortive infection will occur [75]. The p53 tumor suppressor protein similarly plays a pivotal role for human CMV infection, and Luo et al. [76] and Hannemann et al. [77] demonstrated that cells that are p53 wild type support human CMV replication better than cells that are p53 mutant or p53 null. As the IE2-86 protein accumulates, it is capable of binding to p53 and inhibiting its p53 function, although the ramifications of this interaction regarding viral replication remain unclear [78].

Human CMV affects cellular nucleotide metabolism by neutralizing the repressive activity of p107 and pRb (reviewed in refs. 66, 67, 79). Cellular gene array analysis demonstrated that E2F responsive promoters are activated for the expression of cellular enzymes and initiation factors required for cellular DNA synthesis [80]. However, human CMV stops cell cycle progression at the $G_1$/S transition point, and the precursors for DNA synthesis are subverted for use in viral DNA synthesis. The IE2-86 protein of human CMV directly and indirectly controls cell cycle progression. The IE2-86 protein alone can stop the cell cycle of a p53 wild-type cell at the $G_1$/S transition and a p53 null cell at the $G_2$/M transition [81]. The IE2-86 protein also activates the expression of early viral gene products such as UL97 and UL117 that also control cell cycle progression. While UL97 can activate E2F-1 responsive cellular promoters [82], UL117 can prevent MCM loading at cellular origins of DNA replication and consequently inhibit cellular DNA synthesis and movement into the S phase [83]. These properties are unique to human CMV since murine CMV lacks the $G_1$ dependence for MIE gene expression and arrest the cell cycle predominantly in the late $G_2$ compartment [84].

## 12 Conclusion

The last 40 years have seen tremendous progress in our understanding of CMV replication. There are antivirals for the treatment of human CMV-induced disease, but these antivirals have cytotoxic side effects, and drug-resistant viral strains can emerge. Thus, there is a need for the development of new antivirals. Understanding the molecular events involved in replication of the virus may lead to the development of new and unique antivirals. We are just beginning to understand the essential role of viral proteins in regulating late viral gene transcription [85, 86], and the production and function of these late gene products represent additional target for antivirals. We currently do not fully understand the requirements for maintenance of the viral genome during latency or for reactivation of the virus from a latent state. While there has also been tremendous progress in understanding the innate and acquired immune response to CMV and how this persistent virus infection evades the immune response (not reviewed here), a vaccine to prevent CMV infection is still not available. Many of the CMV genes dispensable for replication in fibroblast cells have important roles in the virus–host interaction, and proteomics of viral–host cell protein interactions will tell us more about how the virus navigates from latent to productive replication. Moreover, studies on the potential roles for these genes in latency, reactivation, and pathogenesis will require the use of animal models for

CMV infection. Finally, it has become clear that a combination of molecular, cellular, and immunological approaches will be necessary to understand and control this important human pathogen.

## Acknowledgments

The author regrets not acknowledging all important details and contributions over the last 40 years within the page limitation. The author thanks the National Institutes of Health and the many colleagues and friends for their support.

### References

1. Ho M (1991) Cytomegalovirus: biology and infection, 2nd edn. Plenum Publishing Corp, New York
2. Alford CA, Britt WJ (1990) Cytomegalovirus. In: Fields BN, Knipe DM et al (eds) Virology. Raven Press Ltd, New York, pp 1981–2010
3. Stinski MF (1976) Human cytomegalovirus: glycoproteins associated with virions and dense bodies. J Virol 19:594–609
4. Stinski MF (1977) Synthesis of proteins and glycoproteins in cells infected with human cytomegalovirus. J Virol 23:751–767
5. Stinski MF (1978) Sequence of protein synthesis in cells infected by human cytomegalovirus: early and late virus-induced polypeptides. J Virol 26:686–701
6. Huang E-S, Chen S-T, Pagano JS (1973) Human cytomegalovirus. I. Purification and characterization of viral DNA. J Virol 12:1473–1481
7. Sarov I, Friedman A (1976) Electron microscopy of human cytomegalovirus DNA. Arch Virol 50:343–347
8. Kilpatrick BA, Huang ES (1977) Human cytomegalovirus genome: partial denaturation map and organization of genome sequences. J Virol 24:261–276
9. DeMarchi JM, Blankship ML, Brown GD, Kaplan AS (1978) Size and complexity of human cytomegalovirus DNA. Virology 89:643–646
10. Geelen JLMC, Walig C, Wertheim P, Van der Noordaa J (1978) Human cytomegalovirus DNA. I. Molecular weight and infectivity. J Virol 26:813–816
11. Stinski MF, Mocarski ES, Thomsen DR (1979) DNA of human cytomegalovirus: size heterogeneity and defectiveness resulting from serial undiluted passage. J Virol 31:231–239
12. Tamashiro JC, Spector DH (1980) Molecular cloning of the human cytomegalovirus genome (strain AD169). In: Fields BN, Jaenisch R (eds) Animal virus genetics. Academic, New York, pp 21–37
13. Thomsen DR, Stinski MF (1981) Cloning of the human cytomegalovirus genome as endonuclease XbaI fragments. Gene 16:207–216
14. Fleckenstein B, Muller I, Collins J (1982) Cloning of the complete human cytomegalovirus genome in cosmids. Gene 18:39–46
15. DeMarchi JM, Schmidt CA, Kaplan AS (1980) Patterns of transcription of human cytomegalovirus in permissively infected cells. J Virol 35:277–286
16. LaFemina RL, Hayward GS (1980) Structural organization of the DNA molecules from human cytomegalovirus. In: Fields BN, Jaenisch R (eds) Animal virus genetics. Academic, New York, pp 39–55
17. Weststrate MW, Geelen JLMC, Van der Noordaa J (1980) Human cytomegalovirus DNA: physical maps for the restriction endonuclease BglII, HindIII, and XbaI. J Gen Virol 49:1–21
18. Wathen MW, Thomsen DR, Stinski MF (1981) Temporal regulation of human cytomegalovirus transcription at immediate-early and early times after infection. J Virol 38: 446–459
19. DeMarchi JM (1981) Human cytomegalovirus DNA: restriction enzyme cleavage maps and map location for immediate-early, early, and late RNAs. Virology 124:390–402
20. Stinski MF (1990) Cytomegalovirus and its replication. In: Knipe DM et al (eds) Virology, 2nd edn. Raven, New York, pp 1959–1980
21. Wathen MW, Stinski MF (1982) Temporal patterns of human cytomegalovirus transcription: mapping the viral RNAs synthesized at immediate early, early, and late times after infection. J Virol 41:462–477
22. Stinski MF, Thomsen DR, Wathen MW (1981) Structure and function of the cytomegalovirus

genome. In: Nahmias A, Dowdle WR, Schinazi RF (eds) The human herpesviruses: an interdisciplinary perspective. Elsevier, New York, pp 72–84

23. Stinski MF, Thomsen DR, Stenberg RM, Goldstein LC (1983) Organization and expression of the immediate early genes of human cytomegalovirus. J Virol 46:1–14
24. Stenberg RM, Thomsen DR, Stinski MF (1984) Structural analysis of the major immediate early gene of human cytomegalovirus. J Virol 49:190–199
25. Stenberg RM, Witte PR, Stinski MF (1985) Multiple spliced and unspliced transcripts from human cytomegalovirus immediate-early region 2 and evidence for a common initiation site within immediate-early region 1. J Virol 56:665–675
26. Stinski MF, Stenberg RM, Goins WF (1984) Structure and function of the human cytomegalovirus immediate early genes. In: Rapp F (ed) Herpesvirus. Alan R. Liss, Inc., New York, pp 399–421
27. Stinski MF (1984) The proteins of human cytomegalovirus. In: Plotkin SA, Michelson S, Pagano JS, Rapp F (eds) CMV pathogenesis and prevention of human infection. Alan R. Liss, Inc., New York, pp 49–62
28. Thomsen DR, Stenberg RM, Goins WF, Stinski MF (1984) Promoter-regulatory region of the major immediate early gene of human cytomegalovirus. Proc Natl Acad Sci USA 81:659–663
29. Boshart M, Weber F, Jahn G, Dorsch-Hasler K, Fleckenstein B, Schaffner W (1985) A very strong enhancer is located upstream of an immediate early gene of human cytomegalovirus. Cell 41:521–530
30. Stinski MF, Roehr TJ (1985) Activation of the major immediate early gene of human cytomegalovirus by cis-acting elements in the promoter-regulatory sequence and by virus-specific trans-acting components. J Virol 55:431–441
31. Stinski MF (1999) Cytomegalovirus promoter for expression in mammalian cells. In: Ferandez JM, Hoeffler JP (eds) Gene expression systems: using nature for the art of expression. Academic, San Diego, CA, pp 211–233
32. Chee MA, Bankier AT, Beck S, Bohni R, Brown CM, Cerny R et al (1990) Analysis of the protein-coding content of the sequence of human cytomegalovirus strain AD169. Curr Top Microbiol Immunol 154:125–169
33. Murphy E, Dong Y, Grimwood J, Schmutz J, Dickson M, Jarvis MA et al (2003) Coding potential of laboratory and clinical stains of human cytomegalovirus. Proc Natl Acad Sci USA 100:14976–14981
34. Murphy E, Rigoutsos I, Shibuya T, Shenk T (2003) Reevaluation of human cytomegalovirus coding potential. Proc Natl Acad Sci USA 100:13585–13590
35. Dunn W, Chou C, Hong L, Hai R, Patterson D, Stolc V et al (2003) Functional profiling of a human cytomegalovirus genome. Proc Natl Acad Sci USA 100:14223–14228
36. Gatherer D, Seirafian S, Cunningham C, Holton M, Dargan DJ, Baluchova K et al (2011) High-resolution human cytomegalovirus transcriptome. Proc Natl Acad Sci USA 108:19755–19760
37. Hamzeh FM, Lietman PS, Gibson W, Hayward GS (1990) Identification of the lytic origin of DNA replication in human cytomegalovirus by a novel approach utilizing ganciclovir-induced chain termination. J Virol 64:6184–6195
38. Pari GS, Anders DG (1993) Eleven loci encoding trans-acting factors are required for transient complementation of human cytomegalovirus oriLyt-dependent DNA replication. J Virol 67: 6979–6988
39. Zhu Y, Huang L, Anders DG (1998) Human cytomegalovirus oriLyt sequence requirements. J Virol 72:4989–4996
40. Sarisky RT, Hayward GS (1996) Evidence that the UL84 gene product of human cytomegalovirus is essential for promoting oriLyt-dependent DNA replication and formation of replication compartments in cotransfection assays. J Virol 70:7398–7413
41. Xu Y, Cei SA, Rodriguez HA, Colletti KS, Pari GS (2004) Human cytomegalovirus DNA replication requires transcriptional activation via an IE2- and UL84-responsive bidirectional promoter element within oriLyt. J Virol 78:11664–11677
42. Pari GS (2008) Nuts and bolts of human cytomegalovirus lytic DNA replication. In: Shenk T, Stinski MF (eds) Human cytomegalovirus, Current topics in microbiology and immunology. Springer, Berlin, pp 153–166
43. Prichard MN, Jairath S, Penfold MET, St. Jeor S, Bohlman MC, Pari GS (1998) Identification of persistent RNA-DNA hybrid structures within the origin of replication of human cytomegalovirus. J Virol 72:6997–7004
44. Hermiston TW, Malone CL, Witte PR, Stinski MF (1987) Identification and characterization of the human cytomegalovirus immediate-early region 2 gene that stimulates gene expression from an inducible promoter. J Virol 61:3214–3221

45. Tevethia MJ, Spector DJ, Leisure KM, Stinski MF (1987) Participation of two human cytomegalovirus immediate early gene regions in transcriptional activation of adenovirus promoters. Virology 161(2):276–285
46. Stinski MF, Malone CL, Hermiston TW, Liu B (1991) Regulation of human cytomegalovirus transcription. In: Wagner EK (ed) Herpesvirus transcription and its control. CRC Press, Boca Raton, FL, pp 245–260
47. Stinski MF (1991) Molecular biology of cytomegalovirus replication. In: Ho M (ed) Cytomegalovirus biology and infection, 2nd edn. Plenum Medical Book Co., New York, pp 7–36
48. Stinski MF, Macias MP, Malone CL, Thrower AR, Huang L (1993) Regulation of transcription from the cytomegalovirus major immediate early promoter by cellular and viral proteins. In: Michelson S, Plotkin SA (eds) Multidisciplinary approach to understanding cytomegalovirus. Elsevier Science, Amsterdam, pp 3–12
49. Maul GG (2008) Initiation of cytomegalovirus infection at ND10. In: Shenk T, Stinski MF (eds) Human cytomegalovirus, Current topics in microbiology and immunology. Springer, Berlin, pp 117–132
50. Tavalai N, Papior P, Rechter S, Stamminger T (2008) Nuclear domain 10 components promyelocytic leukemia protein and hDaxx independently contribute to an intrinsic antiviral defense against human cytomegalovirus infection. J Virol 82:126–137
51. Huh YH, Kim YE, Kim ET, Park JJ, Song MJ, Zhu H, Hayward GS, Ahn J-H (2008) Binding STAT2 by the acidic domain of human cytomegalovirus IE1 promotes viral growth and is negatively regulated by SUMO. J Virol 82:10444–10454
52. Castillo JP, Yurochko AD, Kowalik TF (2000) Role of human cytomegalovirus immediate-early proteins in cell growth control. J Virol 74:8028–8037
53. Yee L-F, Lin PL, Stinski MF (2007) Ectopic expression of HCMV IE72 and IE86 proteins is sufficient to induce early gene expression but not production of infectious virus in undifferentiated promonocytic THP-1 cells. Virology 363:174–188
54. Greaves RF, Brown JM, Vieira J, Mocarski ES (1995) Selectable insertion and deletion mutagenesis of the human cytomegalovirus genome using the E. coli guanosine phosphoribosyl transferase (gpt) gene. J Gen Virol 76: 2151–2160
55. Greaves RF, Mocarski ES (1998) Defective growth correlates with reduced accumulation of viral DNA replication protein after low-multiplicity infection by a human cytomegalovirus ie1 mutant. J Virol 72:366–379
56. Meier JL, Stinski MF (1997) Effect of a modulator deletion on transcription of the human cytomegalovirus major immediate-early genes in infected undifferentiated and differentiated cells. J Virol 71:1246–1255
57. Messerle M, Crnkovic I, Hammerschmidt W, Ziegler H, Koszinowski UU (1997) Cloning and mutagenesis of a herpesvirus genome as an infectious bacterial artificial chromosome. Proc Natl Acad Sci USA 94:14759–14763
58. Angulo A, Ghazal P, Messerle M (2000) The major immediate-early gene ie3 of mouse cytomegalovirus is essential for viral growth. J Virol 74:11129–11136
59. Marchini A, Liu H, Ahu H (2001) Human cytomegalovirus with IE-2 (UL122) deleted fails to express early lytic genes. J Virol 75:1870–1878
60. Angulo A, Messerle M, Koszinowski UH, Ghazal P (1998) Enhancer requirement for murine cytomegalovirus growth and genetic complementation by the human cytomegalovirus enhancer. J Virol 72:8502–8509
61. Isomura H, Stinski MF (2003) Effect of substitution of the human cytomegalovirus enhancer or promoter with the murine cytomegalovirus enhancer or promoter on replication in human fibroblast. J Virol 77:3602–3614
62. Meier JL, Pruessner JA (2000) The human cytomegalovirus major immediate-early distal enhancer region is required for efficient viral replication and immediate-early expression. J Virol 74:1602–1613
63. Isomura H, Tatsuya T, Stinski MF (2004) The role of the proximal enhancer of the major immediate early promoter in human cytomegalovirus replication. J Virol 78:12788–12799
64. Isomura H, Stinski MF, Kudoh A, Daikoku T, Shirata N, Tsurumi T (2005) Two Sp1/Sp3 binding sites in the major immediate-early proximal enhancer of human cytomegalovirus have a significant role in viral replication. J Virol 79:9597–9607
65. Lashmit P, Wang S, Li H, Isomura H, Stinski MF (2009) The CREB site in the proximal enhancer is critical for cooperative interaction with the other transcription factor binding sites to enhance transcription of the major intermediate-early genes in human cytomegalovirus-infected cells. J Virol 83: 8893–8904
66. Stinski MF, Meier JL (2007) Immediate-early viral gene regulation and function. In: Arvin A et al (eds) Human herpesviruses biology, therapy, and immunoprophylaxis. Cambridge University Press, New York, pp 241–263

67. Meier JL, Stinski MF (2006) Major immediate-early enhancer and its gene products. In: Reddehase MJ (ed) Cytomegaloviruses molecular biology and immunology. Caister Academic Press, Norfolk, UK, pp 151–166
68. Stinski MF, Isomura H (2008) Role of the cytomegalovirus major immediate early enhancer in acute infection and reactivation from latency. Med Microbiol Immunol 197: 223–231
69. Dittmer D, Mocarski ES (1997) Human cytomegalovirus infection inhibits G1/S transition. J Virol 71:1629–1634
70. Jault FM, Jault J-M, Ruchti F, Fortunato EA, Clark C, Corbeil J et al (1995) Cytomegalovirus infection induces high levels of cyclins, phosphorylated RB, and p53, leading to cell cycle arrest. J Virol 69:6697–6704
71. Lu M, Shenk T (1996) Human cytomegalovirus infection inhibits cell cycle progression at multiple points including the transition from G1 to S. J Virol 70:8850–8857
72. Salvant BS, Fortunato EA, Spector DH (1998) Cell cycle dysregulation by human cytomegalovirus: influence of the cell cycle phase at the time of infection and effects on cyclin transcription. J Virol 72:3729–3741
73. Fortunato E, Sanchez V, Yen JY, Spector DH (2002) Infection of cells with human cytomegalovirus during S phase results in a blockade to immediate-early gene expression that can be overcome by inhibition of the proteasome. J Virol 76:5369–5379
74. Zydek M, Hagemeier C, Wiebusch L (2010) Cyclin-dependent kinase activity controls the onset of the HCMV lytic cycle. PLoS Pathog 6:1–16
75. Du G, Dutta N, Lashmit P, Stinski MF (2011) Alternative splicing of the human cytomegalovirus major immediate-early genes affects infectious-virus replication and control of cellular cyclin-dependent kinase. J Virol 85:804–817
76. Luo MH, Rosenke K, Czornak K, Fortunato EA (2007) Human cytomegalovirus disrupts both ataxia telangiectasia mutated protein (ATM)- and ATM-Rad3-related kinase-mediated DNA damage responses during lytic infection. J Virol 81:1934–1950
77. Hannemann H, Rosenki K, O'Dowd JM, Fortunato EA (2009) The presence of p53 influences the expression of multiple human cytomegalovirus genes at early times postinfection. J Virol 83:4316–4325
78. Tsai HL, Kou GH, Chen SC, Wu CW, Lin YS (1996) Human cytomegalovirus immediate-early protein IE2 tethers a transcriptional repression domain to p53. J Biol Chem 271:3534–3540
79. Stinski MF, Petrik DT (2008) Functional roles of the human cytomegalovirus essential IE86 protein. In: Shenk T, Stinski MF (eds) Human cytomegalovirus, Current topics in microbiology and immunology. Springer, Berlin, pp 133–152
80. Song Y-J, Stinski MF (2002) Effect of the human cytomegalovirus IE86 protein on expression of E2F-responsive genes: a DNA microarray analysis. Proc Natl Acad Sci USA 99:2836–2841
81. Song Y-J, Stinski MF (2004) Inhibition of cell division by the human cytomegalovirus IE86 protein: role of the p53 pathway or cdk1/cyclin B1. J Virol 79:2597–2603
82. Hume AJ, Finkel JS, Kamil JP, Coen DM, Culbertson MR, Kalejta RF (2008) Phosphorylation of retinoblastoma protein by viral protein with cyclin-dependent kinase function. Science 320:797–799
83. Qian Z, Leung-Pineda V, Xuan B, Piwnica-Worms H, Yu D (2010) Human cytomegalovirus protein pUL117 targets the mini-chromosome maintenance complex and suppresses cellar DNA synthesis. PLoS Pathog 6:e1000814
84. Wiebusch L, Neuwirth A, Grabenhenrich L, Voight S, Hagemeier C (2008) Cell cycle-independent expression of immediate-early gene 3 results in G1 and G2 arrest in murine cytomegalovirus-infected cells. J Virol 82: 10188–10198
85. Isomura H, Stinski MF, Murata T, Yamashita Y, Kanda T, Toyokuni S et al (2011) The human cytomegalovirus gene products essential for late viral gene expression assemble into prereplication complexes before viral DNA replication. J Virol 85:6629–6644
86. Perng Y-C, Qian Z, Fehr AR, Xuan B, Yu D (2011) The human cytomegalovirus gene UL79 is required for the accumulation of late viral transcripts. J Virol 85:4841–4852

Chapter 2

# Overview of Human Cytomegalovirus Pathogenesis

Maciej T. Nogalski, Donna Collins-McMillen, and Andrew D. Yurochko

## Abstract

Human cytomegalovirus (HCMV) is a human pathogen that infects greater than 50 % of the human population. HCMV infection is usually asymptomatic in most individuals. That is, primary infection or reactivation of latent virus is generally clinically silent. HCMV infection, however, is associated with significant morbidity and mortality in the immunocompromised and chronic inflammatory diseases in the immunocompetent. In immunocompromised individuals (acquired immune deficiency syndrome and transplant patients, developing children (in utero), and cancer patients undergoing chemotherapy), HCMV infection increases morbidity and mortality. In those individuals with a normal immune system, HCMV infection is also associated with a risk of serious disease, as viral infection is now considered to be a strong risk factor for the development of various vascular diseases and to be associated with some types of tumor development. Intense research is currently being undertaken to better understand the mechanisms of viral pathogenesis that are briefly discussed in this chapter.

**Key words** Human cytomegalovirus, Viral pathogenesis, Immunocompetent, Immunocompromised, Vascular disease, Oncogenesis, Congenital infection, AIDS patients, Transplant patients

## 1 Introduction

Human cytomegalovirus (HCMV) is a prevalent infectious agent affecting the health of the human population. In a simple sense, HCMV pathogenesis can be broken down to that observed in immunocompetent hosts and that observed in immunocompromised hosts [1, 2]. HCMV pathogenesis in immunologically normal individuals is usually considered less severe when compared to the morbidity and mortality seen in immunocompromised individuals. Severe complications such as pneumonia, retinitis, hepatitis, encephalitis, and disseminated HCMV disease with multiorgan involvement are extremely rare in immunologically healthy people [1–4]. The majority of HCMV infections in the immunocompetent are asymptomatic [1]; however, primary infection can result in a mononucleosis-like syndrome [3]. In addition, data from both clinical and experimental studies now define a potentially strong role for HCMV infection in the development and/or severity of inflammatory

Andrew D. Yurochko and William E. Miller (eds.), *Human Cytomegaloviruses: Methods and Protocols*,
Methods in Molecular Biology, vol. 1119, DOI 10.1007/978-1-62703-788-4_2, © Springer Science+Business Media New York 2014

cardiovascular diseases, and HCMV infection has been linked to the development of certain types of cancers [2, 5–8].

In immunocompromised individuals, HCMV infection can result in severe disease [1, 2]. For example, in patients undergoing immunosuppressive therapies, such as in transplant and cancer patients, and in patients with acquired immunodeficiency syndrome (AIDS), HCMV infection is of significant clinical concern. In addition, HCMV is one of the leading infectious agents causing congenital infection [9]. Thus individuals with suppressed or underdeveloped immune systems are prone to severe disease following primary HCMV infection or reactivation of latent virus. This short chapter focuses briefly on some of the consequences of HCMV infection in the immunocompetent and the immunocompromised.

## 2 Pathogenesis in Immunocompetent Hosts

### 2.1 Infectious Mononucleosis

The most common clinical manifestation of HCMV infection in the immunocompetent host is a self-limiting febrile illness that resembles the infectious mononucleosis resulting from infection with Epstein-Barr virus (EBV) [3, 10]. The clinical picture of HCMV mononucleosis is typically indistinguishable from EBV mononucleosis, with the exception that pharyngitis, adenopathy, and splenomegaly occur less commonly with HCMV infection [1, 3, 10, 11]. In addition, HCMV mononucleosis is a heterophile-negative mononucleosis accounting for approximately 10 % of mononucleosis diagnoses [1, 3, 10]. Fever, malaise, myalgia, headache, and fatigue are the most commonly experienced signs and symptoms of the disease [1, 3, 10, 11]. A smaller number of patients can present with splenomegaly, hepatomegaly, adenopathy, and a rash [1, 3, 10, 11]. Laboratory tests commonly reveal lymphocytosis, activated or atypical lymphocytes, and abnormal liver function [1, 3, 12].

### 2.2 Viral Role in Vascular Disease

Mounting evidence suggests that HCMV infection is an etiologic/co-etiologic agent in the development and/or severity of inflammatory cardiovascular diseases [7, 8, 13]. Beginning in the 1980s, several studies established a potential link between HCMV infection and the development of atherosclerosis [14–17]. A prospective study conducted from 1987 to 1992 examined medical records, including anti-HCMV antibody titers taken more than a decade earlier from patients with carotid intimal-medial thickening (IMT) and their age- and gender-matched controls [17]. Investigators observed higher anti-HCMV antibody titers in those patients with IMT than in the control group, consistent with HCMV infection being a possible factor/cofactor in the development of disease [17]. Additional serological studies identified a correlation between HCMV seropositivity and the severity of various cardiovascular diseases, and HCMV nucleic acid and antigen have been detected in tissue sections taken from atherosclerotic lesions [16, 18]. A link has also been established between HCMV infection and coronary restenosis [19, 20]. For example, HCMV seropositivity and higher HCMV IgG titers were shown to correlate with the incidence of restenosis following surgical intervention to treat atherosclerosis [19]. Studies have also linked HCMV infection to transplant vascular sclerosis [8, 21, 22]; and it has been reported that patients with coronary artery disease have higher blood serum levels of C-reactive protein, a marker of the inflammatory response, that correlates with HCMV seropositivity, suggesting that inflammation resulting from HCMV infection may serve as a risk factor for vascular disease [23]. HCMV infects endothelial cells, smooth muscle cells, and monocytes in vivo and in vitro, all of which, when aberrantly activated, can directly contribute to the development of cardiovascular disease [24–30]. As an example, evidence shows that monocytes can migrate into arterial tissue, differentiate into macrophages, and engulf oxidized low-density lipoproteins, becoming the foamy macrophages that accumulate in arteries and atherosclerotic lesions [reviewed in ref. 25]. Because HCMV infection of monocytes has been shown to alter many of these processes [24, 31], it is plausible to hypothesize that viral infection of monocytes and the resulting biological changes in these cells contribute to atherosclerotic disease. The same idea holds true for HCMV infection of vascular endothelial cells and smooth muscle cells. Animal studies in various rodent models have demonstrated (essentially confirming Koch's postulates) that CMV infection is linked to endothelial damage, monocyte infiltration, foam cell accumulation, and vascular disease [30, 32–36]. Thus, both clinical and experimental studies have provided strong evidence that CMV infection promotes vascular disease at almost every stage of the disease process (enhancement of the proinflammatory response, vascular injury, increased migration and proliferation of smooth muscle cells, migration of monocytes into lesions, formation of foamy macrophages, plaque development, and other biological changes consistent with a role in vascular disease).

### 2.3 Possible Viral Role in Oncogenesis

The potential relationship between HCMV infection and cancer has been debated for decades, and during that time, HCMV has been argued to be associated with a variety of malignancies [6]. Seroepidemiological studies, as well as detection of viral nucleic acids and/or antigens in malignant tissues, are suggestive of an etiologic role for HCMV in the development of various types of cancers [6, 37–41]. However, it remains unclear whether HCMV is the causative agent of any of these cancers, as the virus has not been shown to transform normal cells and is not generally thought of as an oncogenic virus in the classical sense [6]. On the other hand, HCMV does possess many of the molecular hallmarks of the small DNA tumor viruses (altering p53 and Rb function, inducing cellular proliferation, enhancing cellular survival, etc.), suggesting at least from a molecular standpoint that it could be an etiologic agent in tumor development and/or progression [42–46]. It has also been suggested that HCMV infection might produce an environment of a "smoldering inflammation," which in turn could mechanistically promote oncogenesis in a similar manner to that defined for the hepatitis viruses [47]. In keeping with this idea, it has been proposed that "oncomodulation" could describe HCMV's effect on tumor cells; in that HCMV could infect tumor cells and enhance their malignancy, thereby promoting tumor progression without being an oncogenic virus per se [48–51]. The body of evidence supporting a possible role for the virus in tumor growth continues to expand, with investigators focusing on both clinical and experimental aspects of HCMV cancer research. HCMV genome and antigen indicative of a persistent low level of replication have been detected in tumor cells, but not in the surrounding tissue, of a variety of malignancies including colorectal cancers [38], malignant gliomas [37, 41, 52, 53], prostate cancers [39], and breast cancers [40]. HCMV has also been proposed to be a co-etiologic agent in the development of certain types of cancers [54]. A recent study aimed at establishing the clinical relevance of HCMV infection in malignant glioblastomas grouped patients based on the level of HCMV-infected tumor cells and uncovered a relationship between the level of infection and the patient's life expectancy: patients with low-level HCMV infection outlived those with higher levels of HCMV infection [5, 55]. There are multiple molecular mechanisms, which are proposed to contribute to HCMV-induced oncomodulation. A recent review by Michaelis et al. provides a comprehensive description of the mechanisms by which viral proteins and noncoding RNAs alter the molecular and functional properties of HCMV-infected tumor cells [6]. Experimental evidence suggests that HCMV alters the cell cycle and inhibits apoptosis in infected cancer cells, thereby promoting proliferation and survival of the cells [6]. HCMV infection also appears to influence invasion, migration, and endothelial adhesion of malignant cells, potentially contributing to metastatic complications in HCMV-infected patients [6]. In addition, HCMV infection has been shown to promote angiogenesis, a process that is central to the initiation and progression of malignancies [6, 26, 56]. Infection with HCMV also diminishes cancer cell immunogenicity and causes chromosomal abnormalities in infected cells [6, 57]. Nevertheless, the relationship between HCMV infection and cancer is unclear/unresolved and remains an important question in the area of HCMV pathogenesis. Additional research is needed to define if and how HCMV infection modulates and/or is an etiologic agent in tumor initiation and/or progression.

## 3 Pathogenesis in Immunocompromised Hosts

### 3.1 Congenital Infection

Congenital HCMV infection occurs when the virus crosses the placental barrier allowing transmission of the virus from mother to baby. This process can occur following primary infection of the mother or from reactivation of latent virus in the mother [1, 9]. It is estimated that in developed countries, up to 7 % of seronegative mothers will develop a primary infection, and from this group, approximately 30–50 % will transmit the virus to the fetus [9, 58–61]. However, only around 10 % of congenitally infected newborns show disease symptoms [9]. If the woman was seropositive before conception, the risk of a newborn being congenitally infected is low and oscillates at about 1 % in developed countries (higher frequencies of infection are reported in developing countries), with a small number of children being severely affected by the virus [9, 60, 62, 63]. Congenitally infected babies can have a multisymptomatic disease affecting many organ systems and ranging from pneumonia, through gastrointestinal and retinal diseases, to central nervous system (CNS) diseases [64, 65]. Congenital HCMV may also manifest as a hematologic disease with, for example, thrombocytopenia and hemolytic anemia being commonly observed abnormalities [1]. Additionally, congenital HCMV infection may result in jaundice, hepatitis, hepatosplenomegaly, petechiae, and thrombocytopenia in the infected neonate. Although symptoms generally subside a few weeks after birth, the disease can be severe for some newborns, even leading to neonatal death in a small percentage of cases [9, 60, 64, 66]. Greater than 50 % of cases of symptomatic congenital HCMV infection present with abnormalities in the CNS [60, 62, 63, 67]. These abnormalities often cause a range of neurological symptoms, such as mental retardation, diminished motor skills, and hearing and/or vision loss [1, 9, 66–68]. CNS sequelae also present (~10 %) in newborns that are asymptomatic at birth for congenital HCMV infection and mainly cause hearing loss [9, 62, 69, 70]. Additionally, it has been documented that symptomatic congenital HCMV infection with a higher severity of disease occurs more often if the mother was exposed to a primary infection during pregnancy [62, 71]. However, because congenital infection can also occur in seropositive mothers [9, 72], it is not entirely clear if seropositivity per se diminishes the incidence of congenital infection. Based on the prevalence and severity of disease, congenital HCMV infection is considered as a leading cause of CNS damage in children [1, 9, 62].

### 3.2 Infection of Infants

An infant's immune system does not generally fully develop until approximately 6 months after birth. Thus, they have a lower ability to mount effective immune responses to pathogens. Nevertheless, maternal antibodies transferred through the placenta and antibodies transferred through breast milk generally serve to protect newborns from infections during their early life.

In regard to HCMV infection, newborns can be infected in utero (as discussed above), via intrapartum transmission, or via consumption of breast milk containing the virus [67]. It is estimated that HCMV is shed in around 10 % of women from the vagina or cervix [1, 73, 74] and that the rate of intrapartum transmission from virus shedding mothers is around 50 % [75]. Human breast milk is also considered to be a common means for mother-to-infant transmission of viruses. The rate of newborn HCMV infection via this route strongly correlates with the length of time that the baby is nursed. It has been thought that approximately 40–60 % of breast-fed infants will be infected by HCMV at the end of their first year of life if fed by seropositive mothers [67, 76]. This high rate of transmission is consistent with studies that have documented that up to ~95 % of tested milk samples are positive for HCMV DNA [67, 77, 78].

Although full-term newborns usually do not present with significant disease if infected early in life, there have been reported cases of hepatomegaly, elevated hepatic enzymes, and inflammation of lung tissue [79–81]. The risk of complications, however, rises in preterm babies. For example, there is an association between HCMV disease manifestation and premature infants of seronegative mothers that received blood from seropositive donors. In these reported cases, symptoms suggested multiorgan dysfunction [82, 83]. In addition, it has been found that preterm neonates are infected at a higher rate than full-term infants if nursed by seropositive mothers [76, 78]. In preterm infants, the risks of complications associated with infection include thrombocytopenia, neutropenia, apnea, liver dysfunction, sepsis syndrome, and a mononucleosis-like disorder [78, 84]. Limiting preterm newborns to blood products from seronegative donors or by eliminating leukocytes from the transfused blood reduces the transfusion-acquired HCMV disease in these infants [83, 85, 86]. Although, the health-related issues resulting from HCMV infection via intrapartum or postnatal transmission are not usually associated with adverse outcomes in full-term babies, the role of infected infants in the epidemiology of HCMV spread is significant, as they can shed the virus for years, thereby enhancing viral transmission [1, 9, 61, 74]. This point may be especially relevant to pregnant seronegative women who have older children in day care, as it increases their risk of exposure to the virus.

### 3.3 Infection of Immunocompromised Hosts

HCMV is considered to be one of the most common opportunistic pathogens seen in immunocompromised patients. These patients are at risk for viral-mediated disease as a result of a primary infection, reinfection (of an already seropositive host), and reactivation of latent virus. It has been documented that the stronger the suppression of the immune system, the greater the risk for HCMV-mediated disease [1]. Allogeneic stem cell transplant patients and AIDS patients are characterized as having the most severe disease manifestations. HCMV infection and disease are also seen in solid organ transplant and cancer patients undergoing immunosuppressive therapy [1]. Clinical manifestations in these patients can range from a short febrile illness to multiple organ system involvement.

Although the investigation of the impact HCMV infection has on immunocompromised patients is complex, studies performed on those patients have allowed a better understanding of HCMV infection, immune control of the virus, and viral-mediated diseases. For example, these studies have provided evidence about the reinfection of patients with new strains of virus [87–89], the importance of humoral and cell-mediated immunity in limiting HCMV infection [9, 90, 91], and evidence that a CMV vaccine may be efficacious [9, 92]. Additionally, the investigation and generation of new antiviral drugs has been influenced by the need for better management of HCMV infection in immunocompromised patients [93].

#### 3.3.1 Infection of Transplant Patients

Complications resulting from HCMV infection of transplant patients significantly increase the overall cost and length of hospitalization for the patient [94, 95]. Viral reactivation is a common cause of HCMV infection in patients receiving solid organ or stem cell transplants [96, 97]. Usually, the virus is detected in the blood or excretions at around 4–8 weeks after transplantation [1, 98]. Nevertheless, primary HCMV infection of transplant patients is usually considered to show more complications than those arising in patients in which active virus is the result of reactivation of latent virus or reinfection (of a seropositive individual) [1, 96, 98, 99]. The most common symptoms of HCMV disease in solid organ transplant patients are fever, leukopenia, malaise, joint pain, and macular rash. However, more severe complications, such as pneumonitis, gastrointestinal ulceration, abnormal liver function, accelerated coronary artery disease, fungal and bacterial infection, and impairment of and/or rejection of the graft, have also been reported [1, 98]. The most common manifestation of HCMV infection in stem cell patients is an interstitial pneumonia that unfortunately has a high mortality (>50 %); mortality is still high in the presence of active antiviral treatment [96, 99–102]. Cases of gastrointestinal disease, hepatitis, encephalitis, and retinitis have also been described [96, 99, 102, 103].

The severity of complications caused by HCMV infection in transplant patients has set the stage for the improved management of HCMV disease. In general, pre-transplant donor and recipient screening as well as posttransplant screening for the presence of HCMV are performed, along with the preemptive and prophylactic administration of antivirals [96, 97, 99, 104, 105]. By having those protocols in place, the rates of HCMV disease and infection-related deaths have been significantly reduced [103]. However, late-onset HCMV disease remains a significant problem in these patients and is responsible for the high mortality in these patients [97, 99, 101, 106–108].

#### 3.3.2 Infection in AIDS Patients

HCMV infection is considered a leading opportunistic pathogen in AIDS patients and has been associated with progression of HIV infection [1, 9, 109–112]. The serological prevalence of HCMV is evident in nearly all adults and about half of the children seropositive for HIV infection [1]. It was estimated that approximately 40 % of adults and about 10 % of children with AIDS showed manifestations of HCMV disease before the introduction of highly active antiretroviral therapy (HAART) [1, 113, 114]. Common manifestations of HCMV disease in AIDS patients are retinitis, esophagitis, and colitis; case reports have also documented encephalitis, neuropathy, polyradiculoneuritis, pneumonitis, gastritis, and liver dysfunction [114]. Because of the use of HAART, incidence of each of these pathologies has significantly decreased in treated patients [9, 115–117]. Nevertheless, there is evidence that HCMV infection remains an independent predictor of morbidity and mortality in AIDS patients [111, 118, 119]. HCMV infection has long been attributed to the progression of HIV infection and morbidity in these patients, although mechanistically it remains unclear how HCMV may affect the outcome of HIV-infected patients (outside of HCMV's role as an opportunistic pathogen). Some examples for how HCMV may alter the course of infection include transactivation of the HIV promoter [109], changes in Fc receptor expression [110], altered immune activation [112], and increased T cell senescence [120]. Regardless, of the mechanism for how HCMV may affect the outcome of HIV-infected patients, it is very clear that these two pathogens possess a partnership and that even in the day of HAART, HCMV remains an important pathogen in AIDS patients.

## 4 Conclusions

HCMV remains an important pathogen of humans because of its high rate of dissemination in the human population and the multitude of disease pathologies caused by or associated with infection. HCMV-mediated disease can loosely be grouped into the diseases observed in immunocompromised individuals and the diseases observed in immunocompetent individuals. In immunocompromised people, HCMV infection can cause severe disease and affect a variety of organ systems. In healthy people, HCMV infection is generally thought of as benign; however, with the association of viral infection with cardiovascular diseases and now with several cancers, this idea of the virus generally being benign needs to be revisited and the virus thought of as a significant pathogen to all hosts. Overall, HCMV is a complex pathogen with an interesting and diverse pathobiology. Additional studies into the basic aspects of the biology of the virus to the specific mechanisms that directly cause disease are needed. In addition, new treatment options and the generation of an effective vaccine are needed to counter the morbidity and mortality associated with infection.

## References


1. Mocarsk ES Jr, Shenk T, Pass RF (2007) Cytomegaloviruses. In: Knipe DM, Howley PM (eds) Fields virology. Lippincott Williams & Wilkins, Philadelphia, PA, pp 2701–2772
2. Britt W (2008) Manifestations of human cytomegalovirus infection: proposed mechanisms of acute and chronic disease. In: Stinksi MF, Shenk T (eds) Human cytomegaloviruses. Springer, Berlin, pp 417–470
3. Bravender T (2010) Epstein-Barr virus, cytomegalovirus, and infectious mononucleosis. Adolesc Med State Art Rev 21:251–264, ix–64, ix
4. Eddleston M, Peacock S, Juniper M, Warrell DA (1997) Severe cytomegalovirus infection in immunocompetent patients. Clin Infect Dis 24:52–56
5. Söderberg-Nauclér C (2008) HCMV microinfections in inflammatory diseases and cancer. J Clin Virol 41:218–223
6. Michaelis M, Doerr HW, Cinatl J (2009) The story of human cytomegalovirus and cancer: increasing evidence and open questions. Neoplasia 11:1–9
7. Caposio P, Orloff SL, Streblow DN (2011) The role of cytomegalovirus in angiogenesis. Virus Res 157:204–211
8. Streblow DN, Dumortier J, Moses AV, Orloff SL, Nelson JA (2008) Mechanisms of cytomegalovirus-accelerated vascular disease: induction of paracrine factors that promote angiogenesis and wound healing. Curr Top Microbiol Immunol 325:397–415
9. Manicklal S, Emery VC, Lazzarotto T, Boppana SB, Gupta RK (2013) The "silent" global burden of congenital cytomegalovirus. Clin Microbiol Rev 26:86–102
10. Klemola E, Von Essen R, Henle G, Henle W (1970) Infectious-mononucleosis-like disease with negative heterophile agglutination test. Clinical features in relation to Epstein-Barr virus and cytomegalovirus antibodies. J Infect Dis 121:608–614
11. Jordan MC, Rousseau W, Stewart JA, Noble GR, Chin TD (1973) Spontaneous cytomegalovirus mononucleosis. Clinical and laboratory observations in nine cases. Ann Intern Med 79:153–160
12. Wreghitt TG, Teare EL, Sule O, Devi R, Rice P (2003) Cytomegalovirus infection in immunocompetent patients. Clin Infect Dis 37: 1603–1606
13. Streblow DN, Dumortier J, Moses AV, Orloff SL, Nelson JA (2008) Mechanisms of cytomegalovirus-accelerated vascular disease: induction of paracrine factors that promote angiogenesis and wound healing. In: Stinksi MF, Shenk T (eds) Human cytomegaloviruses. Springer, Berlin, pp 397–416
14. Adam E, Melnick JL, Probtsfield JL, Petrie BL, Burek J, Bailey KR, McCollum CH, DeBakey ME (1987) High levels of cytomegalovirus antibody in patients requiring vascular surgery for atherosclerosis. Lancet 2: 291–293
15. Hendrix MG, Salimans MM, van-Boven CP, Bruggeman CA (1990) High prevalence of latently present cytomegalovirus in arterial walls of patients suffering from grade III atherosclerosis. Am J Pathol 136:23–28
16. Melnick JL, Petrie BL, Dreesman GR, Burek J, McCollum CH, DeBakey ME (1983) Cytomegalovirus antigen within human arterial smooth muscle cells. Lancet 2:644–647
17. Nieto FJ, Adam E, Sorlie P, Farzadegan H, Melnick JL, Comstock GW, Szklo M (1996) Cohort study of cytomegalovirus infection as a risk factor for carotid intimal-medial thickening, a measure of subclinical atherosclerosis. Circulation 94:922–927
18. Gyorkey F, Melnick JL, Guinn GA, Gyorkey P, DeBakey ME (1984) Herpesviridae in the endothelial and smooth muscle cells of the proximal aorta in atherosclerotic patients. Exp Mol Pathol 40:328–339
19. Zhou YF, Leon MB, Waclawiw MA, Popma JJ, Yu ZX, Finkel T, Epstein SE (1996) Association between prior cytomegalovirus infection and the risk of restenosis after coronary atherectomy. N Engl J Med 335: 624–630

20. Speir E, Modali R, Huang ES, Leon MB, Shawl F, Finkel T, Epstein SE (1994) Potential role of human cytomegalovirus and p53 interaction in coronary restenosis. Science 265:391–394
21. Hussain T, Burch M, Fenton MJ, Whitmore PM, Rees P, Elliott M, Aurora P (2007) Positive pretransplantation cytomegalovirus serology is a risk factor for cardiac allograft vasculopathy in children. Circulation 115:1798–1805
22. Söderberg-Naucĺer C, Emery VC (2001) Viral infections and their impact on chronic renal allograft dysfunction. Transplantation 71:SS24–SS30
23. Zhu J, Quyyumi AA, Norman JE, Csako G, Epstein SE (1999) Cytomegalovirus in the pathogenesis of atherosclerosis: the role of inflammation as reflected by elevated C-reactive protein levels. J Am Coll Cardiol 34:1738–1743
24. Chan G, Nogalski MT, Stevenson EV, Yurochko AD (2012) Human cytomegalovirus induction of a unique signalsome during viral entry into monocytes mediates distinct functional changes: a strategy for viral dissemination. J Leukoc Biol 92:743–752
25. Hansson GK, Libby P (2006) The immune response in atherosclerosis: a double-edged sword. Nat Rev Immunol 6:508–519
26. Bentz GL, Yurochko AD (2008) HCMV infection of endothelial cells induces an angiogenic response through viral-binding to the epidermal growth factor receptor and the $\beta_1$ and $\beta_3$ integrins. Proc Natl Acad Sci USA 105:5531–5536
27. Falk E (2006) Pathogenesis of atherosclerosis. J Am Coll Cardiol 47:C7–C12
28. Lusis AJ (2000) Atherosclerosis. Nature 407:233–241
29. Streblow DN, Soderberg-Naucler C, Vieira J, Smith P, Wakabayashi E, Ruchti F, Mattison K, Altschuler Y, Nelson JA (1999) The human cytomegalovirus chemokine receptor US28 mediates vascular smooth muscle cell migration. Cell 99:511–520
30. Streblow DN, Orloff SL, Nelson JA (2001) Do pathogens accelerate atherosclerosis? J Nutr 131:2798–2804
31. Chan G, Nogalski MT, Yurochko AD (2012) Human cytomegalovirus stimulates monocyte-to-macrophage differentiation via the temporal regulation of caspase 3. J Virol 92:10714–10723
32. Streblow DN, Kreklywich CN, Smith P, Soule JL, Meyer C, Yin M, Beisser P, Vink C, Nelson JA, Orloff SL (2005) Rat cytomegalovirus-accelerated transplant vascular sclerosis is reduced with mutation of the chemokine-receptor R33. Am J Transplant 5:436–442
33. Orloff SL, Hwee YK, Kreklywich C, Andoh TF, Hart E, Smith PA, Messaoudi I, Streblow DN (2011) Cytomegalovirus latency promotes cardiac lymphoid neogenesis and accelerated allograft rejection in CMV naïve recipients. Am J Transplant 11:45–55
34. Streblow DN, Kreklywich CN, Andoh T, Moses AV, Dumortier J, Smith PP, Defilippis V, Fruh K, Nelson JA, Orloff SL (2008) The role of angiogenic and wound repair factors during CMV-accelerated transplant vascular sclerosis in rat cardiac transplants. Am J Transplant 8:277–287
35. Gombos RB, Brown JC, Teefy J, Gibeault RL, Conn KL, Schang LM, Hemmings DG (2013) Vascular dysfunction in young, mid-aged and aged mice with latent cytomegalovirus infections. Am J Physiol Heart Circ Physiol 304:H183–H194
36. Tang-Feldman YJ, Lochhead SR, Lochhead GR, Yu C, George M, Villablanca AC, Pomeroy C (2013) Murine cytomegalovirus (MCMV) infection upregulates P38 MAP kinase in aortas of Apo E KO mice: a molecular mechanism for MCMV-induced acceleration of atherosclerosis. J Cardiovasc Transl Res 6:54–64
37. Bhattacharjee B, Renzette N, Kowalik TF (2012) Genetic analysis of cytomegalovirus in malignant gliomas. J Virol 86:6815–6824
38. Harkins L, Volk AL, Samanta M, Mikolaenko I, Britt WJ, Bland KI, Cobbs CS (2002) Specific localisation of human cytomegalovirus nucleic acids and proteins in human colorectal cancer. Lancet 360:1557–1563
39. Samanta M, Harkins L, Klemm K, Britt WJ, Cobbs CS (2003) High prevalence of human cytomegalovirus in prostatic intraepithelial neoplasia and prostatic carcinoma. J Urol 170:998–1002
40. Harkins LE, Matlaf LA, Soroceanu L, Klemm K, Britt WJ, Wang W, Bland KI, Cobbs CS (2010) Detection of human cytomegalovirus in normal and neoplastic breast epithelium. Herpesviridae 1:8
41. Cobbs CS, Harkins L, Samanta M, Gillespie GY, Bharara S, King PH, Nabors LB, Cobbs CG, Britt WJ (2002) Human cytomegalovirus infection and expression in human malignant glioma. Cancer Res 62:3347–3350
42. Hwang ES, Zhang Z, Cai H, Huang DY, Huong SM, Cha CY, Huang ES (2009) Human cytomegalovirus IE1-72 protein

interacts with p53 and inhibits p53-dependent transactivation by a mechanism different from that of IE2-86 protein. J Virol 83: 12388–12398

43. Shen Y, Zhu H, Shenk T (1997) Human cytomegalovirus IE1 and IE2 proteins are mutagenic and mediate "hit-and-run" oncogenic transformation in cooperation with the adenovirus E1A proteins. Proc Natl Acad Sci USA 94:3341–3345
44. E X, Pickering MT, Debatis M, Castillo J, Lagadinos A, Wang S, Lu S, Kowalik TF (2011) An E2F1-mediated DNA damage response contributes to the replication of human cytomegalovirus. PLoS Pathog 7:e1001342
45. Castillo JP, Frame FM, Rogoff HA, Pickering MT, Yurochko AD, Kowalik TF (2005) Human cytomegalovirus IE1-72 activates ataxia telangiectasia mutated kinase and a p53/p21-mediated growth arrest response. J Virol 79:11467–11475
46. Hume AJ, Kalejta RF (2009) Regulation of the retinoblastoma proteins by the human herpesviruses. Cell Div 4:1
47. Soroceanu L, Cobbs CS (2011) Is HCMV a tumor promoter? Virus Res 157:193–203
48. Cinatl J Jr, Cinatl J, Vogel JU, Rabenau H, Kornhuber B, Doerr HW (1996) Modulatory effects of human cytomegalovirus infection on malignant properties of cancer cells. Intervirology 39:259–269
49. Barami K (2010) Oncomodulatory mechanisms of human cytomegalovirus in gliomas. J Clin Neurosci 17:819–823
50. Söderberg-Nauclér C, Johnsen JI (2012) Cytomegalovirus infection in brain tumors: a potential new target for therapy? Oncoimmunology 1:739–740
51. Maussang D, Verzijl D, van Walsum M, Leurs R, Holl J, Pleskoff O, Michel D, van Dongen GA, Smit MJ (2006) Human cytomegalovirus-encoded chemokine receptor US28 promotes tumorigenesis. Proc Natl Acad Sci USA 103:13068–13073
52. Mitchell DA, Xie W, Schmittling R, Learn C, Friedman AD, McLendon RE, Sampson JH (2008) Sensitive detection of human cytomegalovirus in tumors and peripheral blood of patients diagnosed with glioblastoma. Neuro Oncol 10:10–18
53. Scheurer ME, Bondy ML, Aldape KD, Albrecht T, El-Zein R (2008) Detection of human cytomegalovirus in different histological types of gliomas. Acta Neuropathol 116:79–86
54. Shen CY, Ho MS, Chang SF, Yen MS, Ng HT, Huang E-S, Wu CW (1993) High rate of concurrent genital infections with human cytomegalovirus and human papillomaviruses in cervical cancer patients. J Infect Dis 168:449–452
55. Rahbar A, Stragliotto G, Orrego A, Peredo I, Taher C, Willems J, Söderberg-Naucler C (2012) Low levels of human cytomegalovirus infection in glioblastoma multiforme associates with patient survival—a case-control study. Herpesviridae 3:3
56. Dumortier J, Streblow DN, Moses AV, Jacobs JM, Kreklywich CN, Camp D, Smith RD, Orloff SL, Nelson JA (2008) Human cytomegalovirus secretome contains factors that induce angiogenesis and wound healing. J Virol 82:6524–6535
57. Fortunato EA, Spector DH (2003) Viral induction of site-specific chromosome damage. Rev Med Virol 13:21–37
58. Adler SP (2005) Congenital cytomegalovirus screening. Pediatr Infect Dis J 24:1105–1106
59. Adler SP (2011) Screening for cytomegalovirus during pregnancy. Infect Dis Obstet Gynecol 2011:1–9
60. Nigro G, Adler SP (2011) Cytomegalovirus infections during pregnancy. Curr Opin Obstet Gynecol 23:123–128
61. Marshall BC, Adler SP (2009) The frequency of pregnancy and exposure to cytomegalovirus infections among women with a young child in day care. Am J Obstet Gynecol 200:163.e1–163.e5
62. Fowler KB, Stagno S, Pass RF, Britt WJ, Boll TJ, Alford CA (1992) The outcome of congenital cytomegalovirus infection in relation to maternal antibody status. N Engl J Med 326:663–667
63. Ross SA, Boppana SB (2005) Congenital cytomegalovirus infection: outcome and diagnosis. Semin Pediatr Infect Dis 16:44–49
64. Vancikova Z, Dvorak P (2001) Cytomegalovirus infection in immunocompetent and immunocompromised individuals—a review. Curr Drug Targets Immune Endocr Metabol Disord 1:179–187
65. Mussi-Pinhata MM, Yamamoto AY, Moura Brito RM, de Lima Isaac M, de Carvalho Oliveira PF, Boppana S, Britt WJ (2009) Birth prevalence and natural history of congenital cytomegalovirus infection in a highly seroimmune population. Clin Infect Dis 49:522–528
66. Boppana SB, Pass RF, Britt WJ, Stagno S, Alford CA (1992) Symptomatic congenital cytomegalovirus infection: neonatal morbidity and mortality. Pediatr Infect Dis J 11:93–99
67. Nassetta L, Kimberlin D, Whitley R (2009) Treatment of congenital cytomegalovirus

infection: implications for future therapeutic strategies. J Antimicrob Chemother 63: 862–867

68. Bale JF Jr, Blackman JA, Sato Y (1990) Outcome in children with symptomatic congenital cytomegalovirus infection. J Child Neurol 5:131–136
69. Williamson WD, Percy AK, Yow MD, Gerson P, Catlin FI, Koppelman ML, Thurber S (1990) Asymptomatic congenital cytomegalovirus infection. Audiologic, neuroradiologic, and neurodevelopmental abnormalities during the first year. Am J Dis Child 144:1365–1368
70. Saigal S, Lunyk O, Larke RP, Chernesky MA (1982) The outcome in children with congenital cytomegalovirus infection. A longitudinal follow-up study. Am J Dis Child 136:896–901
71. Stagno S, Pass RF, Cloud G, Britt WJ, Henderson RE, Walton PD, Veren DA, Page F, Alford CA (1986) Primary cytomegalovirus infection in pregnancy. Incidence, transmission to fetus, and clinical outcome. JAMA 256:1904–1908
72. Boppana SB, Fowler KB, Britt WJ, Stagno S, Pass RF (1999) Symptomatic congenital cytomegalovirus infection in infants born to mothers with preexisting immunity to cytomegalovirus. Pediatrics 104:55–60
73. Stagno S, Pass RF, Dworsky ME, Alford CA Jr (1982) Maternal cytomegalovirus infection and perinatal transmission. Clin Obstet Gynecol 25:563–576
74. Cannon MJ, Hyde TB, Schmid DS (2011) Review of cytomegalovirus shedding in bodily fluids and relevance to congenital cytomegalovirus infection. Rev Med Virol 21:240–255
75. Reynolds DW, Stagno S, Hosty TS, Tiller M, Alford CA Jr (1973) Maternal cytomegalovirus excretion and perinatal infection. N Engl J Med 289:1–5
76. Dworsky M, Yow M, Stagno S, Pass RF, Alford C (1983) Cytomegalovirus infection of breast milk and transmission in infancy. Pediatrics 72:295–299
77. Jim WT, Shu CH, Chiu NC, Kao HA, Hung HY, Chang JH, Peng CC, Hsieh WS, Liu KC, Huang FY (2004) Transmission of cytomegalovirus from mothers to preterm infants by breast milk. Pediatr Infect Dis J 23:848–851
78. Vochem M, Hamprecht K, Jahn G, Speer CP (1998) Transmission of cytomegalovirus to preterm infants through breast milk. Pediatr Infect Dis J 17:53–58
79. Chiba S, Hori S, Kawamura N, Nakao T (1975) Primary cytomegalovirus infection and liver involvement in early infancy. Tohoku J Exp Med 117:143–151
80. Stagno S, Brasfield DM, Brown MB, Cassell GH, Pifer LL, Whitley RJ, Tiller RE (1981) Infant pneumonitis associated with cytomegalovirus, Chlamydia, Pneumocystis, and Ureaplasma: a prospective study. Pediatrics 68:322–329
81. Kumar ML, Nankervis GA, Cooper AR, Gold E (1984) Postnatally acquired cytomegalovirus infections in infants of CMV-excreting mothers. J Pediatr 104:669–673
82. Ballard RA, Drew WL, Hufnagle KG, Riedel PA (1979) Acquired cytomegalovirus infection in preterm infants. Am J Dis Child 133:482–485
83. Yeager AS, Grumet FC, Hafleigh EB, Arvin AM, Bradley JS, Prober CG (1981) Prevention of transfusion-acquired cytomegalovirus infections in newborn infants. J Pediatr 98:281–287
84. Hamprecht K, Maschmann J, Vochem M, Dietz K, Speer CP, Jahn G (2001) Epidemiology of transmission of cytomegalovirus from mother to preterm infant by breastfeeding. Lancet 357:513–518
85. Eisenfeld L, Silver H, McLaughlin J, Klevjer-Anderson P, Mayo D, Anderson J, Herson V, Krause P, Savidakis J, Lazar A, Rosenkrantz T, Pisciotto P (1992) Prevention of transfusion-associated cytomegalovirus infection in neonatal patients by the removal of white cells from blood. Transfusion 32:205–209
86. Gilbert GL, Hayes K, Hudson IL, James J (1989) Prevention of transfusion-acquired cytomegalovirus infection in infants by blood filtration to remove leucocytes. Neonatal Cytomegalovirus Infection Study Group. Lancet 1:1228–1231
87. Chou SW (1986) Acquisition of donor strains of cytomegalovirus by renal-transplant recipients. N Engl J Med 314:1418–1423
88. Drew WL, Sweet ES, Miner RC, Mocarski ES (1984) Multiple infections by cytomegalovirus in patients with acquired immunodeficiency syndrome: documentation by southern blot hybridization. J Infect Dis 150:952–953
89. Grundy JE, Lui SF, Super M, Berry NJ, Sweny P, Fernando ON, Moorhead J, Griffiths PD (1988) Symptomatic cytomegalovirus infection in seropositive kidney recipients: reinfection with donor virus rather than reactivation of recipient virus. Lancet 2:132–135
90. Snydman DR, Werner BG, Heinze-Lacey B, Berardi VP, Tilney NL, Kirkman RL, Milford EL, Cho SI, Bush HL Jr, Levey AS, Strom TB,

Carpenter CB, Levey RH, Harmon WE, Zimmerman CE II, Shapiro ME, Steinman T, LoGerfo F, Idelson B, Schröter GPJ, Levin MJ, McIver J, Leszczynski J, Grady GF (1987) Use of cytomegalovirus immune globulin to prevent cytomegalovirus disease in renal-transplant recipients. N Engl J Med 317:1049–1054

91. Walter EA, Greenberg PD, Gilbert MJ, Finch RJ, Watanabe KS, Thomas ED, Riddell SR (1995) Reconstitution of cellular immunity against cytomegalovirus in recipients of allogeneic bone marrow by transfer of T-cell clones from the donor. N Engl J Med 333:1038–1044
92. Plotkin SA, Smiley ML, Friedman HM, Starr SE, Fleisher GR, Wlodaver C, Dafoe DC, Friedman AD, Grossman RA, Barker CF (1984) Towne-vaccine-induced prevention of cytomegalovirus disease after renal transplants. Lancet 1:528–530
93. Erice A, Gil-Roda C, Perez JL, Balfour HH Jr, Sannerud KJ, Hanson MN, Boivin G, Chou S (1997) Antiviral susceptibilities and analysis of UL97 and DNA polymerase sequences of clinical cytomegalovirus isolates from immunocompromised patients. J Infect Dis 175:1087–1092
94. Legendre CM, Norman DJ, Keating MR, Maclaine GD, Grant DM (2000) Valaciclovir prophylaxis of cytomegalovirus infection and disease in renal transplantation: an economic evaluation. Transplantation 70:1463–1468
95. Mauskopf JA, Richter A, Annemans L, Maclaine G (2000) Cost-effectiveness model of cytomegalovirus management strategies in renal transplantation. Comparing valaciclovir prophylaxis with current practice. Pharmacoeconomics 18:239–251
96. Boeckh M, Geballe AP (2011) Cytomegalovirus: pathogen, paradigm, and puzzle. J Clin Invest 121:1673–1680
97. Kowalsky S, Arnon R, Posada R (2013) Prevention of cytomegalovirus following solid organ transplantation: a literature review. Pediatr Transplant 17:499–509
98. Eid AJ, Razonable RR (2010) New developments in the management of cytomegalovirus infection after solid organ transplantation. Drugs 70:965–981
99. Ljungman P, Hakki M, Boeckh M (2011) Cytomegalovirus in hematopoietic stem cell transplant recipients. Hematol Oncol Clin North Am 25:151–169
100. Boeckh M, Bowden R (1995) Cytomegalovirus infection in marrow transplantation. Cancer Treat Res 76:97–136
101. Boeckh M, Leisenring W, Riddell SR, Bowden RA, Huang ML, Myerson D, Stevens-Ayers T, Flowers ME, Cunningham T, Corey L (2003) Late cytomegalovirus disease and mortality in recipients of allogeneic hematopoietic stem cell transplants: importance of viral load and T-cell immunity. Blood 101:407–414
102. Boeckh M, Nichols WG, Papanicolaou G, Rubin R, Wingard JR, Zaia J (2003) Cytomegalovirus in hematopoietic stem cell transplant recipients: current status, known challenges, and future strategies. Biol Blood Marrow Transplant 9:543–558
103. Ljungman P (1996) Cytomegalovirus infections in transplant patients. Scand J Infect Dis Suppl 100:59–63
104. Hebart H, Kanz L, Jahn G, Einsele H (1998) Management of cytomegalovirus infection after solid-organ or stem-cell transplantation. Current guidelines and future prospects. Drugs 55:59–72
105. Prentice HG, Kho P (1997) Clinical strategies for the management of cytomegalovirus infection and disease in allogeneic bone marrow transplant. Bone Marrow Transplant 19:135–142
106. Akalin E, Sehgal V, Ames S, Hossain S, Daly L, Barbara M, Bromberg JS (2003) Cytomegalovirus disease in high-risk transplant recipients despite ganciclovir or valganciclovir prophylaxis. Am J Transplant 3:731–735
107. Limaye AP, Bakthavatsalam R, Kim HW, Kuhr CS, Halldorson JB, Healey PJ, Boeckh M (2004) Late-onset cytomegalovirus disease in liver transplant recipients despite antiviral prophylaxis. Transplantation 78:1390–1396
108. Razonable RR, Rivero A, Rodriguez A, Wilson J, Daniels J, Jenkins G, Larson T, Hellinger WC, Spivey JR, Paya CV (2001) Allograft rejection predicts the occurrence of late-onset cytomegalovirus (CMV) disease among CMV-mismatched solid organ transplant patients receiving prophylaxis with oral ganciclovir. J Infect Dis 184:1461–1464
109. Barry PA, Pratt-Lowe E, Peterlin BM, Luciw PA (1990) Cytomegalovirus activates transcription directed by the long terminal repeat of human immunodeficiency virus type 1. J Virol 64:2932–2940
110. McKeating JA, Griffiths PD, Weiss RA (1990) HIV susceptibility conferred to human fibroblasts by cytomegalovirus-induced Fc receptor. Nature 343:659–661
111. Griffiths PD (2006) CMV as a cofactor enhancing progression of AIDS. J Clin Virol 35:489–492
112. Ostrowski MA, Krakauer DC, Li Y, Justement SJ, Learn G, Ehler LA, Stanley SK, Nowak M, Fauci AS (1998) Effect of immune activation

on the dynamics of human immunodeficiency virus replication and on the distribution of viral quasispecies. J Virol 72:7772–7784

113. Gallant JE, Moore RD, Richman DD, Keruly J, Chaisson RE (1992) Incidence and natural history of cytomegalovirus disease in patients with advanced human immunodeficiency virus disease treated with zidovudine. The Zidovudine Epidemiology Study Group. J Infect Dis 166:1223–1227

114. Cheung TW, Teich SA (1999) Cytomegalovirus infection in patients with HIV infection. Mt Sinai J Med 66: 113–124

115. Deayton J, Mocroft A, Wilson P, Emery VC, Johnson MA, Griffiths PD (1999) Loss of cytomegalovirus (CMV) viraemia following highly active antiretroviral therapy in the absence of specific anti-CMV therapy. AIDS 13:1203–1206

116. O'Sullivan CE, Drew WL, McMullen DJ, Miner R, Lee JY, Kaslow RA, Lazar JG, Saag MS (1999) Decrease of cytomegalovirus replication in human immunodeficiency virus infected-patients after treatment with highly active antiretroviral therapy. J Infect Dis 180: 847–849

117. Jacobson MA, Schrier R, McCune JM, Torriani FJ, Holland GN, O'Donnell JJ, Freeman WR, Bredt BM (2001) Cytomegalovirus (CMV)-specific CD4+ T lymphocyte immune function in long-term survivors of AIDS-related CMV end-organ disease who are receiving potent antiretroviral therapy. J Infect Dis 183:1399–1404

118. Spector SA, Hsia K, Crager M, Pilcher M, Cabral S, Stempien MJ (1999) Cytomegalovirus (CMV) DNA load is an independent predictor of CMV disease and survival in advanced AIDS. J Virol 73:7027–7030

119. Deayton JR, Sabin CA, Johnson MA, Emery VC, Wilson P, Griffiths PD (2004) Importance of cytomegalovirus viraemia in risk of disease progression and death in HIV-infected patients receiving highly active antiretroviral therapy. Lancet 363:2116–2121

120. Dock JN, Effros RB (2011) Role of CD8 T cell replicative senescence in human aging and in HIV-mediated immunosenescence. Aging Dis 2:382–397

# Chapter 3

# Distinct Properties of Human Cytomegalovirus Strains and the Appropriate Choice of Strains for Particular Studies

**Giada Frascaroli and Christian Sinzger**

## Abstract

Human cytomegalovirus is routinely isolated by inoculating fibroblast cultures with clinical specimens suspected of harboring HCMV and then monitoring the cultures for cytopathic effects characteristic of this virus. Initially, such clinical isolates are usually strictly cell associated, but continued propagation in cell culture increases the capacity of an HCMV isolate to release cell-free infectious progeny. Once cell-free infection is possible, genetically homogenous virus strains can be purified by limiting dilution infections. HCMV strains can differ greatly with regard to the titers that can be achieved, the tropism for certain cell types, and the degree to which nonessential genes have been lost during propagation. As there is no ideal HCMV strain for all purposes, the choice of the most appropriate strain depends on the requirements of the particular experiment or project. In this chapter, we provide information that can serve as a basis for deciding which strain may be the most appropriate for a given experiment.

**Key words** Cytomegalovirus, Isolate, Strain, Tropism, Titer

## 1 Introduction

In the 50 years since the discovery of the original HCMV strain (AD169) [19], numerous different strains have been propagated and used for HCMV research in various cell culture systems. With increasing knowledge about phenotypic and genotypic differences between these strains, it has become clear that the choice of the virus strains used for experiments will greatly influence the results of a certain project. It may be particularly relevant to decide whether recent clinical HCMV isolates or established laboratory strains should be used, which is the most suitable strain for a given cell type, or whether a BAC-cloned HCMV strain that can be easily modified by genetic manipulation will be advantageous. This chapter is designed to facilitate this decision-making process by describing the distinct features of the various HCMV strains that are available and discussing their advantages and disadvantages for certain experimental approaches.

Andrew D. Yurochko and William E. Miller (eds.), *Human Cytomegaloviruses: Methods and Protocols*, Methods in Molecular Biology, vol. 1119, DOI 10.1007/978-1-62703-788-4_3, © Springer Science+Business Media New York 2014

## 2 Materials

1. Plastic ware: sterile 1.0 ml cryovials, sterile 1.5 ml microtubes, sterile 15 ml conical tubes, 96-well flat bottom cell culture plates, 6-well cell culture plates, 25 $cm^2$ tissue culture flasks, sterile syringe filters (pore size 0.45 μm).
2. Solutions and culture media: cell culture grade $H_2O$, Dulbecco's Modified Eagle Medium with 5 % fetal bovine serum, 2.4 mmol/l glutamine and 50 mg/l gentamicin (DMEM5), Minimum Essential Medium with 5 % fetal bovine serum, 2.4 mmol/l glutamine and 50 mg/l gentamicin (MEM5), 2× MEM with 10 % fetal bovine serum, 4.8 mmol/l glutamine and 100 mg/l gentamicin, RPMI 1640 medium containing 10 % human serum (CMV seronegative), 50 mg/l gentamicin, 5 IU/ml heparin, and 50 μg/ml endothelial cell growth supplement (RPMI10), Trypsin/EDTA (with 0.05 % trypsin).
3. Other chemicals: cell culture grade agarose, mitomycin C.
4. Cells: human foreskin fibroblasts (HFF), human umbilical vein endothelial cells (HUVEC).

## 3 Methods

### 3.1 HCMV Strains

#### 3.1.1 Clinical Isolates

HCMV can be isolated from various specimens of patients with acute infection, including urine, throat washings, bronchoalveolar lavage, leukocytes, and biopsies from affected organs. This is a routine procedure in most virological laboratories, and human fibroblasts are the cell of choice for these isolations. Virus recovered from a human specimen and passaged in culture is regarded as a clinical isolate until phenotypic alteration occurs (e.g., release of cell-free progeny is an indicator of adaptation to cell culture; *see* Subheading 3.1.2).

*Isolation of HCMV from Clinical Specimens (e.g., urine)*

1. Dilute 2 ml of the urine sample with 2 ml of DMEM5 in a 15 ml conical tube (*see* **Note 1**).
2. Centrifuge for 10 min at 3,000 × *g*.
3. Filter supernatant through a 0.45 μm filter to remove bacteria and fungi.
4. Inoculate multiple subconfluent HFF cultures (e.g., duplicates or triplicates) with the filtered supernatant (e.g., 100 μl per 15,000 cells in a 96-well microculture plate).
5. Centrifuge plates at 300 × *g* for 30 min.
6. Incubate for additional 30 min at 37 °C with 5 % $CO_2$ in an incubator.

7. Remove inoculum and wash cells with 200 μl of DMEM5.
8. Add 200 μl of DMEM5 and incubate cells at 37 °C with 5 % $CO_2$ in an incubator until a focal cytopathic effect becomes detectable under the inverted light microscope.
9. Detach cells of infected cultures with Trypsin/EDTA and coculture with uninfected fibroblasts for passaging of the isolate.

Advantages of clinical isolates are that (1) they are phenotypically and genotypically closely related to the viruses replicating within patients including maintenance of wild-type cell tropism, (2) they are well suitable for correlative studies between viral genetic polymorphisms and clinical symptoms, and (3) they may serve as sources for the identification of new viral genes or gene variants lost in laboratory strains due to the selective pressure in cell culture. (4) Additionally, clinical isolates are an ideal material for analyses of natural genetic variability present in HCMV.

On the other hand, it may be disadvantageous that isolates are not plaque purified, hence are not genetically pure [17] and do not usually yield high-titer cell-free viral stocks. Low passage clinical isolates usually grow strictly cell associated with almost no infectivity released into the supernatant [26, 33]. This means that the virus has to be propagated by coculture of infected and uninfected cells, and synchronous infection of cultured cells with cell-free virus preparations is not possible. Thus, these low passage clinical isolates may not be useful for downstream techniques that require synchronous infection such as analysis of gene expression kinetics in cell lysates by immunoblotting. To obtain cell-free infectivity from recent clinical isolates, release of "intracellular virus" from infected cells by ultrasonication, freeze–thaw lysis, or mechanical disruption has been suggested. However, reasonable virus titers can only be achieved by these methods when the isolate is capable of releasing at least small amounts of infectious virus progeny into the supernatant (unpublished observation).

A further drawback is that clinical isolates are genetically unstable due to the selective pressure for replicative fitness under certain cell culture conditions. Viral genes necessary for interaction with the host immune system or for replication in specialized cell types and tissues may be lost if they are not beneficial in the cell type used during isolation. For example, clinical isolates rapidly lose the replication inhibitor RL13 during propagation in various cell types. Isolates that have lost RL13 acquire the increased capacity to release infectious progeny into the supernatant (*see* **Note 2**) [6].

#### *3.1.2 Established Laboratory Strains with Restricted Tropism*

It is commonly assumed that once an HCMV isolate is fully adapted for growth in fibroblast cell culture and there is no further selective pressure against any of the viral genes, the genome will remain relatively stable during continued propagation. This is supposedly the case when—after an initial increment of progeny virus titers during cell-free passages in fibroblasts—no further increase occurs. At this point, a genetically homogenous HCMV strain may be generated by plaque purification, which relies on the assumption that a single viral genome can cause the formation of a focus of infected cells or "plaque" in an otherwise uninfected monolayer. To increase the chance of picking a single genome and hence the chance for selection of a homogenous strain, the procedure is repeated three times, before the resulting strain is considered "purified" (*see* **Note 3**). A number of highly productive HCMV strains now widely used as standard strains by many researchers were established in this way. The most frequently used of these "laboratory strains" are AD169 and Towne [16, 19]. Other laboratory strains like Toledo or TR [14, 18, 27] have also been proven to be very useful for HCMV research.

*Plaque Purification*

1. Seed HFFs in a 6-well plate at a cell density of 150,000 cells per well in MEM5.
2. The next day, add serial tenfold dilutions of virus to the cell cultures (e.g., $10^{-3}$ to $10^{-8}$ fold dilutions of virus in medium) and incubate for 1 h.
3. Prepare 0.6 % agarose solution by adding 0.6 g agarose to 100 ml $H_2O$. Boil the solution in a microwave oven and be sure to account for evaporative loss of fluid. Let cool to 60 °C and add 100 ml of twofold concentrated cell culture medium (e.g., 2× MEM with 10 % fetal bovine serum, 4.8 mmol/l glutamine and 100 mg/l gentamicin). Let the agarose-medium solution cool to 37 °C and maintain at this temperature in a water bath.
4. When the incubation with virus from **step 1** is finished, remove supernatants and replace with 2 ml of the agarose medium. Incubate for 5–12 days or until plaques are easily visible. The amount of time required for this incubation is variable depending on the growth properties of the virus being used.
5. The day before harvesting plaques, seed HFFs in 25 cm² tissue culture flasks at a density of 300,000 cells per flask.
6. Under a phase contrast microscope, identify plaques of infected cells, scratch them with a pipette tip, and aspirate 20–30 μl. Transfer aspirate immediately to 1 ml of MEM5 and add to HFFs in 25 cm² flasks.
7. Incubate for 7–14 days until plaques have formed in the inoculated culture and passage infected cultures by detaching cells with Trypsin/EDTA solution and reseeding them in the appropriate medium.

8. When >90 % of HFFs show a late stage cytopathic effect (CPE) (*see* **Note 4**), wait two additional days, harvest plaque-purified virus by clarifying the supernatants from cell debris. Store aliquots of virus at −80 °C until further use.
9. Repeat **steps 1–8** twice to increase the chance of yielding a genetically pure "clonal" virus.

Besides highly efficient growth yielding high titers of viral progeny, another important advantage of these strains is that they have been widely distributed allowing one to compare experimental results with published data obtained under similar conditions. For example, in the case of the AD169 strain, many viral genes have been cloned from this strain and subsequently analyzed in isolation, yielding an enormous amount of information useful for various aspects of molecular HCMV research. It is also advantageous that the genomic sequences are available for many of these established strains allowing for genotype–phenotype comparison, cloning of viral genes, etc. One caveat to consider is that derivatives of established strains may exist, such as the genotypically and phenotypically distinct variants varATCC, varUK, and varUC of strain AD169 and varS and varL of strain Towne [5].

An obvious disadvantage of these well-established strains is the loss of genomic material due to extensive adaptation for growth in fibroblasts, which almost inevitably results in the potential loss of important functions not required for optimal growth in fibroblasts. HCMV strains such as AD169 and Towne that have been extensively propagated in fibroblasts have a restricted host cell range, i.e., besides fibroblasts these strains can infect smooth muscle cells, hepatocytes, and trophoblast cells but they can only inefficiently infect endothelial cells, epithelial cells, macrophages, and dendritic cells. This phenotypic change occurs after about 20 passages and is due to functional disruption of the UL128 gene locus, i.e., at least one of the three proteins pUL128, pUL130, and pUL131A is deleted or expressed in a nonfunctional form [3, 6, 7, 10, 26, 30]. As a consequence, strains AD169 and Towne are inadequate for infection of endothelial cells, epithelial cells, and cells derived from peripheral blood monocytes.

#### *3.1.3 Established Laboratory Strains with Extended Tropism*

The host cell restriction of HCMV strains observed during extensive passaging in fibroblasts can be prevented by propagation in endothelial cell cultures, which have the capability to preserve the natural broad cell tropism of HCMV [30]. Although there is considerable interstrain variation, almost all clinical HCMV isolates have an extended cell tropism represented by their ability to form foci both in fibroblast and in endothelial cell cultures [26]. During further propagation in fibroblasts, the strict cell association is usually lost around passage number 10, presumably due to disruption of the replication inhibitor RL13. Disruption of RL13 allows the resulting virus strains to infect a broad range of cell types in a cell-free mode for a limited number of passages until the abovementioned disruption of the UL128 locus occurs, restricting the host cell range [6, 26]. A convenient way to preserve the extended cell tropism of newly isolated virus strains is to "transfer" the virus to endothelial cell cultures shortly after isolation.

Transfer of cell-associated HCMV isolates from fibroblasts to endothelial cells

1. Propagate the virus isolate in HFFs until 10 % of the cells exhibit a CPE (*see* **Note 4**).
2. To induce terminal differentiation and abolish mitotic activity, treat fibroblast cultures with $2 \times 10^{-6}$ mol/mitomycin C for 48 h. Wash cultures and incubate in medium without mitomycin C for 48 h. Repeat treatment with mitomycin C (*see* **Note 5**).
3. Coculture aliquots of the terminally differentiated isolate-infected HFF cultures with HUVECs at an HFF/HUVEC ratio of 1/10 in RPMI10. Propagate cultures until the desired CPE is reached.

For an alternative approach, in case the isolate releases cell-free infectivity, *see* **Note 6**.

Typical strains that have been adapted to endothelial cell cultures shortly after primary isolation and thus have preserved the extended tropism are VHL/E [30] and TB40/E [26]. Both have the additional advantage that they produce high titers (*see* **Note 7**), in contrast to other endotheliotropic HCMV strains like VR1814 (Figs. 1 and 2). The extended tropism also includes high infectivity for epithelial cells and monocyte-derived macrophages and dendritic cells [23].

#### *3.1.4 BAC-Cloned Strains*

To facilitate mutation of viral genes in the context of replicating virus, HCMV genomes of strains AD169, VR1814, TR, PH, Toledo, and TB40/E have been cloned into F1 plasmid vectors that allow for amplification, mutation, and selection of the viral DNA as bacterial artificial chromosomes (BACs) in bacteria [2, 4, 9, 14, 24]. Briefly, in a first step, HCMV genomes replicating in infected fibroblasts are recombined with a cassette containing the bacterial origin of replication, a resistance gene for selection in fibroblasts and a resistance gene for selection in bacteria. In a second step, the selected recombined viral genomes are purified and transferred into bacteria, where they can then be amplified and mutated. In a third step, the mutated replication-competent viral genomes are transfected into fibroblasts, and successfully transfected fibroblasts will then reconstitute the BAC-cloned virus. More detailed protocols regarding isolation and genetic manipulation of viral BACs can be found in other chapters in this volume. As the size of the HCMV genome is already at the packaging limit, insertion of additional DNA, such as the BAC cassette, will reduce the replicative fitness of

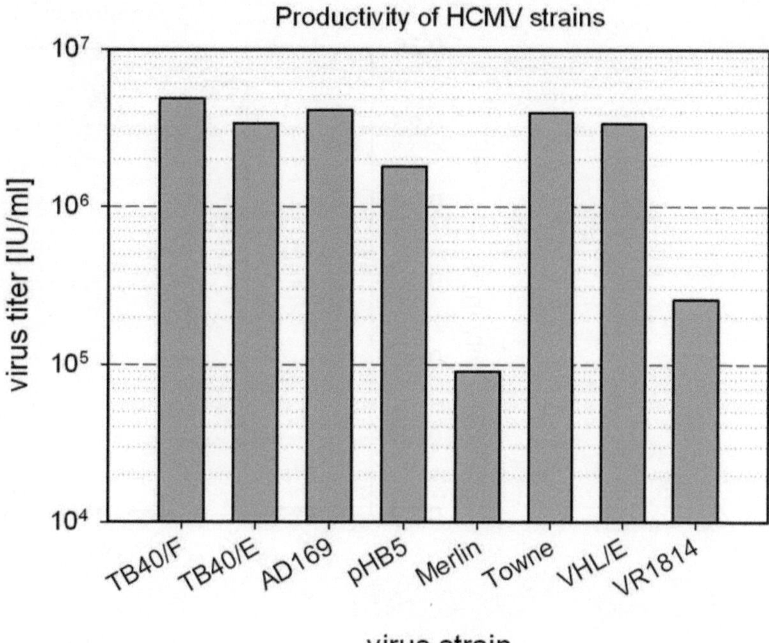


**Fig. 1** HCMV strains differ with regard to their productivity. Fibroblast cultures were infected with the respective strain and washed extensively on day 3 after infection to remove residual input virus. Progeny virus was collected on day 7 after infection and stored at −80 °C until determination of virus titers. Titers of infectivity were determined by infection of fibroblast cultures with dilution series of the various virus preparations and detection of viral IE antigens by indirect immunofluorescence 24 h after infection. The result of a representative experiment is shown

the virus [34] unless viral genes are deleted. In BAC clones generated according to the method developed by Borst et al., the viral open reading frames US2–6, known to be nonessential for replication in cell culture, are replaced with the BAC cassette [4]. The resulting genomes are still about 4.8 kb larger than the wild-type genomes, and this may be one reason for the slightly decreased titers obtained by BAC-derived virus when compared to wild-type virus (compare AD169 and pHB5 in Fig. 1). In BAC clones of strain Towne, generated with the method by Marchini et al., IRS1 and US12 are truncated and US1 to US11 are deleted [12].

Attempts have been made to generate HCMV-BAC clones without deletion of viral sequences [34]. In this approach, a BAC cloning cassette flanked by LoxP sites was introduced between the US28 and US29 open reading frames of HCMV by homologous recombination in fibroblasts, and the successfully recombined genomes were then transferred into *Escherichia coli* where mutagenesis techniques can be applied. If desired, the BAC cassette can be excised during reconstitution of virus in fibroblasts by

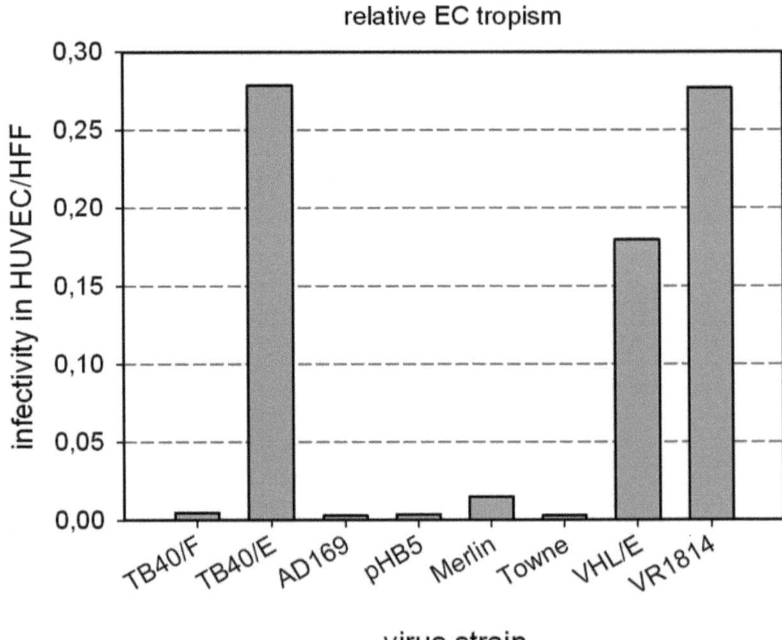


**Fig. 2** HCMV strains differ with regard to their endothelial cell (EC) tropism. Fibroblast cultures were infected with the respective strain and washed extensively on day 3 after infection to remove residual input virus. Progeny virus was collected on day 7 after infection and stored at −80 °C until determination of virus titers. The relative EC tropism of the various virus preparations was determined by simultaneous infection of human umbilical vein endothelial cell (HUVEC) cultures and fibroblast (HFF) cultures with dilution series of the various virus preparations and detection of viral IE antigens by indirect immunofluorescence 24 h after infection. The relative EC tropism was calculated as the infection efficiency in HUVECs/HFFs. The result of a representative experiment is shown

cotransfection of Cre recombinase expression plasmid, resulting in a mostly intact HCMV carrying only the desired mutations. The genome of the reconstituted virus can however not be retransferred into bacteria as it no longer contains the F1-plasmid sequences. A further caveat comes from the possibility that introduction of additional sequences during the initial homologous recombination step in fibroblasts may lead to undesired deletions as a compensation of the increased genome size.

When working with HCMV strains reconstituted from BACs, it is essential to consider which genes are present or missing due to the respective BAC cloning procedure or due to strain-specific polymorphisms in the genetic backbone (*see* Table 1).

#### 3.1.5 Genetically Repaired BAC-Derived Strains

While many cell culture-adapted HCMV strains are available as BAC clones, BAC-cloned clinical isolates (*see* **Note 2**) have not been reported. Obviously, the first step of the BAC cloning procedure, i.e., recombination of viral genomes with the BAC cassette in

**Table 1**
**Properties of selected HCMV strains and their BAC clones**

| Strain | Productivity | Tropism | Sequence | BAC clone | Sequence | Genes known to be disrupted/deleted | References |
|---|---|---|---|---|---|---|---|
| AD169varUC | High | Restricted | FJ527563.1 | | | RL5A, RL13<br>UL131A<br>UL140–144 | [5] |
| AD169varUK | High | Restricted | X17403.1<br>BK000394 | | | RL5A, RL13<br>(UL42–43)<br>UL131A<br>UL133–150 | [5] |
| AD169 | High | Restricted | | pHB5 | | BAC: US2–6 | [4] |
| AD169varATCC | High | Restricted | | BADwt | AC146999 | RL5A, RL13<br>(UL42–43)<br>UL131A<br>UL133–150 | [5, 34] |
| Towne | High | Restricted | FJ616285.1 | $\text{Towne}_{\text{BAC}}$ | AY315197.2 | UL133–146<br>BAC: IRS1; US1–12 | [8, 12] |
| TB40/F | High | Restricted | – | No | – | n.a. | |
| TB40/E | High | Extended | – | TB40-BAC4 | EF999921.1 | RL6, RL13, (UL141)<br>BAC: IRS1, US1–6 | [24] |
| VHL/E | High | Extended | – | No | – | n.a. | |
| VR1814 | Low | Extended | GU179289.1 | FIX-BAC | AC146907.1 | BAC: IRS1, US1–6 | [9] |
| Merlin | Low | Restricted | AY446894.2<br>NC_006273.2 | pAL1128 | GU179001 | RL13, UL128 | [6, 28] |

infected fibroblasts, will put a strong selective pressure against cell-associated genomes with intact RL13, thus promoting selection of BAC clones with mutations in RL13, which can spread more efficiently due to the capacity to release higher titers of infectious virus.

To overcome this limitation, RL13 has been repaired in an existing BAC clone of the HCMV strain Merlin originally derived from the urine of a congenitally infected infant. The rapid appearance of disabling RL13 mutations already during the reconstitution of repaired genomes in fibroblasts emphasized how strong is the abovementioned selective pressure against RL13 [28]. Only when expression of the intact RL13 gene was conditionally repressed, it was possible to reconstitute BAC-cloned viruses with wild-type RL13 sequences [28]. This new strain combines many of the advantages of BAC-cloned HCMV strains mentioned before such as genetic stability and purity with an isolate-like phenotype, i.e., strictly cell-associated growth (*see* **Note 2**). Moreover, it allows investigators to test targeted mutations in short-term infection experiments in a system where the reconstituted Merlin viruses largely represent the genetic equivalents of an HCMV clinical isolate. The key here is the short term as after a few rounds of replication, RL13 will again be disrupted. One drawback of this system is that the virus can only be maintained in an immortalized cell line expressing the Tet repressor and high titers of virus progeny have not yet been obtained.

### 3.2 Choice of Strains for Particular Projects

As described in the previous section, many different HCMV strains are available, with quite distinctive features. There is no such thing as the ideal strain serving all purposes. The appropriate choice of an HCMV strain depends on the specific experimental design and requirements. The following paragraphs are meant as a guide for decision making in the planning phase of a project.

#### 3.2.1 Experiments Addressing Genetic Variability Between HCMV Isolates

Although herpesviral genomes are regarded as rather stable when compared to other viruses, genetic polymorphisms have been reported for many HCMV genes when different laboratory strains or isolates from different patients have been compared [14]. For example, surface glycoproteins gN and gO show a high degree of polymorphism, probably reflecting the selective pressure of neutralizing antibodies [13–15].

It is important to employ clinical HCMV strains for experiments in which naturally occurring polymorphisms within HCMV genes are to be considered. Examples are investigations into (1) strain specificity of neutralizing antibodies, (2) strain susceptibility to newly developed antiviral agents, and (3) strain evolution in patients that have been unsuccessfully treated with experimental or prescribed pharmaceuticals and in which the emergence of drug-resistant strains is suspected.

The drawback of using newly generated clinical isolates is that these viruses remain cell associated and are not useful when infections with defined infection multiplicities using virus-containing supernatants are required. Protocols for focus expansion assays and plaque reduction assays have been developed to overcome this limitation and to allow a certain degree of standardization when using clinical isolates.

Alternatively, the use of established cell culture-adapted HCMV strains can reveal the influence of genetic polymorphisms on gene function, as indicated by the interstrain comparison of HCMV glycoproteins [24].

#### 3.2.2 Experiments Depending on High Infection Multiplicities

High virus titers are a prerequisite for all approaches where synchronous infection of a high percentage (>95 %) of the cells is desired. For example, HCMV-induced downregulation of cellular factors may go undetected if too many uninfected cells are present in the sample. On the other hand, upregulation of cellular factors in an infected culture can be due to modifications in uninfected bystander cells, and again infection rates approaching 100 % are desired to exclude this possibility and attribute the observed effect directly to infected cells.

The fraction of infected cells ($F_{inf}$) depends on the infection multiplicity (MOI, *see* **Note 8**) and is calculated according to the principles of Poisson distribution: $F_{inf} = 1 - (e^{-MOI})$. This means that about 63 % of cells are infected with an MOI of 1 infectious unit (IU)/cell and about 99 % of cells are infected with an MOI of 5 IU/cell. High MOIs are desired for analyses of viral attachment and entry, especially when the number of virions per infected cell needs to be counted after infection with fluorescent-labeled virus or after immunofluorescence staining of virus particles. High MOIs are also desired when virion proteins or nucleic acids from input particles need to be detected by blotting techniques in infected cell lysates. Even more input virus is necessary for electron microscopic analyses of entry events: to detect only one virion per cell in an ultrathin section with conventional transmission EM, virus loads of >100 particles per cell are required.

Cell culture-adapted HCMV strains like AD169, Towne, VHL/F, and TB40/F are particularly suitable for these purposes. Titers of $>10^6$ IU/ml can usually be achieved, allowing for experiments with MOIs of >10 with native virus preparations. Higher virus titers ($>10^7$ IU/ml) can be achieved by ultracentrifugation of large volumes of native virus preparations and resuspension of the virion pellet in small volumes of medium (*see also* **Note 9** on the problem of particle/infectivity ratios and **Note 10** on avoidance of dense bodies in virus preparations). Such enriched virus suspensions are often necessary for ultrastructural analyses of virus entry. If the cell type of interest cannot be infected by fibroblast-adapted strains, EC-adapted strains like VHL/E and TB40/E may be useful as they combine broad cellular tropism with relatively high titer.

*3.2.3 Infection of Endothelial Cells, Epithelial Cells, or Professional Antigen Presenting Cells*

For analyses of HCMV infection of endothelial cells, epithelial cells or professional antigen presenting cells, several options exist. Newly isolated clinical strains are a viable option as they have a naturally broad cellular tropism and are usually able to infect endothelial, epithelial, and myeloid cells. However, their strict cell-associated growth phenotype prevents the generation of virus stocks containing high titers of cell-free progeny virus [6, 26]. This greatly limits the repertoire of possible experiments as one is limited to approaches based on coculture of infected and uninfected cells. Alternatively, cell culture-adapted RL13–/UL128+ derivatives of newly isolated clinical strains can be used if synchronous infection by cell-free supernatants is required. Still, virus titers produced by such strains ($10^3$–$10^4$ IU/ml) may be too low for a highly efficient infection of such cell cultures [6]. However, at least two well-established highly endotheliotropic HCMV strains (TB40/E and VHL/E) are available that produce high viral titers in cell-free supernatants and allow for high infection rates (Fig. 2). Alternatively, genetically repaired variants of AD169 with a functional UL131A ORF are also available [1, 31].

In addition to the choice of the appropriate strain, the choice of the producer cell type for generation of virus stocks is important. Surprisingly, progeny virus produced by infected fibroblast cultures is more endotheliotropic than progeny virus produced by infected endothelial cells [21]. Therefore, to achieve maximum infection rates in endothelial cells, epithelial cells, and antigen presenting cells, virus stocks should be produced in fibroblasts. One must be careful to use fibroblasts only to amplify a high-titer stock as repeated passage of endotheliotropic viruses in fibroblasts can lead to loss of UL128–UL131A, as discussed previously.

*3.2.4 Genetic Manipulation of HCMV Genomes*

When a targeted mutation in viral genes in the context of replication-competent virus is desired, BAC-cloned viruses are the appropriate choice for most projects. Once the viral genome is available as a replicating bacterial artificial chromosome in *E. coli*, many tools for genetic engineering of plasmids can be applied in order to modify the viral genes contained in this vector. The tailored virus will then be recovered from fibroblasts that have been transfected with the mutated BACs. Importantly, the UL128 gene locus, which confers endothelial cell tropism, is usually not disrupted during reconstitution in fibroblasts. In principle, HCMV-BACs may also be reconstituted by transfection into cell types other than fibroblasts (e.g., epithelial cells) [28].

Many of the relevant HCMV strains are available as BAC-cloned viruses. They differ from each other regarding the viral genes that were replaced with the BAC cassette and they carry the genetic characteristics of their parental strain, which will influence the choice depending on the planned project. In order to have a valid wild-type control for a desired mutation, the chosen BAC clone should be intact regarding the gene of interest. Knock-in mutants restoring the gene of interest are an alternative strategy. In addition to the integrity of the gene of interest, tropism and titer production may also influence the choice of a certain strain.

#### *3.2.5 Virus Host–Defense Interactions*

Virus genes that counteract antiviral actions of CD8 T cells, CD4 T cells, NK cells, and antibodies are usually nonessential for growth in cell culture and may therefore be counterselected during propagation of the virus in cell culture. For example, the NK evasion genes UL141 and UL142 are deleted in the genome of AD169varATCC (genetic defects of various strains are listed in Table 1). This has to be considered when studying virus host–defense interactions: clinical isolates can be assumed to encode the full set of HCMV genes, whereas "nonessential" genes may have been lost in cell culture-adapted strains. None of the available HCMV strains can guarantee genetic completeness. On the other hand, a certain gene usually is not disrupted in all strains (*see* Table 1). Therefore, inclusion of several strains reduces the risk of missing a viral immune evasion gene in a project addressing yet undefined immune evasive functions of HCMV.

With regard to well-established immune evasive functions of HCMV (e.g., US2, 3, 6, and 11 for MHC class I mediated antigen presentation; UL16, 18, 40, 141 and 142 for NK cell function) [32], an appropriate strain can be chosen according to the available sequence information, in order to ensure that the viral genes of interest are intact.

#### *3.2.6 Cell-Associated Spread*

The fact that newly isolated clinical strains are able to grow in fibroblast cultures with almost no detectable virus in the culture supernatant suggests that cell-to-cell spread rather than cell-free transmission of virions is the default mode of viral growth in vivo. Recent clinical isolates are appropriate to study the underlying mechanisms. Infected cells have to be cocultured with uninfected indicator cells before the extent of cell-to-cell transmissions can be analyzed by quantitative measurement of focus size or number [25]. The limits of such an approach are mentioned in Subheading 3.1.1.

To analyze the effect of targeted mutations on cell-to-cell spread, cell-associated derivatives of BAC-cloned HCMV strains may under some circumstances serve as viable surrogates for clinical isolates. For example, deletion of UL99 or UL74 almost completely abrogates release of cell-free infectivity but still allows cell-associated focal spread of the virus [11, 22]. Such an experimental setup is advantageous in that the viral genome is known and can be further manipulated and that the cell-associated spread phenotype is apparently stable (i.e., escape mutants releasing cell-free infectivity have not been reported). The disadvantage of such as experimental setup is that the mechanisms of virus transfer may differ from clinical isolates in which the UL99 and/or UL74 genes are intact. Therefore, results from such studies have to be interpreted carefully.

The BAC-cloned version of the Merlin strain which conditionally represses RL13 in the presence of the Tet repressor expressed from stably transfected fibroblasts [28] (*see* Subheading 3.1.5) can help to overcome many limitations of clinical isolates, i.e., allow targeted mutation of genes of interest, ensure a high genetic stability, and allow for synchronous infection at the beginning of the experiment before derepressing the RL13 gene.

### 3.3 Conclusions

It has become evident that HCMV strains can differ genetically and phenotypically either due to natural interstrain polymorphisms or due to alterations occurring during extended propagation in cell culture. In particular, differences concern the titers that can be achieved, the tropism for certain cell types, and the degree to which nonessential genes have been lost during propagation. As there is no ideal HCMV strain for all purposes, the choice of the most appropriate strain depends on the requirements of the particular experiment or project. Recent clinical isolates are ideal for correlative studies between viral genetic polymorphisms and clinical symptoms and may serve as a source for the identification of new viral genes or gene variants. The well-established fibroblast-adapted HCMV strains combine the advantage of high-titer production with a plethora of reference data available in the literature, however at the cost of a restricted host cell tropism. For experiments in endothelial cells, epithelial cells, and leukocyte-derived cells, HCMV strains with unrestricted cell tropism are preferable. For genetic manipulation, BAC-cloned HCMV strains are the best choice as they can be easily engineered in *E. coli.* Many of the BAC-cloned viruses will however grow to lower titers than their parental strains and have genetic defects due to insertion of the BAC cassette into the viral genome.

## 4 Notes

1. For isolation of HCMV from throat washing specimens directly start with **step 2** (*see* Subheading 3.1.1) using 4 ml of the specimen.
2. "Clinical isolates" usually grow focally in a cell-associated manner and do not release virus progeny into the culture supernatant. As soon as the virus starts to release significant amounts of infectivity into the supernatant (usually within the first ten passages), the isolate becomes presumably adapted to growth in cell culture (including loss of RL13) and is therefore not "clinical isolate" in the true sense of the definition. This means that VHL/E, Toledo, TB40/E, VR1814, and similar viruses are not "clinical isolates," but cell culture-adapted "strains." To avoid confusion, we do not use the term "clinical strain" and we call a fresh patient isolate "clinical isolate" only as long as it grows cell associated.

3. As demonstrated for strain TB40/E, plaque purification is no guarantee for genetic homogeneity [7, 24, 29]. The best way to achieve genetic homogeneity is by cloning the viral genome into a BAC and reconstituting replicating virus by transfection of the cloned genome into a permissive cell culture.
4. When a cytopathic effect (CPE) is mentioned in the protocols of this chapter, the detection of nuclear inclusions in phase contrast microscopy is meant, as this will indicate that the cell culture is producing virus progeny. Besides this CPE indicative of the late phase of viral replication, HCMV can cause also other CPEs like cell rounding in the early phase of infection, syncytia formation in the late stage of infection, and cell destruction in the final stage of infection. While the time course and extent of the latter CPEs may vary depending on the virus strain, a nuclear inclusion will regularly appear the latest on day 4 after infection in each productively infected HFF cell.
5. To prevent that during the coculture of partially infected fibroblasts with uninfected endothelial cells, the latter ones are overgrown by the faster replicating fibroblasts, mitosis in the fibroblast culture can be irreversibly inhibited with mitomycin C.
6. Alternatively, the recent isolate can be transferred by cell-free transmission onto endothelial cell monolayers directly after the appearance of cell-free virus, given that sufficient virus progeny is released into the supernatant. An attempt to isolate HCMV directly in endothelial cell cultures is also possible.
7. When fibroblasts and endothelial cells are infected with the same preparation of an HCMV strain with extended tropism, the tropism of progeny virus differs to some degree depending on the producer cell type. Progeny from fibroblast cultures has a higher endothelial cell tropism than progeny from endothelial cells [21]. High-titer stocks of "EC-tropic" virus should therefore be produced in fibroblasts.
8. MOI in this sense means the average number of infectious particles that have bound per cell in a given experiment. By definition, this MOI can only be determined restrospectively, e.g., in the same cells infected in parallel with a dilution series of the same virus preparation. Often the MOI is estimated in advance as an extrapolation from previous experiments. This "estimated MOI" may differ from the "actual MOI" to some extent.
9. Particle/infectivity ratio is an important issue when a method basically detects the physical particle rather than the biologically active infectious unit. This concerns detection of virions under the electron microscope, fluorescent virus particles in live microscopy, immunofluorescence detection of capsids, and also the determination of DNA copies with quantitative DNA detection methods. Usually, not all particles are infectious.

Inevitably, a virus preparation will contain enveloped capsids lacking DNA or packed with damaged DNA. In addition, those particles that are infectious immediately after harvest have a biological half-life of about 1 day (unpublished data), which means that supernatant collected over a period of several days will contain a majority of inactive particles. Hence, the fraction of non-active particles can be limited by using supernatants produced over a short time period, e.g., over night after a medium exchange. Unfortunately, procedures like freezing the virus and even ultracentrifugation will also affect the biological activity of HCMV virions. This has to be taken into account, when interpreting the localization of particles in EM or the behavior of particles in fluorescence assays. If necessary, the particle/infectivity ratio can be determined by counting the number of capsids per cell and comparing this to the MOI.

A particular source of misunderstanding is the determination of a "particle/infectivity" ratio by comparison of the genome copies in a given supernatant with its ability to infect cell cultures. The genome copy number will reflect the total number of particles in the preparation, whereas only a small fraction of infectious particles in the inoculum is absorbed by a cell culture during the incubation. Such "particle/infectivity" ratio will reflect the absorption rate rather than the biological potency of the virions. Even if all virions were biologically active, only few will bind to and infect a cell, resulting in a very high "particle/infectivity" ratio.

10. If the presence of dense bodies needs to be avoided, virus preparations have to be gradient purified. However, the procedure of ultracentrifugation through a density gradient will increase the number of noninfectious particles in the purified virion fraction. Fresh preparations of pp65-deletion mutants are a good alternative in this situation as they do not produce dense bodies and can therefore be used directly without the need of gradient purification [20].

We apologize for not mentioning certain HCMV strains or not describing them in detail. We invite suggestions from colleagues who have experience with strains not commented here in order to improve future versions of this chapter.

## References

1. Adler B, Scrivano L, Ruzcics Z, Rupp B, Sinzger C, Koszinowski U (2006) Role of human cytomegalovirus UL131A in cell type-specific virus entry and release. J Gen Virol 87:2451–2460
2. Adler H, Messerle M, Koszinowski UH (2003) Cloning of herpesviral genomes as bacterial artificial chromosomes. Rev Med Virol 13: 111–121
3. Akter P, Cunningham C, McSharry BP, Dolan A, Addison C, Dargan DJ, Hassan-Walker AF, Emery VC, Griffiths PD, Wilkinson GW, Davison AJ (2003) Two novel spliced genes in human cytomegalovirus. J Gen Virol 84:1117–1122

4. Borst EM, Hahn G, Koszinowski UH, Messerle M (1999) Cloning of the human cytomegalovirus (HCMV) genome as an infectious bacterial artificial chromosome in Escherichia coli: a new approach for construction of HCMV mutants. J Virol 73:8320–8329
5. Bradley AJ, Lurain NS, Ghazal P, Trivedi U, Cunningham C, Baluchova K, Gatherer D, Wilkinson GW, Dargan DJ, Davison AJ (2009) High-throughput sequence analysis of variants of human cytomegalovirus strains Towne and AD169. J Gen Virol 90:2375–2380
6. Dargan DJ, Douglas E, Cunningham C, Jamieson F, Stanton RJ, Baluchova K, McSharry BP, Tomasec P, Emery VC, Percivalle E, Sarasini A, Gerna G, Wilkinson GW, Davison AJ (2010) Sequential mutations associated with adaptation of human cytomegalovirus to growth in cell culture. J Gen Virol 91:1535–1546
7. Dolan A, Cunningham C, Hector RD, Hassan-Walker AF, Lee L, Addison C, Dargan DJ, McGeoch DJ, Gatherer D, Emery VC, Griffiths PD, Sinzger C, McSharry BP, Wilkinson GW, Davison AJ (2004) Genetic content of wild-type human cytomegalovirus. J Gen Virol 85:1301–1312
8. Dunn W, Chou C, Li H, Hai R, Patterson D, Stolc V, Zhu H, Liu F (2003) Functional profiling of a human cytomegalovirus genome. Proc Natl Acad Sci U S A 100:14223–14228
9. Hahn G, Khan H, Baldanti F, Koszinowski UH, Revello MG, Gerna G (2002) The human cytomegalovirus ribonucleotide reductase homolog UL45 is dispensable for growth in endothelial cells, as determined by a BAC-cloned clinical isolate of human cytomegalovirus with preserved wild-type characteristics. J Virol 76:9551–9555
10. Hahn G, Revello MG, Patrone M, Percivalle E, Campanini G, Sarasini A, Wagner M, Gallina A, Milanesi G, Koszinowski U, Baldanti F, Gerna G (2004) Human cytomegalovirus UL131-128 genes are indispensable for virus growth in endothelial cells and virus transfer to leukocytes. J Virol 78:10023–10033
11. Jiang XJ, Sampaio KL, Ettischer N, Stierhof YD, Jahn G, Kropff B, Mach M, Sinzger C (2011) UL74 of human cytomegalovirus reduces the inhibitory effect of gH-specific and gB-specific antibodies. Arch Virol 156:2145–2155
12. Marchini A, Liu H, Zhu H (2001) Human cytomegalovirus with IE-2 (UL122) deleted fails to express early lytic genes. J Virol 75:1870–1878
13. Mattick C, Dewin D, Polley S, Sevilla-Reyes E, Pignatelli S, Rawlinson W, Wilkinson G, Dal Monte P, Gompels UA (2004) Linkage of human cytomegalovirus glycoprotein gO variant groups identified from worldwide clinical isolates with gN genotypes, implications for disease associations and evidence for N-terminal sites of positive selection. Virology 318:582–597
14. Murphy E, Yu D, Grimwood J, Schmutz J, Dickson M, Jarvis MA, Hahn G, Nelson JA, Myers RM, Shenk TE (2003) Coding potential of laboratory and clinical strains of human cytomegalovirus. Proc Natl Acad Sci U S A 100:14976–14981
15. Pignatelli S, Dal Monte P, Landini MP (2001) gpUL73 (gN) genomic variants of human cytomegalovirus isolates are clustered into four distinct genotypes. J Gen Virol 82:2777–2784
16. Plotkin SA, Furukawa T, Zygraich N, Huygelen C (1975) Candidate cytomegalovirus strain for human vaccination. Infect Immun 12:521–527
17. Prichard MN, Penfold ME, Duke GM, Spaete RR, Kemble GW (2001) A review of genetic differences between limited and extensively passaged human cytomegalovirus strains. Rev Med Virol 11:191–200
18. Quinnan GV Jr, Delery M, Rook AH, Frederick WR, Epstein JS, Manischewitz JF, Jackson L, Ramsey KM, Mittal K, Plotkin SA et al (1984) Comparative virulence and immunogenicity of the Towne strain and a nonattenuated strain of cytomegalovirus. Ann Intern Med 101:478–483
19. Rowe WP, Hartley JW, Waterman S, Turner HC, Huebner RJ (1956) Cytopathogenic agent resembling human salivary gland virus recovered from tissue cultures of human adenoids. Proc Soc Exp Biol Med 92:418–424
20. Schmolke S, Kern HF, Drescher P, Jahn G, Plachter B (1995) The dominant phosphoprotein pp 65 (UL83) of human cytomegalovirus is dispensable for growth in cell culture. J Virol 69:5959–5968
21. Scrivano L, Sinzger C, Nitschko H, Koszinowski UH, Adler B (2011) HCMV spread and cell tropism are determined by distinct virus populations. PLoS Pathog 7:e1001256
22. Silva MC, Yu QC, Enquist L, Shenk T (2003) Human cytomegalovirus UL99-encoded pp 28 is required for the cytoplasmic envelopment of tegument-associated capsids. J Virol 77:10594–10605
23. Sinzger C, Digel M, Jahn G (2008) Cytomegalovirus cell tropism. Curr Top Microbiol Immunol 325:63–83
24. Sinzger C, Hahn G, Digel M, Katona R, Sampaio KL, Messerle M, Hengel H,

Koszinowski U, Brune W, Adler B (2008) Cloning and sequencing of a highly productive, endotheliotropic virus strain derived from human cytomegalovirus TB40/E. J Gen Virol 89:359–368

25. Sinzger C, Knapp J, Plachter B, Schmidt K, Jahn G (1997) Quantification of replication of clinical cytomegalovirus isolates in cultured endothelial cells and fibroblasts by a focus expansion assay. J Virol Methods 63:103–112
26. Sinzger C, Schmidt K, Knapp J, Kahl M, Beck R, Waldman J, Hebart H, Einsele H, Jahn G (1999) Modification of human cytomegalovirus tropism through propagation in vitro is associated with changes in the viral genome. J Gen Virol 80(Pt 11):2867–2877
27. Smith IL, Taskintuna I, Rahhal FM, Powell HC, Ai E, Mueller AJ, Spector SA, Freeman WR (1998) Clinical failure of CMV retinitis with intravitreal cidofovir is associated with antiviral resistance. Arch Ophthalmol 116:178–185
28. Stanton RJ, Baluchova K, Dargan DJ, Cunningham C, Sheehy O, Seirafian S, McSharry BP, Neale ML, Davies JA, Tomasec P, Davison AJ, Wilkinson GW (2010) Reconstruction of the complete human cytomegalovirus genome in a BAC reveals RL13 to be a potent inhibitor of replication. J Clin Invest 120:3191–3208
29. Tomasec P, Wang EC, Davison AJ, Vojtesek B, Armstrong M, Griffin C, McSharry BP, Morris RJ, Llewellyn-Lacey S, Rickards C, Nomoto A, Sinzger C, Wilkinson GW (2005) Downregulation of natural killer cell-activating ligand CD155 by human cytomegalovirus UL141. Nat Immunol 6:181–188
30. Waldman WJ, Roberts WH, Davis DH, Williams MV, Sedmak DD, Stephens RE (1991) Preservation of natural endothelial cytopathogenicity of cytomegalovirus by propagation in endothelial cells. Arch Virol 117:143–164
31. Wang D, Shenk T (2005) Human cytomegalovirus UL131 open reading frame is required for epithelial cell tropism. J Virol 79: 10330–10338
32. Wilkinson GW, Tomasec P, Stanton RJ, Armstrong M, Prod'homme V, Aicheler R, McSharry BP, Rickards CR, Cochrane D, Llewellyn-Lacey S, Wang EC, Griffin CA, Davison AJ (2008) Modulation of natural killer cells by human cytomegalovirus. J Clin Virol 41:206–212
33. Yamane Y, Furukawa T, Plotkin SA (1983) Supernatant virus release as a differentiating marker between low passage and vaccine strains of human cytomegalovirus. Vaccine 1:23–25
34. Yu D, Smith GA, Enquist LW, Shenk T (2002) Construction of a self-excisable bacterial artificial chromosome containing the human cytomegalovirus genome and mutagenesis of the diploid TRL/IRL13 gene. J Virol 76: 2316–2328

# Chapter 4

# Use of Diploid Human Fibroblasts as a Model System to Culture, Grow, and Study Human Cytomegalovirus Infection

Elizabeth A. Fortunato

## Abstract

Primary human diploid fibroblasts are used routinely to study host/pathogen interactions of human cytomegalovirus (HCMV). Fibroblasts' ease of culture and tremendous permissiveness for infection allow the study of all facets of infection, an abbreviated list of which includes ligand/receptor interactions, activation of cell signaling responses, and dysregulation of the cell cycle and DNA repair processes. Another advantage to fibroblasts' permissiveness for HCMV is the capability to grow high titer stocks of virus in them. This chapter will discuss the production of viral stocks of HCMV in primary human fibroblasts, commencing with culturing and infection of cells and continuing through harvest, titration (determining the infectious capacity of a particular virus preparation), and storage of viral stocks for use in downstream experiments.

**Key words** Human cytomegalovirus, Culture of primary fibroblasts, Preparation and storage of virus stocks, Titrating viral stocks, Growth of virus in culture

## 1 Introduction

HCMV has a wide range of permissiveness in vivo [1]. Although fibroblasts may not be the first cells that come to mind as clinically relevant to the drastic ramifications seen during congenital infection or transplant rejection, they provide a useful tool for the study of a fully permissive infection in the context of a tissue culture environment. Several compelling reasons to use fibroblasts include the following: (1) they are quite easy to culture and thrive for many passages, (2) they synchronize easily, (3) they can be grown in large quantities with relative ease, and (4) they display all the characteristics of a fully permissive infection, producing the full range of viral antigens and (with laboratory-adapted virus strains) producing high titer stocks of cell-free virus, which is readily harvested from the supernatant of infected cells.

Andrew D. Yurochko and William E. Miller (eds.), *Human Cytomegaloviruses: Methods and Protocols*,
Methods in Molecular Biology, vol. 1119, DOI 10.1007/978-1-62703-788-4_4, © Springer Science+Business Media New York 2014

Fibroblasts are not identical and different sources can be used. In general, primary cells (referred to as normal diploid fibroblasts) are utilized in most experiments. Cells can be obtained from cell banks such as ATCC or from local hospital sources and can be derived either from adult or fetal/newborn tissue. Cells from many different areas of the body, including lung, kidney, and foreskin, have been successfully used for HCMV studies.

Fibroblasts have historically served as a "prototype" for lytic infection in vitro. An attractive characteristic of relatively high multiplicity of infection (MOI) HCMV infections in synchronized (via serum starvation or confluence arrest [2, 3]) fibroblasts is a "synchronous" infection (*see* **Note 7** for explanation of MOI). Following a high MOI infection, it is relatively easy to follow the entire lytic life cycle in these cells. Initial interactions begin with the receptor and continue through wholesale necrosis of the cells after 5–7 days (depending upon the virus strain used). In addition, fibroblasts have traditionally been the "vehicle" of choice for growing virus stocks in the lab. Due to tissue culture "adaptation" (as was described in Chapter 3) in fibroblasts, HCMV replicates to high levels and is shed in large quantities. Fibroblasts are also frequently used in virus titration (determining the infectious capacity of virions shed from infected cells). The latter is true even when clinical strains are utilized or if an experiment is performed in another cell type. This is due to the ease with which all HCMV strains infect fibroblasts and produce assayable plaques. All HCMV strains possess the proteins necessary for interaction with and entry into fibroblasts (as described in Chapter 8). In essence, virus titration in fibroblasts can "level the playing field" for different virus strains when assessing functional virion output.

It is noteworthy that this last characteristic of fibroblasts, namely, that all fibroblasts are permissive for all HCMV strains, permits quick propagation of large-scale virus preparations. As was noted in Chapter 3, long-term passage of clinical isolates within fibroblast is not advisable, since these viruses can "lose" the important ULb′ cassette/suite of viral genes important for infection and growth in vivo (and in certain distinct tissue types). Clinical strains are highly cell associated (when compared to the laboratory-adapted Towne and AD169 strains) and require different culturing and collection conditions in order to obtain high titer virus (*see* **Notes 4, 15** and **22** for brief discussion of these differences). Therefore, the bulk of this chapter focuses on growth of the laboratory-adapted strains that are utilized by the large majority of laboratories working on HCMV for the bulk of their experiments.

This chapter covers the steps needed to produce high titer viral stocks of HCMV for use in experiments. It will cover culturing fibroblasts (including the critical parameters important for maintaining healthy cells), the low multiplicity infection conditions necessary for the production of high titer stocks, harvesting viral supernatants, proper freezing and storage conditions, and the titer assay for calculating the number of infectious virions/ml of supernatant. Also discussed will be other options for measuring antigen positivity in viral cultures when circumstances preclude the use of plaque assays during an experiment.

## 2 Materials

All procedures described in this chapter will be carried out in a biological safety cabinet to maintain sterile conditions and biohazard abatement. HCMV is a biosafety level 2 pathogen and adequate precautions and disposal techniques must be utilized during all handling of the virus and virus-infected cultures. Unless they are presterilized by the manufacturer, all solutions should be either filter sterilized or autoclaved before use.

### 2.1 Cells and Culture Media

1. Human diploid fibroblasts: As mentioned in Subheading 1, these cells can be obtained from several sources, the most common being American Type Culture Collection (ATCC). Many labs derive cells from tissue obtained from neonatal foreskins.
2. Culture media for fibroblasts: Eagle's minimal essential medium (MEM, Gibco BRL) supplemented with final concentrations of 10 % fetal bovine serum (FBS), penicillin (200 U/ml), streptomycin (200 μg/ml), L-glutamine (2 mM), and Amphotericin B, i.e., Fungizone, (1.5 μg/ml) (*see* **Note 1**).
3. Parameters for maintenance of cells in culture: In general, most tissue culture cells are maintained at 37 °C in a humidified atmosphere containing 5 % $CO_2$.

### 2.2 Additional Solutions

1. Phosphate-buffered saline (PBS): For 1 l add 8 g NaCl, 0.2 g KCl, 1.44 g $Na_2HPO_4$, and 0.24 g $NaH_2PO_4$ to 950 ml distilled $H_2O$. Adjust the pH to 7.4, bring to volume, and then autoclave.
2. 2.5 % Trypsin: Dilute tenfold in sterile PBS to a working 0.25 % stock (store refrigerated at 4 °C when not in use). Undiluted stock should be stored at −20 °C.
3. DMSO for frozen storage of virus stocks: Add stock to final concentration of 1 % in supernatant (*see* **Note 2**).
4. Agarose (2 %) diluted in water for overlays: Agarose is added to water and then autoclaved to sterilize (*see* **Note 3**).

### 2.3 Plasticware

The following plasticware is needed (and should either be purchased sterile from the manufacturer or autoclaved):

1. Large format tissue culture flasks (~ 185 $cm^2$ growth area).
2. 24-well tissue culture plates.

3. Conical tubes for spinning and freezing viral stocks.
4. Micro-centrifuge tubes for serial dilutions.
5. Cotton-plugged barrier pipette tips for serial dilutions.
6. Serological pipettes for transferring media and virus supernatants.

### 2.4 Other Necessary Equipment

The following equipment is also necessary to carry out these experiments:

1. A clinical centrifuge.
2. A hemocytometer for counting cells.
3. A tissue culture incubator for culturing cells and growing stocks.
4. A biological safety cabinet for all culturing of cells and handling of virus.
5. A –80 °C freezer for storing stocks.

## 3 Methods

### 3.1 Growing Viral Stocks on Fibroblasts

1. Trypsinize and seed actively dividing, low passage human fibroblasts onto T185 flasks (~185 cm$^2$ seeding area). Seed approximately $3.5 \times 10^6$ cells into each flask. In order to yield roughly 200 ml of supernatant per harvest, between 10 and 15 flasks should be seeded. Seed cells the night before they are to be infected. This allows the cells a chance to settle and adhere (*see* **Notes 4–6**).
2. The following morning infect the cells at a low MOI (0.02 is suggested). This low MOI allows ample time for cell-to-cell spread of the propagating virus (*see* **Note 7**).
3. Remove the inoculation media ~8–10 h postinfection (hpi). Each flask should be refed with 17–18 ml of fresh media (*see* **Note 8**).
4. Observe the monolayer each day, refeeding the cells every 2–3 days as necessary (*see* **Note 9**).
5. Watch for when the monolayer displays approximately 80 % cytopathic effect (CPE). CPE is defined as enlarged/rounded cells containing clear virus replication centers visualized by light microscopy. At this point, refeed the cells again with 17 ml/flask in preparation for your first harvest.
6. After 3–4 days harvest the supernatants from the flasks and refeed your cells. If the monolayers have remained intact and there are only a small number of lysed cells in the flasks (*see* **Note 10**), another harvest should be possible in 2–3 days. Note that because a significant proportion of the cells are starting to die at this point, the titer from the second stock will likely be lower than the first stock.

7. For laboratory-adapted virus stocks, the supernatant collected from the infected cells can be cleared by centrifugation in a tabletop centrifuge. Supernatant can be dispersed equally in 50 ml conical tubes and spun at approximately 1,500 rpm (500 × *g*) for 5–10 min to pellet any cellular debris that might be present. If debris is still present after a first clearing spin, the supernatants can be transferred to new tubes and spun again for an additional 5 min.
8. Transfer the cleared supernatant to a T185 flask in order to pool the entire stock, being careful to avoid the cellular debris. Determine how many mls are in the flask.
9. Add 10 % of a 10 % DMSO stock made up in growth media to the supernatant; for example, if you have 200 ml of supernatant, add 20 ml of your 10 % DMSO stock. This yields a final DMSO concentration of 1 % in the virus stock (*see* **Note 11**).
10. Once the DMSO is added, cap the flask and mix well. Aliquot the supernatant into freezer-safe conical tubes (15 ml conical tubes are suggested) and store at −80 °C (*see* **Notes 12–15**).

### 3.2 Titration of Virus Stocks

After virus stocks have been prepared and before they can be used in experiments, the number of functional virions (or plaque-forming units—pfu) per ml must be determined (*see* **Note 16**). The following steps should be followed to determine the pfu/ml:

1. The day before beginning the titration process, seed a total of $1.8 \times 10^6$ fibroblasts into a 24-well plate (for a final concentration of $7.5 \times 10^4$ cells per well). The cells should be resuspended in 24 ml of media, and each well should receive 1 ml (*see* **Note 17**).
2. The following morning quickly thaw a 1 ml aliquot of frozen virus stock in a 37 °C water bath. In complete media, dilute this stock serially, using tenfold dilutions (*see* **Note 18**).
3. Aspirate the media from the wells of the 24-well plate seeded the previous day. Pipette 200–250 μl of each dilution into a separate well of the plate. Record and use a logical pattern when plating the dilutions (*see* **Note 19**). Rock the plate gently to ensure even coverage of the monolayers in each well.
4. After the dilutions are plated, replace the 24-well plate back into the incubator. Allow this inoculum to adsorb for 4–6 h (*see* **Note 20**).
5. Check the monolayers after 4–6 h of incubation; they should all be firmly attached.
6. Place an agarose overlay onto the wells. Final concentration of the agarose should be ~0.25 % in media (diluted from a 2 % stock made up in water; 3 ml agarose + 21 ml media) (*see* **Note 21**).
7. Place 1 ml of overlay into each well (place directly over the supernatant already in the wells, DO NOT aspirate!).

Allow the agarose to solidify (this usually takes 5–10 min in the safety hood) and replace the plate into the incubator.

8. Monitor the plate for plaques every few days. Plaques should start appearing clearly on the lower dilutions between 4 and 5 days. Count the plaques twice between days 7 and 11 post plating (*see* **Note 22**). Ideally between 20 and 50 plaques per well will be counted in order to make accurate calculations. Calculate the titer based upon the number of plaques, the dilution, and the amount of supernatant you plated (*see* **Notes 23** and **24**).

When are titer/plaque assays performed? Scenarios include determining the capability of mutant viruses to grow in culture or to determine if the replication capacity of a given virus varies in different cell types. There are two accepted methods for performing growth curves, single-round and multistep curves. The first technique is to perform single-round growth curves, in which all cells are infected at the outset of the experiment. In single-round growth curves, there are two methods of harvesting and assessing the output from the cells that can be used. The first method determines "cumulative" yields (i.e., taking a small aliquot each day and leaving the media on for the entire experiment). The second method assesses how much virus the cells are releasing "each day." In this method the entire supernatant is harvested each day and calculated for a total output each day of the experiment. The second technique for performing growth curves is multistep curves. In these experiments the initial infection is done at a low MOI. This second method allows for the determination of whether the initial virus is capable of replicating and releasing additional virus that can infect another cell.

## 4 Notes

1. Different laboratories use variations of this media including changes in base media (some use Dulbecco's modified Eagle's media), amount of serum added (anywhere from 5 to 20 %), addition (or not) of antibiotics, and supplementation (or not) of extra glutamine. The use of heat-inactivated serum is recommended, since the complement present in serum can be detrimental to certain cell types in culture. It should be noted that some cell types do not grow well in the presence of amphotericin B. Amphotericin B can be omitted, but cells must be monitored more closely for fungal contamination.
2. A stock of 10 % DMSO solution in growth media can be made just prior to freezing. This solution can then be diluted tenfold (to a final concentration of 1 %) into collected supernatants.

3. When not in use the 2 % agarose stock can be allowed to solidify and stored at room temperature (RT). Simply heat the stock in a microwave to liquefy and cool (in a 56 °C water bath) a minimum of 1 h before use. Overlaying of a 24-well plate requires 24 ml of overlay medium (3 ml of melted 2 % agarose mixed with 21 ml media for a ~0.25 % final concentration of agarose).
4. The cell type in which to grow virus stocks should be carefully considered. Generally, HCMV laboratory-adapted virus strains are grown on primary fibroblasts. Other primary cells might be preferable for growing a clinical strain (i.e., endothelial cells to maintain the integrity of the ULb′ region). Clinical isolates can be grown for a few passages on fibroblasts but care must be used, as many clinical isolates very quickly lose the ULb′ cassette. In some strains this loss of ULb′ integrity has been reported to occur as early as 1–2 passages after infection of fibroblasts. PCR assays should be performed to ensure that this cassette is still intact within any clinical virus preparation. Although highly unlikely, any new cells being used for experiments should be screened to look for viral antigen positivity prior to use (either by immunofluorescence analysis and/or by PCR amplification of a region of the viral genome). Lastly, although this might seem obvious to the experienced herpes virologist, the cytomegaloviruses are highly species restricted, and only human cells can be used for culturing HCMV.
5. The condition of the cells in culture is extremely important to obtain high titer stocks. In essence, treat the cells delicately, affectionately, and well! Maintaining cell cultures at subconfluence is important to the vigor of the cells. DO NOT let cultures remain confluent for more than 1 or 2 days. Confluent cells are not exceptionally healthy and will yield suboptimal results. Cells should be refed 2–3×/week, and split 1:3 approximately once a week or when they near confluence.
6. The number of cells seeded per flask should be on the high side (approximately $3.5 \times 10^6$/flask). In addition, anecdotal evidence has found that lower passage cells yield higher titer stocks. This is likely linked to the slowdown in kinetics as primary cells age. Lower passage cells divide more regularly and are healthier in general.
7. What is an MOI? Multiplicity of infection refers to the plaque-forming units per cell. A synchronized infection is more likely to be produced by using a higher MOI (in general between 2 and 10, depending upon the cells). A high MOI ensures that all the cells are infected and produce viral antigens simultaneously. This is particularly the case if the cells have been previously synchronized in $G_0$ versus an asynchronous infection [4, 5].

Producing high titer stocks requires starting the infection at a quite low MOI. A low MOI offers the virus a chance to "simmer" and replicate multiple times before the entire monolayer is lysed. If a high MOI were used to initiate the stock preparation (as might well be used during an actual experiment), only one round of infection will occur without any cell-to-cell spread of the virus. High MOI infections generally produce high particle to pfu ratios (*see* **Note 16** for an explanation), which is not necessarily desirable in a stock preparation.

8. Leaving the inoculum on overnight can sometimes lead to the infection proceeding too quickly with subsequent lower virus yields. Maintaining a relatively low volume of media in flasks produces a more concentrated stock.
9. The infection should proceed relatively slowly at first, with the first harvest occurring between 7 and 10 days pi. Keep an eye on the monolayer. After a few days distinct "foci" of infected cells will be seen. As the infection proceeds these foci will coalesce, and eventually the entire monolayer will exhibit strong cytopathic effect.
10. Timing of the initial harvest is critical. If the supernatant is harvested too early, the virus yield will be low. In addition, while waiting 3–4 days before the first harvest may seem like a long time, the virus in the supernatant will not suffer at 37 °C. The best titer yields are produced when this first harvest is allowed to "simmer" on the cells.
11. Use care when adding the DMSO. Pipette slowly into the flask of supernatant while swirling the flask to avoid "flashing" the DMSO and creating "hotspots" of high concentrations, which could damage the virions.
12. Freeze 1, 5, and 10 ml aliquots so that appropriate quantity aliquots are available depending on requirements for future experiments. It is not acceptable to freeze and thaw virus stocks more than once. The virus loses significant viability following repeated freeze/thaw cycles rendering MOI calculations invalid.
13. "Mock-infected supernatant" for use as a control for infection can be created in parallel with virus stock preparation. Collect "spent" media that has been incubated for 2–4 days from an equivalent set of flasks of uninfected fibroblasts in culture. This "mock-infected supernatant" will have serum and a significant component of the chemokines/cytokines that are normally secreted by growing cells. This is referred to as "conditioned media" and is added in equivalent amounts to mock-infected cultures as virus is added to infected cultures.
14. Experiments that are sensitive to the effects of serum in virus stocks can benefit by creating virus stocks in serum-free

medium. In this case the cells are washed several times in warm PBS and then incubated in serum-free media prior to collecting virions; however, this is a suboptimal condition for cell growth. An alternative approach is to pellet virions from the supernatant with a high-speed spin in an ultracentrifuge (23,000 rpm (90,000×*g*) for 75 min at 10 °C). The pelleted virions can be washed in PBS to remove any residual serum and then respun under the same conditions.

Some labs pellet these virions through a 25 % sucrose "cushion" made in PBS. The theory is to cushion the envelope from the blows of multiple rounds of pelleting. The sucrose cushion may also serve to separate less dense cellular debris (which will float) from the more dense pelleted particles. Once pelleted and washed, these particles can be resuspended either in an equivalent amount of serum-free media (or PBS) or, if more concentrated stocks are required, less volume can be used to resuspend. Note that this pelleting will not separate full/infectious virions from the other particles secreted by permissive cells, which include noninfectious enveloped particles (NIEPS) and dense bodies (DBs). If pure virions are required, particles must be layered onto a gradient (e.g., glycerol tartrate gradients), which will isolate the three distinct bands of particles [6]. Purity of the fractions from these gradients should be assessed before use.

15. Unlike laboratory-adapted strains, clinical isolates are very cell associated. Additional steps must be taken when harvesting clinical virus stocks. Rather than harvesting only the supernatant, the cells must be harvested as well. First, the supernatant should be harvested and spun as described above. The cellular debris from this spin should be saved. In addition, trypsinize and harvest the cells on the flask. Spin these harvested cells and collect the pelleted cells. Add this cell pellet to the other cells derived from the supernatant clearing spin by resuspending in a small volume of media (5 ml or less). Sonicate all these cells (2 min on ice at an amplitude of 30 % in a cup horn sonicator). Spin out the debris in a tabletop centrifuge at ~1,500 rpm (500×*g*) for 10 min. Pipette off the 5 ml of media and add it to the reserved supernatant, being careful to avoid the debris at the bottom. Respin if the supernatant still looks cloudy. When clear, pellet all the particles out of the supernatant in an ultracentrifuge (as described in **Note 14**) and resuspend in 1/10th to 1/20th the initial volume. This stock can now be titrated on fibroblasts.
16. A plaque-forming unit (pfu) is defined as a functional virion that is capable of infecting a cell and producing enough virus within that cell to infect the cells that are surrounding it (i.e., capable of creating an infected plaque in a monolayer of uninfected cells). Pfu is not a measurement of the number of viral

particles released into the supernatant; the infectious virions are a small proportion of the total particles in the supernatant. There are many other particles interacting with and penetrating infected cells; these additional particles could elicit a response. These particles include NIEPs and DBs. The particle to pfu ratio for HCMV can be as high as 100:1 [7].

17. Be aware that not all multi-well plates are the same. Occasionally, it appears that plates are unevenly coated or that there may be no coating on several wells. Corning plates seem to be the most consistent. Be careful when you are seeding cells on these small circular well plates. Gently rock the plates; do not swirl in a circular motion, as doing so will clump the cells into the middle of the well. The recommended "magic shake" consists of holding the plate in front of your body and gently shaking it back and forth away from your body a few times followed by gently shaking it back and forth across your body a few more times (sometimes a little dance does not hurt either!).
18. Make $10^{-1}$ to $10^{-6}$ dilutions in 1 ml each. It is very important to mix each dilution thoroughly before removing the next aliquot. Always change tips between tubes to avoid cross contamination.
19. As a general rule of thumb, when plating dilutions aspirate media from only half of the plate at a time in order to avoid the cells drying out. It is advisable to use at least duplicate (if not triplicate) wells for each dilution. This improves the accuracy, as the counts from the multiple wells of a given dilution will be averaged.
20. DO NOT allow the cells to dry out during this incubation. Since this is a relatively small amount of media, rock the plate every hour to ensure cells are continuously covered.
21. After completing the dilution series, melt the agarose stock in a microwave and then place into a 56 °C water bath to cool. This will ensure it will have cooled enough to add to the cells later without causing them harm. Work quickly making the agarose dilution in media. If it is too cool, it will solidify in the pipette. Carboxymethyl cellulose is an alternate choice of overlay material.
22. Titrating clinical isolates may require a longer period of incubation. These strains often plaque slowly and only produce small plaques.
23. A sample calculation:
    Count plaques on the $10^{-5}$ dilution wells. Three wells have counts of 18, 22, and 20 plaques, respectively. The average number of plaques is therefore 20. The per ml titer is calculated as follows: 20 plaques $\times 10^5 \times 4$ (because only 250 μl were plated) $= 8 \times 10^6$ pfu/ml.

24. An alternative to performing plaque assays to assess the titer of a virus stock is to assess the immediate early protein positivity ($IE^{+}$) of cells after several hours of incubation. It should be noted that this does not establish whether the virion that has entered a cell has the capacity to proceed through an entire infectious cycle and produce more virions. However, this type of assay can be utilized when a virus is incapable of actually producing plaques (e.g., mutant viruses that do not shed virus).

## References


1. Sinzger C, Jahn G (1996) Human cytomegalovirus cell tropism and pathogenesis. Intervirology 39:302–319
2. Casavant NC, Luo MH, Rosenke K, Winegardner T, Zurawska A, Fortunato EA (2006) Potential role for p53 in the permissive life cycle of human cytomegalovirus. J Virol 80:8390–8401
3. Fortunato EA, Spector DH (1998) p53 and RPA are sequestered in viral replication centers in the nuclei of cells infected with human cytomegalovirus. J Virol 72:2033–2039
4. Fortunato EA, Sanchez V, Yen JY, Spector DH (2002) Infection of cells with human cytomegalovirus during S phase results in a blockade to immediate early gene expression that can be overcome by inhibition of the proteasome. J Virol 76(11):5369–5379
5. Salvant BS, Fortunato EA, Spector DH (1998) Cell cycle dysregulation by human cytomegalovirus: influence of the cell cycle phase at the time of infection and effects on cyclin transcription. J Virol 72:3729–3741
6. Irmiere A, Gibson W (1983) Isolation and characterization of a noninfectious virion-like particle released from cells infected with human strains of cytomegalovirus. Virology 130:118–133
7. Benyesh-Melnick M, Probstmeyer F, McCombs R, Brunschwig JP, Vonka V (1966) Correlation between infectivity and physical virus particles in human cytomegalovirus. J Bacteriol 92:1555–1561

# Chapter 5

# Use of Recombinant Approaches to Construct Human Cytomegalovirus Mutants

**Iryna Dekhtiarenko, Luka Čičin-Šain, and Martin Messerle**

## Abstract

To fully understand the function of cytomegalovirus (CMV) genes, it is imperative that they be studied in the context of infection. Therefore, the targeted deletion of individual viral genes and the comparison of loss of function viral mutants to the wild-type virus allow the identification of the relevance and role for a particular gene in the viral replication cycle. Targeted CMV mutagenesis has made huge advances over the past 15 years. The cloning of CMV genomes into (*E. coli*) as bacterial artificial chromosomes (BAC) allows not only quick and efficient deletion of viral genomic regions, individual genes, or single nucleotide exchanges in the viral genome but also the insertion of heterologous genetic sequences for gain of function approaches. The conceptual advantage of this strategy is that it overcomes the restrictions of recombinant technologies in cell culture systems. Namely, recombination in infected cells occurs only in a few clones, and their selection is not possible if the targeted genes are relevant for virus replication and are not able to compete for growth against the unrecombined viruses. On the other hand, BAC mutagenesis enables the selection for antibiotic resistance in *E. coli*, allowing a selective growth advantage to the recombined genomes. Here we describe the methods used for the generation of a CMV BAC, targeted mutagenesis of BAC clones, and transfection of human cells with CMV BAC DNA in order to reconstitute the viral infection process.

**Key words** Bacterial artificial chromosome (BAC), Targeted mutagenesis, Homologous recombination, Antibiotic selection

## 1 Introduction

The study of viral gene function in the context of CMV infection requires the comparison of phenotypes elicited by the wild-type virus and a viral mutant lacking the gene of interest. Recombinant DNA technologies in virus infected cells allowed the generation of CMV mutants with site-directed deletions of genes of interest [1]. However, site-directed manipulation of cytomegalovirus genomes in tissue culture involves difficult methods that often yield suboptimal results. This method works moderately well when applied to genes that are not necessary for virus replication, but even in this case, laborious techniques such as plaque purification or limiting dilution are needed to separate a mutant from the wild-type virus.

Andrew D. Yurochko and William E. Miller (eds.), *Human Cytomegaloviruses: Methods and Protocols*,
Methods in Molecular Biology, vol. 1119, DOI 10.1007/978-1-62703-788-4_5, © Springer Science+Business Media New York 2014

If genes that contribute to in vitro replication of the virus are deleted, this leads to decreased replication of the recombined viruses, making the isolation of recombinant clones from the unrecombined wild-type viruses nearly impossible.

Human cytomegalovirus (HCMV) is characterized by extremely slow replication in tissue culture and by the fact that it possesses one of the largest genomes among clinically relevant viral pathogens [2–4]. Furthermore, passaging of virus isolates in tissue culture results in genetic drift, loss of tropism for defined cell types, and positive selection of strains adapted to tissue-culture conditions [5, 6]. Hence, even successful recombination events could not exclude the possibility of the occurrence of concomitant uncharacterized mutations elsewhere in the genome and the equally arduous generation of virus revertant viruses.

These problems have been largely mitigated by the cloning of the HCMV genome into *E. coli,* within a bacterial artificial chromosome (BAC) vector [7]. The propagation of viral genomes as BACs in *E. coli* strains lacking recombination enzymes allows for genome stability and selection of recombined BAC clones by antibiotic resistance. Large amounts of HCMV BAC DNA are easily isolated from bacterial cultures and can be used for transfection into eukaryotic cells, whereupon viral gene expression and replication restarts. The method has been successfully applied to the generation of cloned genomes of laboratory strains [7–9] and clinical isolates [10–13]. Here we will describe the methodology for (1) the cloning of HCMV isolates as BACs, (2) state-of-the-art methods for directed mutagenesis, and (3) common methodology for the reconstitution of infectious HCMV from its cloned BAC DNA.

## 2 Materials

Standard materials and equipment for molecular biology and cell culture work is required (e.g., a table top centrifuge, incubators for bacterial and eukaryotic cells, a refrigerator or cold room, freezers (−20 and −70 °C), pipets, microcentrifuge tubes, Erlenmeyer flasks, Petri dishes, cell culture flasks, and plates).

### 2.1 Cloning of HCMV Isolates as Bacterial Artificial Chromosomes (BACs)

1. Restriction endonucleases.
2. BAC vector plasmid (e.g., pEB1097 [7], available from Martin Messerle, or pBeloBAC11 [#E4154S; New England Biolabs]). Please *see* **Note 1**.
3. 100 % ethanol (store at −20 °C).
4. 70 % ethanol.
5. Sterile double-distilled water (dd$H_2O$).
6. 3 M sodium acetate in $H_2O$. Adjust pH to 5.2 with glacial acetic acid.

7. 1 M Tris–HCl stock solution in $H_2O$. Adjust pH to 8.0 with HCl and autoclave.
8. 0.5 M ethylenediaminetetraacetic acid (EDTA) in $H_2O$, adjust to pH 8.0 with NaOH, and autoclave.
9. TE buffer: 10 mM Tris–HCl, 1 mM EDTA, pH 8.0. Prepare from stock solutions using sterile dd$H_2O$.
10. Human fibroblasts (e.g., MRC5 cells ATCC# CCL-171).
11. Cell culture medium (e.g., DMEM, Opti-MEM from Gibco-BRL).
12. Fetal calf serum (FCS).
13. Trypsin for cell culture: 0.25 % trypsin, 1 mM EDTA.
14. Mycophenolic acid—add to the cell culture medium (final concentration—100 μM).
15. Xanthine (Sigma)—add to the cell culture medium (final concentration—25 μM).
16. Phosphate-buffered saline (PBS) for cell culture (Gibco-BRL).
17. 20 mM EDTA, pH 8.0. Prepare from the stock solution.
18. 10 % (w/v) sodium dodecyl sulfate (SDS) stock solution in dd$H_2O$.
19. 1.2 % SDS. Prepare from stock solution with sterile dd$H_2O$.
20. Autoclaved 5 M NaCl.
21. Phenol–chloroform.
22. Isopropanol.
23. RNase A (stock solution 10 mg/mL in TE; final concentration, 5–10 ng/mL).
24. *E. coli* strains DH10B and GS1783 [14] (or any other *E. coli* strain that either lacks or displays only low constitutive recombination activity). DH10B are commercially available from several sources and GS1783 are available upon request from Gregory A. Smith, Northwestern University, Chicago, IL. Store at −70 °C as glycerol stocks, or generate electroporation competent bacteria by using the method detailed in Subheading 3.2.
25. Dialysis membrane (0.25 μm pore size; e.g., Millipore VSWP04700).
26. LB medium: mix 10 g Bacto-Tryptone, 5 g Bacto yeast extract, and 10 g NaCl in 900 mL $H_2O$; adjust pH to 7.0 with NaOH; adjust volume to 1 L with $H_2O$. Autoclave and store at 4 °C.
27. LB agar plates: add 15 g Bacto-Agar to 1 L LB medium and sterilize by autoclaving. Cool down to 50 °C before adding appropriate amount of desired antibiotic (e.g., chloramphenicol). Pour medium into sterile plates; allow the medium to solidify and store the dishes at 4 °C in inverted position.

28. Chloramphenicol. Stock solution: 34 mg/mL in 70 % ethanol. Store at −20 °C.
29. Electroporation cuvettes (2 mm wide for the electroporation of bacteria and 4 mm wide for the electroporation of eukaryotic cells).
30. Electroporator machine [e.g., Gene Pulser (Bio-Rad)].

### 2.2 En Passant Mutagenesis of HCMV Genomes

In addition to the materials from the previous section, several other reagents are necessary at this step:

1. PCR template: plasmid pGP704 I-SceIKan is a derivative of the plasmid pGP704, in which the I-SceI restriction site was inserted in front of the sequence for the kanamycin resistance gene, and both were introduced into the polylinker of pGP704, in the EcoRI site. The insertion of the I-SceI sequence into the plasmid has allowed the design of PCR primers that do not contain this restriction site and hence may carry longer sequences to be inserted into the viral genome. pGP704 I-SceIKan can be obtained from Iryna Dekhtiarenko and Luka Cicin-Sain.
2. High-Fidelity Taq DNA Polymerase.
3. DpnI restriction enzyme.
4. QIAquick Gel Extraction Kit.
5. Ethidium bromide.
6. TAE (Tris–acetate–EDTA): first make a concentrated (50×) stock solution of TAE by weighing out 242 g Tris base and dissolving in approximately 750 mL deionized water. Carefully add 57.1 mL glacial acetic acid and 100 mL of 0.5 M EDTA (pH 8.0). If necessary adjust pH to 8.3 and adjust the solution to a final volume of 1 L. This stock solution can be stored at room temperature. Sterilize by autoclaving. The working solution of 1× TAE buffer is made by diluting the stock solution in deionized water. Final solution concentrations are 40 mM Tris acetate and 1 mM EDTA.
7. TBE (Tris–borate–EDTA): first make a concentrated (10×) stock solution of TBE by weighing 108 g Tris base and 55 g boric acid and dissolving both in 900 mL deionized water. Add 40 mL of 0.5 M EDTA (pH 8.0), adjust pH to 8.0 and adjust the solution to a final volume of 1 L. This solution can be stored at room temperature, but a precipitate will form in older solutions. Store the buffer in glass bottles and discard if a precipitate has formed. The working solution of 0.5× TBE buffer is made by simply diluting the stock solution in deionized water.
8. 1 % agarose TAE gel: boil 1 g of agarose in 100 mL of 1× TAE buffer until agarose is completely dissolved; cool down to approximately 50 °C and add 5 μL of a 10 mg/mL stock solution of ethidium bromide (EtBr) for DNA visualization.

Pour solution into the electrophoresis chamber and allow to solidify at room temperature. Depending on the size of your chamber, you may have gel material for 1 or 2 chambers.

9. 0.8 % agarose TBE gel: boil 2.4 g of agarose in 300 mL of 0.5× TBE buffer and follow the protocol in the previous point, adjusting the volume of EtBr to 15 μL. Use large gel chamber (20 cm length) and pour 300 mL of gel.
10. Tissue paper.
11. Kanamycin sulfate. Stock solution: 50 mg/mL in $ddH_2O$. Store at −20 °C.
12. Zeocin. Stock solution: 100 mg/mL in $ddH_2O$. Light sensitive. Store in dark, at −20 °C.
13. 10 % v/v glycerol in $ddH_2O$. Autoclave, store at 4 °C.
14. MangoMix PCR Kit (Bioline) (*see* **Note 2**).
15. 2 % solution (w/v) of L-arabinose in LB medium (do not autoclave, use sterile filtration).
16. LB agar plates with 1 % L-arabinose (should be always made fresh): prepare liquid LB agar, autoclave it, and let it cool down. In parallel prepare a 10 % (w/v) L-arabinose solution in LB medium, sterile filter, and mix it 1:10 with liquid LB agar when it cools to 50 °C. Add 25 μg/mL chloramphenicol.
17. Water bath shaker.

#### *2.3 Transfection of HCMV BACs into Human Cells and Reconstitution of HCMV Infection*

In addition to the materials from the previous sections, the following materials are also necessary for HCMV reconstitution from BAC DNA:

1. EndoFree Plasmid Maxi Kit.
2. FuGENE HD (Promega).
3. Polystyrene tubes.
4. 6-well cell culture plates.
5. Tissue-culture grade penicillin–streptomycin solution.
6. Tissue-culture grade L-Glutamine solution.
7. Fully supplemented DMEM: add 50 mL of FCS, 5 mL of penicillin–streptomycin, and 5 mL of L-Glutamine to 500 mL of DMEM. Mix thoroughly and store at +4 °C for up to 6 weeks.

## 3 Methods

#### *3.1 Cloning of HCMV Isolates as Bacterial Artificial Chromosomes (BACs)*

The cloning of an HCMV genome in *E. coli* requires its physical connection with a bacterial replicon. BAC vectors derived from the *E. coli* F-factor offer properties (large cloning capacity, low copy number, high stability of the inserted sequences) that make them the first choice for this purpose. Theoretically, it should be possible

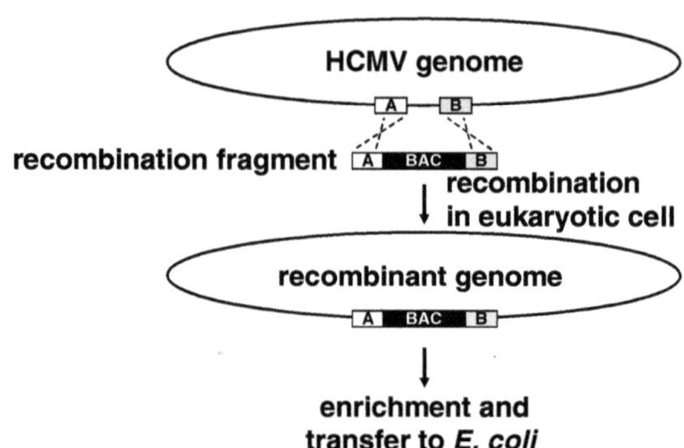


**Fig. 1** Generation of an HCMV BAC. First, a recombination plasmid is constructed carrying the BAC vector sequences (*black*) flanked by viral sequences (*A* and *B*) homologous to the desired integration site in the viral genome. The recombination fragment is isolated from the plasmid and transfected into HCMV infected cells. Homologous recombination between the DNA fragment and the viral genome occurs spontaneously, leading to a recombinant genome carrying the BAC replicon. Following enrichment of the recombinant virus, circular genomes are isolated from infected cells and transferred to a suitable *E. coli* strain for further propagation and maintenance

to directly ligate linear HCMV genomes isolated from virus particles to such a BAC vector. This approach has, however, not been reported, probably due to the large size of the HCMV DNA genome and its vulnerability to shearing. Therefore, the connection of the HCMV genome with the BAC vector has been performed by traditional recombination techniques in infected, eukaryotic cells (*see* Fig. 1 and **Note 1**). One needs to consider that this is the most laborious and technically challenging step of the mutagenesis procedure. Fortunately, this step only needs to be performed once for each HCMV isolate to be cloned. Moreover, BAC clones have already been generated for many CMV laboratory strains, as well as for strains with properties of clinical CMV isolates. These CMV BACs are available for research purposes, and the generation of new CMV BACs may therefore only be considered for strains with additional interesting properties.

In order to facilitate the identification and enrichment of a CMV recombinant carrying the BAC vector, we recommend the insertion of appropriate screening and selection markers into the CMV genome together with the BAC vector sequences. *Gfp* and *gpt* (guanosine phosphoribosyl transferase) markers have been employed successfully for this purpose. Recombination between the BAC vector and the viral genome is a rare event, resulting in a mixture of recombinant and parental wild-type viruses, with the latter one prevailing. Recombinant CMV can be amplified using plaque purification or selection with mycophenolic acid and hypoxanthine, employing the markers mentioned above. Once the recombinant CMV has been enriched to a substantial amount, circular genomes are isolated from infected cells and transferred to suitable *E. coli* strains (for instance, the recombination-deficient *E. coli* strain DH10B) that support stable propagation of the BAC. Careful characterization of the resulting BAC by restriction analysis and preferably by sequencing is recommended to guarantee that the cloned genome carries the full set of intact ORFs. If needed, mutated or disrupted ORFs can be repaired by mutagenesis in *E. coli* [13]. The known CMV BACs displayed remarkable stability upon propagation in *E. coli.* Nevertheless, we recommend the application of good microbiological standards in manipulating the BAC clones, such as the storage of master stocks of the original *E. coli* clones carrying the CMV BAC. These master stocks will then serve as the basis for further mutagenesis steps.

#### 3.1.1 Generation of a CMV Recombinant Carrying the BAC Vector

The initial cloning step requires the generation of a recombination plasmid carrying the BAC vector sequences flanked by viral sequences that are homologous to the target site in the CMV genome (*see* Fig. 1). The generation of this plasmid is easily attainable with standard molecular-cloning protocols [15]; specific examples have been described in Stanton et al. [13], Borst et al. [7], and Yu et al. [13], while a general overview about this cloning strategy is already available in this series [16]. Therefore, the generation of a recombination plasmid is outside the scope of this chapter and we will only point out the general considerations about its design. The BAC vector should be targeted towards a nonessential genomic region and it may even be helpful to replace some nonessential genes by the BAC vector sequences (*see* **Note 1**). PCR fragments of 300–500 bp length, representing the sequences for homologous recombination, are generated using appropriate primers and CMV DNA as template. One should design primers with unique restriction sites at their 5′ ends, to facilitate their cloning into the plasmid, but also the release of the recombination fragment from the plasmid backbone by treatment with the respective restriction enzymes. For the following procedure, it is assumed that the recombination plasmid is already available:

1. Incubate 5–30 μg of the recombination plasmid with 5–20 units of the appropriate restriction enzymes and buffer in a total volume of 50–100 μL for 1–2 h at 37 °C. Check an aliquot of the sample by agarose gel electrophoresis to see if cleavage of the plasmid is complete.
2. Add 1/10 volume of 3 M sodium acetate, pH 5.2, and 2.5× volumes of 100 % ethanol, mix, and collect precipitated DNA fragments by centrifugation (~10 min at 18,000 × *g* in a table top centrifuge).
3. Wash the DNA with 70 % ethanol (~0.5–1 mL), dry, and resuspend in a small volume (20–50 μL) of $ddH_2O$ or TE buffer.

4. Grow human fibroblasts on approximately three culture dishes (diameter 10 cm) to at least 50 % confluency (~$10 \times 10^6$ cells in total) using suitable cell culture medium (e.g., DMEM with 10 % FCS).
5. Detach cells from culture dishes by trypsin/EDTA treatment, collect cells by low speed centrifugation ($300 \times g$, 5 min), wash with culture medium, and resuspend in Opti-MEM1 medium at ~$3 \times 10^6$ cells/400 μL.
6. Transfect three samples of the cells with 5, 10, and 20 μg of the prepared recombination plasmid DNA by electroporation using a Bio-Rad electroporator or equivalent and the settings: 280 V and 1,500 μF (*see* **Note 3**).
7. Seed the cells in suitable cell culture dishes (e.g., 10-cm dishes) and incubate with cell culture medium overnight at 37 °C and 5 % $CO_2$.
8. On the next day infect the cultures at a multiplicity of infection of 1–3 PFU/ cell with human CMV by adding the viral inoculate in a small volume of culture medium (~3 mL); incubate for ~1 h gently shaking at 15, 30, and 45 min post infection; and finally replace the inoculum with culture medium (10 mL).
9. Incubate the dishes until a complete cytopathic effect (CPE) is observed (all cells are infected and dying).
10. Collect the supernatant, remove cell debris by centrifugation, and store the supernatant at −70 °C in 1-mL aliquots.

#### *3.1.2 Enrichment of the CMV Recombinant Carrying the BAC Vector*

The viruses obtained in the supernatant from the step described above are a mixture of the parental and the recombinant virus. In almost all cases, the amount of the parental virus will usually be found in excess. Enrichment of the recombinant virus is therefore mandatory. In our hands, amplification by utilizing the *gpt* selection marker of the recombinant virus is usually successful, but other procedures such as plaque purification or limiting dilution will also work.

1. Grow several dishes (e.g., 10 cm dishes) of human fibroblasts to near confluency (~90 %). Replace the cell culture medium with medium containing 100 μM mycophenolic acid and 25 μM xanthine. Incubate cells at 37 °C and 5 % $CO_2$ overnight (*see* **Note 4**).
2. Infect the cultures with the supernatant obtained from the samples of **step 1** (*see* **Note 5**).
3. Incubate the cells until complete CPE occurs (*see* **Note 6**). Harvest the supernatant and subject aliquots to a second round of enrichment by repeating **steps 1–3**.

Usually, at least three rounds of enrichment are required to get about 50 % (or more) recombinant viruses in the progeny. Since transfer of the circular viral genomes to *E. coli* in the next step is poorly efficient and hence rate limiting, it is necessary that the recombinant genomes be present in substantial numbers.

#### 3.1.3 Transfer of the Circular Virus Genomes to E. coli

1. Infect ~$3 \times 10^6$ human fibroblasts (cells in a 10-cm dish) with the virus preparation (supernatant) obtained in the previous step. Incubate for approximately 5 days or until CPE becomes visible (*see* **Note 7**).
2. Harvest the infected cells by trypsin/EDTA digestion; wash them once with medium containing 10 % FCS and once with PBS.
3. Resuspend the cells in 500 μL 20 mM EDTA pH 8.0 and lyse them by adding the same amount of 1.2 % SDS and gently inverting the tubes.
4. Add 660 μL of 5 M NaCl and mix. This will lead to the precipitation of proteins and large chromosomal and linear DNA molecules. Circular and small DNA molecules will remain in solution (Hirt extract).
5. Incubate the sample at 4 °C (on ice) for 4 h or better overnight.
6. Transfer the clear aqueous phase obtained after centrifugation in a table top centrifuge (30 min at 18,000 × *g* and 4 °C) to new tubes and extract the DNA by adding an equal volume of phenol–chloroform.
7. After a second centrifugation step, transfer the aqueous phase to new tubes, and precipitate the DNA by adding 0.8 vol of isopropanol.
8. Centrifuge the DNA immediately in a table top microcentrifuge (20–30 min at 18,000 × *g*). Do not incubate at −20 °C as this may result in precipitation of excess amounts of sodium chloride.
9. Wash the DNA pellet with 70 % ethanol to get rid of residual sodium chloride. Repeat the washing step.
10. Air-dry the DNA pellet and resuspend it in ~100 μL TE containing 5–10 ng/mL RNase A.
11. Prior to electroporation of *E. coli* bacteria (*see* **Note 8**), the DNA sample needs to be dialyzed against TE buffer to get rid of any remaining salt impurities. Add ~20 μL of the DNA sample onto a dialysis membrane, floating on ~10 mL TE buffer in a 10-cm Petri dish, and incubate for 30 min at room temperature.
12. A 50-μL aliquot of electrocompetent *E. coli* prepared according to standard protocols [15] is thawed on ice and transferred to a cooled electroporation cuvette (1 or 2 mm wide), before ~5 μL of the dialyzed DNA sample is added.

13. Electroporate *E. coli* at 200 Ω, 25 μF, and 2,500 V in a Bio-Rad gene pulser. If using another electroporation device, follow the manufacturer´s recommendations for electroporation of *E. coli* (*see* **Note 9**).
14. Immediately add 500 μL of prewarmed LB medium to the bacteria, followed by incubation at 37 °C for 1 h.
15. Spread bacteria on LB agar plates containing 17–34 μg/mL chloramphenicol (or appropriate antibiotic based on the selectable marker present in the BAC vector), and incubate overnight at 37 °C.
16. Grow overnight cultures from single bacterial clones in LB medium with chloramphenicol (17–34 μg/mL)
17. Isolate BAC DNA from 10 mL of the overnight cultures by alkaline lysis following standard procedures [15].
18. Approximately 2 μg of BAC DNA is obtained from 10 mL cultures, providing sufficient DNA for 2–3 restriction analyses with different enzymes. We recommend sequencing of the full-length BAC at this point. This ensures that the BAC contains all the appropriate genetic material prior to engaging in mutagenesis procedures and characterization of recombinant viral phenotypes. *E. coli* clones carrying CMV BACs with the expected restriction profile are stored as glycerol stocks at −70 °C.

### 3.2 En Passant Mutagenesis of HCMV Genomes

En passant mutagenesis is a method of choice for the generation of HCMV recombinants with point mutations, targeted deletions, or insertions of DNA sequences without introducing superfluous additional genetic alterations [14, 17]. It relies on inducible redαβγ-mediated recombination of BAC DNA with linear DNA fragments carrying a selection marker [18–20], but it is characterized by a second step in which BACs are first cut by a meganuclease and then rejoined by recombination of homologous sequences flanking the selection marker. The major benefit of this method is the exclusion of potential and undesired effects of superfluous noncoding sequences and the ability to sequentially manipulate the viral genome.

#### 3.2.1 Generation of Linear DNA Products for Mutagenesis by Homologous Recombination

1. Design primers for amplification of the I-SceIKan cassette to create the construct for the en passant mutagenesis (*see* **Note 10**).
2. Prepare PCR reaction (use High-Fidelity Taq DNA polymerase) in a total volume 100 μL (*see* **Note 11**). Perform 35 PCR cycles using the following conditions:
   - 30 s 94 °C/30 s 64–46 °C (decrease 1 °C per cycle)/2 min 68 °C for 18 cycles.
   - 30 s at 94 °C/30 s at 45 °C/2 min 68 °C for 17 cycles.

3. Treat PCR product with 30–40 U of DpnI in 70 μL of total volume (buffer provided by NEB along with enzyme) to remove template DNA (overnight, at 37 °C).
4. (Optional step) Run PCR product on a preparative 1 % agarose TAE gel; cut the band out and purify using a commercial DNA purification kit and manufacturer instructions (*see* **Note 12**).

#### *3.2.2 Preparation of Recombination- and Electroporation-Competent GS1783*

Before starting make sure that you prepare the following:

- 5 mL of overnight cultures of GS1783 *E. coli* containing the respective HCMV BAC clones (grown in 5 mL of LB medium with 25 μg/mL chloramphenicol in a bacterial shaker at 32 °C).
- Turn on shaking water bath to warm to 42 °C.
- Prepare ice-cold dd-$H_2O$ and ice-cold glycerol (10 % v/v solution of glycerol in dd-$H_2O$).
- Cool down the table top centrifuge (e.g., Heraeus Multifuge 3SR+with the rotor #3057 and adaptors for 50 mL Falcon tubes or equivalent) to +1 °C.
- Prepare liquid nitrogen (in case you want to store electrocompetent cells and perform the electroporation step at a later time—*see* **Note 13**).

1. Add 1 mL of overnight culture to 50 mL LB supplemented with chloramphenicol in an Erlenmeyer flask. Incubate at 32 °C to an $OD_{600nm}$ of 0.55–0.6 (*see* **Notes 14** and **15**).
2. Immediately transfer the flask with the bacterial culture to the shaking (220 rpm) water bath and heat-shock the culture for exactly 15 min at 42 °C (*see* **Note 15**).
3. Transfer bacteria directly on ice and allow to cool for 5 min. Transfer the cultures to chilled 50 mL tubes (it is better to proceed with the following washing steps shortly after transferring bacteria to the tubes rather than incubating bacteria on ice for an extended period of time).
4. Spin for 5 min at 3,500×*g*/+1 °C. Remove all supernatant and resuspend the bacterial pellet in 1 mL of ice-cold dd-$H_2O$ by pipetting. Add 45 mL of ice-cold dd-$H_2O$.
5. Repeat the previous **step**, but use a 10 % v/v solution of ice-cold glycerol in dd-$H_2O$.
6. Spin for 5 min at 5,000×*g*/+1 °C. Pour off all the supernatant and wipe the remaining supernatant from the tube wall with tissue paper. Keep the bacterial pellet on ice and check the volume of the suspension. It should be between 240 and 300 μL (*see* **Note 16**).

7. Make 60–70 μL aliquots in microcentrifuge tubes (you need 60–70 μL per electroporation, so one batch contains enough electrocompetent *E. coli* for ~4 transformations). Keep electrocompetent bacteria on ice until electroporation. Cells not used may be snap-frozen in liquid $N_2$ and stored at −70 °C for up to 4 weeks.

#### *3.2.3 Electroporation and First Red Recombination Step*

1. Add 200–500 ng of the purified PCR product to a bacterial aliquot. Transfer competent cells and recombination DNA to electroporation cuvettes (2 mm wide). The same volume of dd $ddH_2O$ with electrocompetent cells may be used in a control transformation reaction. Pulse the competent cell–DNA mixes with 2,500 V, 200 Ω, and 25 μF in an electroporation machine (Gene Pulser from Bio-Rad or equivalent), and add immediately 1 mL of 32 °C prewarmed LB medium without antibiotics.
2. Transfer the bacteria into a 1.5 mL microcentrifuge tube. Incubate at 32 °C for 1–3 h in a shaker (*see* **Note 17**).
3. Plate 100 μL of bacteria on an LB agar plate with 25 μg/mL chloramphenicol and 30 μg/mL kanamycin (or 20 μg/mL Zeocin, if using Zeo selection). Centrifuge the remainder of bacteria at 2,400 × *g* for 5 min in a table top microcentrifuge, and resuspend the pellet in 100 μL LB medium and plate on another plate with the same selection. Incubate plates for approximately 24–48 h at 32 °C.
4. Verify that the PCR fragment is integrated into the HCMV BAC by restriction fragment analysis, colony PCR, sequencing, or any other appropriate technique (*see* **Note 18**).

#### *3.2.4 Resolution of Co-integrates*

The selection of clones mutated by the targeted approach is achieved by antibiotic resistance, yet the antibiotic resistance gene needs to be excluded from the final construct to avoid unwanted cis effects of its sequence on the neighboring genes. This is achieved by a two-step process: First the expression of the I-SceI endonuclease is induced by the addition of L+-arabinose, resulting in a double-stranded DNA break of the BAC at the unique restriction site adjacent to the antibiotic resistance gene. In the second step, the sites of homology which flank the antibiotic resistance gene and the I-SceI sequence are recombined by the induction of the heat-sensitive promoter driving the expression of the Red recombinase. Consequently, the BAC is recircularized, without the antibiotic selection gene inserted in the first Red recombination step, but with the mutated sequence. Selection of the bacteria carrying BAC clones with successful resolution is achieved by growing them in the presence of chloramphenicol, and correct recombination is confirmed by excluding that bacteria grow in the presence of kanamycin [17].

1. Inoculate bacteria harboring positive co-integrates into 1 mL of LB medium with 25 μg/mL chloramphenicol. Shake for 1–2 h at 32 °C and 220 rpm until the solution becomes faintly clouded (*see* **Note 19**).
2. Add 1 mL of prewarmed LB medium with 25 μg/mL chloramphenicol and 2 % L-ARABINOSE (*see* **Note 19**).
3. Shake 1 h at 32 °C, 220 rpm.
4. Transfer immediately to 42 °C water bath shaker and shake for another 30 min at 220 rpm.
5. Transfer culture to 32 °C and shake for 2–3 h at 220 rpm.
6. Take 1 mL of the culture to measure $OD_{600}$ and plate 5–10 μL of a 1:100 ($OD_{600} < 0.5$) or a 1:1,000 ($OD_{600} > 0.5$) dilution on an LB agar plate with 25 μg/mL chloramphenicol and 1 % L-arabinose.
7. Incubate at 32 °C for 1–2 days until average size bacteria colonies are grown.
8. Optional: Pick replicas of the bacterial colonies from the arabinose-containing agar plates and transfer in parallel to LB agar plates with 25 μg/mL chloramphenicol and to plates with 25 μg/mL chloramphenicol plus 30 μg/mL kanamycin (or 20 μg/mL Zeocin when using it as selection marker). The colonies that are resistant to chloramphenicol but not to kanamycin (or Zeocin) contain the BACs with resolved co-integrates.
9. Confirm positive clones with restriction fragment analysis, colony PCR, sequencing, or any other appropriate technique.

### 3.3 Transfection of HCMV BACs into Human Cells and Reconstitution of HCMV Infection

Once the recombinant viral clones have been generated and their identity verified, one may proceed to reconstitute infectious virus out of the viral DNA. Viral DNA is best isolated from approximately 200 mL of bacteria by means of a commercial kit for column purification of BAC DNA (several kits may be used here, for instance, the EndoFree Plasmid Maxi Kit from Qiagen). This results in a typical yield of 50 μL of DNA preparation with a concentration of DNA at 200–500 ng/μL. The BAC DNA may be then used for transfection of human cells, upon which the viral genes start to get expressed and virus restarts its infectious cycle. There are several commercial transfection reagents available, including Metafectene Pro (Biontex, Germany), Lipofectamine2000 (Invitrogen), or FuGENE HD (Promega, Germany). Essentially all of them allow for the reconstitution of the infectious process, provided that DNA preparations are of high purity and concentration. We will describe here a protocol adapted for FuGENE HD-mediated transfection of BAC DNA into MRC-5 cells, a commonly used fibroblast cell line.

1. Grow MRC-5 cells in tissue culture flasks and transfer them to 6-well plates on the day prior to transfection. Distribute cells at 50 % of maximum confluency (approximately $10^5$ cells per well).
2. On the day of transfection, inspect cells to make sure that cells are spread uniformly across the plate. Allow FuGENE HD Reagent, DNA, and Opti-MEM to adjust to room temperature and vortex all reagents except BAC DNA prior to use.
3. Mix 2 μg BAC DNA with Opti-MEM in a polystyrene tube (*see* **Note 20**). Adjust total volume to 100 μL.
4. Add transfection reagent to the diluted DNA in a dropwise manner. Avoid touching the plastic of the tube.
5. Mix contents thoroughly, by pipetting up and down ten times.
6. Incubate mixture for 15 min at room temperature.
7. Meanwhile, replace the cell medium with Opti-MEM medium with no additives.
8. Add transfection complex to cells in a dropwise manner and swirl plate to allow distribution over the entire plate surface.
9. At 3–8 h (optimally at 6 h) after transfection, replace Opti-MEM with fully supplemented DMEM.
10. Following transfection, incubate cells for 48–72 h.
11. Split cells of each transfected well to two T25 flasks. Incubate cells further and monitor by microscopy for areas of viral CPE starting at 7 days post transfection.
12. Once cells become confluent, combine the cells from the T25 flasks into a T175 flask. You will likely need to split the flask again a few days later (*see* **Note 21**).
13. Supernatant of completely infected and lysed cells can be used for virus passaging and virus stock generation.
14. Supernatant which is not used immediately can be stored at −70 °C.

## 4 Notes

1. pEB1097 allows a ready-to-go approach for the generation of novel HCMV BAC, because it already flanks the BAC vector with sites of HCMV homology to ORFs US2 and US6, respectively. However, this vector results in recombinants that omit the US2–US6 ORFs, which may be suboptimal for some applications. Therefore, the insertion of target HCMV sequences in a novel BAC vector, like the pBeloBAC11, allows the generation of CMV recombinants that contain the US2–US6 region. In generating such mutants, it is important to consider that the replacement of some nonessential genes of

the CMV genomes by the BAC vector sequences is favorable in order to prevent overlength of the resulting recombinant CMV genome. Once cloned in *E. coli*, the missing sequences could be easily re-added to the BAC. One integration site commonly used was between the ORFs US2 and US6 [7, 8]. Other chosen insertion sites were between US28 and US29 [21] or between US28 and US34 [13].

2. The MangoMix is not a critical reagent, yet it is a very useful reagent for colony PCR screening, because it is affordable and already contains loading buffer in the reaction mix, which is time-saving when 10–20 colonies are screened by PCR.
3. We recommend performing the protocol at least in triplicate, using different amounts of the digested recombination plasmid. It is hard to predict if recombination takes place, with which frequency, and which conditions are optimal. There is probably a lot of randomness in this approach. Consequently, one should not rely on just one attempt.
4. Please note that the additives will lead to a growth arrest of the cells. It is advisable to first establish suitable conditions to determine how fast the growth arrest occurs with the particular passage of fibroblasts that the lab is using. Moreover, the presence of a sufficient amount of cells is required to allow spread of the viruses to neighboring cells in order to generate a sufficient amount of progeny virus.
5. Since it is hard to predict which transfection experiment was successful and how much of the recombinant virus may be present in the supernatant, we recommend performing **step 2** with supernatant from the independent attempts. Furthermore, it is advisable to add different amounts of the supernatants. As a rule of thumb, we recommend using substantial amounts for the second step (~1–5 mL).
6. This time point does not seem to be critical; however, we recommend waiting until the late phase of the infection cycle to guarantee that sufficient amounts of circular genomes are present.
7. Depending how much virus is present in the inoculum, it can take 2–3 weeks until complete CPE occurs.
8. It is critical to use highly competent *E. coli* preparations. One useful strain is DH10B.
9. Transformation of chemically competent bacteria by heat-shock treatment was never successful in our hands as it does not yield sufficient transformation efficiency.
10. We recommend the use of Vector NTI software (Invitrogen), as it allows the design of large annotated BAC sequences and in silico analysis of prospective restriction patterns of the complete BACs, but other noncommercial or Web-based tools

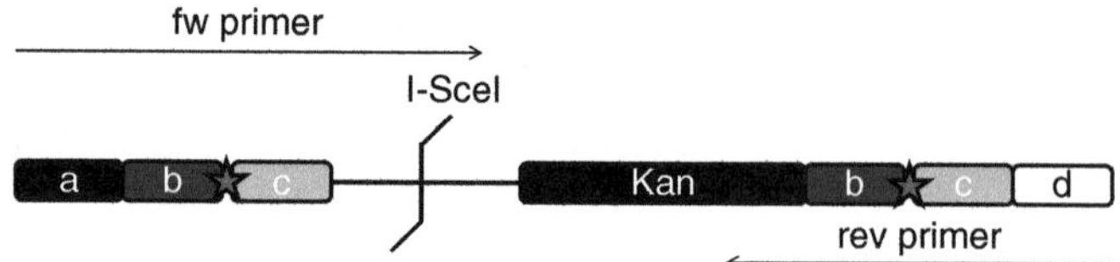


**Fig. 2** *a* and *d*—Sequences of homology to the viral genome, which are used in the first step of the recombination (each ca. 20 bp). *b* and *c*—Sequences of homology to the viral genome which are used in the second step of the recombination (each ca. 20 bp). The *red star symbol* indicates the sequence that should be changed. This may include a point mutation, a DNA sequence for a peptide, or gaps in the viral sequence for the deletion of genes of interest. *I-SceI*-I-SceI restriction sequence (*see* **Note 24**), *Kan*-kanamycin resistance gene

may also be used for primer design (e.g., Genestream or BioEdit). In the primer design, consider the scheme of the final PCR product (*see* Fig. 2). The primers are indicated in the figure and should be designed to harbor the following sequences:

- Forward: sequences on the 5′-end are identical to the segments a–c and contain the mutation between the fragments b and c. This is followed by the sequence that allows the annealing to the I-SceI restriction site (for pGP704 I-SceIKan: 5′-TAGGGATAACAGGGTAATCGAT-3′).
- Reverse: sequences on the 5′-end are the reverse complementary sequence of the d–b sequences and the 3′-end contains the sequence for the annealing to the marker cassette (for pGP704 I-SceIKan: 5′-GTGTTACAACC AATTAACCAAT-3′).

Primers up to 110 bp are commercially available (e.g., from Metabion, Martinsried, Germany; Invitrogen; or Sigma). This allows the insertion of sequences of up to 30 bp using one PCR amplification step. If longer sequences (30–130 bp) need to be inserted, a second set of primers can be used. In this case, the first primer set will contain the homology to the pGP704 I-SceI plasmid on the 3′-end (sequence shown above) and up to 80 bp of the sequence to be inserted on the 5′-end of the each primer. The second primer pair will carry the regions of homology to the virus on their 5′-end (a and b in the case of the fw and rev primer, respectively), a sequence homologous to the 5′-end of the first primer pair on their 3′-end (minimum of 20 bp), and up to 50 bp of additional insert size. The sequence of the PCR product and the strategy for the primer design is shown (*see* Fig. 3).

Alternatively, for inserts larger than 130 bp, one may order synthesized DNA constructs (in this case, it may be preferable to use Zeocin resistance for selection, as the Zeo resistance

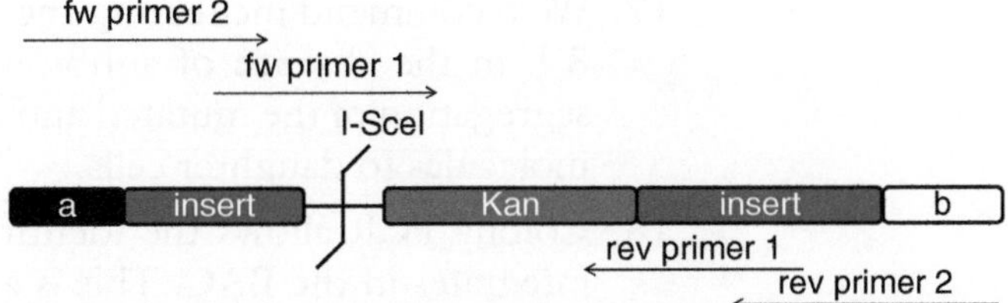


**Fig. 3** *a* and *b*—Sequences of homology to the viral genome flanking the construct and are used in the first recombination step (each ca. 50 bp). *Insert* inserted part (as it is used for the second step of recombination; it should not be less than 40 bp. If its length is less, you may consider introducing appropriate parts of the viral genome flanking the insert). *I-SceI*-I-SceI restriction sequence (*see* **Note 22**), *Kan*-kanamycin resistance gene

gene is shorter than Kan. Sequences of both Zeo and Kan are found in **Notes 22** and **23**).

11. Two 50 μL reactions work more efficiently than one 100 μL reaction, probably due to the lower thermic inertia of the reaction when performed in a smaller volume. For PCR constructs above 2 kbp, use always two PCR reactions of 50 μL each.
12. DpnI treatment alone may be sufficient, as it will degrade the methylated DNA of the template plasmid, but not the unmethylated DNA of the PCR product. However, we found it useful in difficult mutagenesis procedures to use both DpnI digestion and gel purification steps, as these mutagenesis steps will be difficult for the less experienced scientist. One may also routinely purify bands from gels instead of DpnI treatment, but this is usually more labor intensive than DpnI-mediated degradation.
13. Electrocompetent and recombination-proficient *E. coli* preparations can be stored for several weeks at −70 °C (up to 3 months in our hands) without losing the ability for recombination.
14. Times until completion may vary. To catch the cells in the right growth phase, start to measure OD after 2 h in 20 min intervals.
15. It is necessary to grow GS1783 bacteria at 32 °C, to keep the promoter driving the expression of the Red recombinase in its off state. Conversely, growing them at 42 °C for 15 min activates the promoter and leads to Red recombinase expression, making the bacteria ready to recombine any homologous DNA sequences. This induction should be brief to minimize nontargeted recombination events.
16. Do not dry by inverting the tube; the pellet may be really loose at this time and may detach. If you obtain less than 250 μL, adjust volume with ice-cold 10 % glycerol.

17. We recommend incubating the transformed bacteria for up to 3 h in the absence of antibiotic selection, in order to allow segregation of the mutated and non-mutated copy of the BAC molecules to daughter cells.

18. Colony PCR allows the identification of clones carrying co-integrates in the BAC. This is achieved with diagnostic primers, which generate PCR products of different lengths in the recombined and in the unrecombined BAC. We suggest the use of MangoMix PCR Kit and 25 μL volume per reaction. Before setting the PCR reaction, prepare 3 mL aliquots of LB medium with an adequate amount of selective antibiotic for each tested colony. Pick single colonies with a clean toothpick or yellow tip and resuspend them in the PCR reaction mix. Then put some material from the same colonies in the LB medium. Run the PCR reaction with an appropriate program. Liquid cultures incubate overnight at 32 °C, 220 rpm. Liquid cultures from positive clones may be used for DNA extraction and glycerol stocks preparation.

    Restriction fragment analysis allows the identification of restriction patterns that are specific for the recombined BAC while simultaneously allowing insight into the overall stability of the entire HCMV genome. Therefore, while more laborious than the colony PCR, it offers discrete advantages as a method. In order to culture enough bacteria to isolate DNA, pick clones from the plate with a sterile loop and inoculate antibiotic-supplemented LB medium. You may use mini (12 mL culture) or maxi (200 mL culture) preparation protocols. Digest half of the miniprep DNA or up to 2 μg of the maxiprep DNA with appropriate restriction enzyme (3 h, 37 °C, amount of enzyme 5–20 units). Prepare a large (20 cm long) 0.8 % agarose gel in 0.5× TBE (Tris-borate-EDTA) buffer. Load all restriction mixture with loading dye on gel. Run the gel at 60–70 V for approximately 16 h. Please note that you need to use the TBE buffer and gel, as the buffering capacity of the TAE gel is not sufficient for overnight runs.

19. Use the incubation time to prepare the L-arabinose solutions in LB medium. It should always be freshly prepared, sterile filtered, and not autoclaved.

20. It is essential to use Opti-MEM devoid of additives, like fetal calf serum or antibiotics, and tubes made of polystyrene and not polypropylene, as this may decrease the transfection efficiency.

21. It will likely take 2–3 weeks before you may be able to observe CPE. It is not uncommon for this amount of time to pass before significant CPE is observed, so it is imperative to be patient and continue passaging confluent cell cultures until the CPE is evident.

22. The sequence of the Kan resistance cassette is below:
CGATTTATTCAACAAAGCCACGTTGTGTCTC
AAAATCTCTGATGTTACATTGCACAAGATAAAAATATA
TCATCATGAACAATAAAACTGTCTGCTTACATAAACA
GTAATACAAGGGGTGTTATGAGCCATATTCAACGGG
AAACGTCTTGCTCGAGGCCGCGATTAAATTCCAACA
TGGATGCTGATTTATATGGGTATAAATGGGCTCGCG
ATAATGTCGGGCAATCAGGTGCGACAATCTATCGATT
GTATGGGAAGCCCGATGCGCCAGAGTTGTTTCTG
AAACATGGCAAAGGTAGCGTTGCCAATGATGTTAC
AGATGAGATGGTCAGACTAAACTGGCTGACGGAA
TTTATGCCTCTTCCGACCATCAAGCATTTTATCC
GTACTCCTGATGATGCATGGTTACTCACCACTGC
GATCCCCGGGAAAACAGCATTCCAGGTATTAGAA
GAATATCCTGATTCAGGTGAAAATATTGTTGATG
CGCTGGCAGTGTTCCTGCGCCGGTTGCATTCGA
TTCCTGTTTGTAATTGTCCTTTTAACAGCGATCG
CGTATTTCGTCTCGCTCAGGCGCAATCACGAATG
AATAACGGTTTGGTTGATGCGAGTGATTTTGATG
ACGAGCGTAATGGCTGGCCTGTTGAACAAGTCT
GGAAAGAAATGCATAAGCTTTTGCCATTCTCACC
GGATTCAGTCGTCACTCATGGTGATTTCTCACTT
GATAACCTTATTTTTGACGAGGGGAAATTAATAGG
TTGTATTGATGTTGGACGAGTCGGAATCGCAGACC
GATACCAGGATCTTGCCATCCTATGGAACTGCCTC
GGTGAGTTTTCTCCTTCATTACAGAAACGGCTTTTT
CAAAAATATGGTATTGATAATCCTGATATGAATAAA
TTGCAGTTTCATTTGATGCTCGATGAGTTTTTCTAA
TCAGAATTGGTTAATTGGTTGTAACAC

23. The sequence of the Zeo resistance cassette is below:
GTAAGAGGTTCCAACTTTCACCATAATGAAA
TAAGATCACTACCGGGCGTATTTTTTGAGTTATCGA
GATTTTCAGGAGCTAAGGAAGCTAAAATGGCCAAG
TTGACCAGTGCCGTTCCGGTGCTCACCGCGCGCG
ACGTCGCCGGAGCGGTCGAGTTCTGGACCGACCGG
CTCGGGTTCTCCCGGGACTTCGTGGAGGACGACTT
CGCCGGTGTGGTCCGGGACGACGTGACCCTGTTCA
TCAGCGCGGTCCAGGACCAGGTGGTGCCGGACAAC
ACCCTGGCCTGGGTGTGGGTGCGCGGCCTGGACGA
GCTGTACGCCGAGTGGTCGGAGGTCGTGTCCACGA
ACTTCCGGGACGCCTCCGGGCCGGCCATGACCGAG
ATCGGCGAGCAGCCGTGGGGGCGGGAGTTCGCCCT
GCGCGACCCGGCCGGCAACTGCGTGCACTTCGTGG
CCGAGGAGCAGGACTGATTTTTTTAAGGCAGTTATT

24. The sequence of the I-SceI restriction site is below:
TAGGGATAACAGGGTAATCGATTT

## Acknowledgments

I.D. is supported by a stipend from the Helmholtz International Graduate School for Infection Research; L.C.S. and M.M. are supported by the Helmholtz Virtual Institute "Viral Strategies of Immune Evasion" (VH-VI-424).

### References

1. Spaete RR, Mocarski ES (1987) Insertion and deletion mutagenesis of the human cytomegalovirus genome. Proc Natl Acad Sci U S A 84:7213–7217
2. Chee MS, Bankier AT, Beck S, Bohni R, Brown CM, Cerny R, Horsnell T, Hutchison CA 3rd, Kouzarides T, Martignetti JA et al (1990) Analysis of the protein-coding content of the sequence of human cytomegalovirus strain AD169. Curr Top Microbiol Immunol 154:125–169
3. Murphy E, Rigoutsos I, Shibuya T, Shenk TE (2003) Reevaluation of human cytomegalovirus coding potential. Proc Natl Acad Sci U S A 100:13585–13590
4. Davison AJ, Dolan A, Akter P, Addison C, Dargan DJ, Alcendor DJ, McGeoch DJ, Hayward GS (2003) The human cytomegalovirus genome revisited: comparison with the chimpanzee cytomegalovirus genome. J Gen Virol 84:17–28
5. Dargan DJ, Douglas E, Cunningham C, Jamieson F, Stanton RJ, Baluchova K, McSharry BP, Tomasec P, Emery VC, Percivalle E, Sarasini A, Gerna G, Wilkinson GW, Davison AJ (2010) Sequential mutations associated with adaptation of human cytomegalovirus to growth in cell culture. J Gen Virol 91:1535–1546
6. Sinzger C, Schmidt K, Knapp J, Kahl M, Beck R, Waldman J, Hebart H, Einsele H, Jahn G (1999) Modification of human cytomegalovirus tropism through propagation in vitro is associated with changes in the viral genome. J Gen Virol 80(Pt 11):2867–2877
7. Borst EM, Hahn G, Koszinowski UH, Messerle M (1999) Cloning of the human cytomegalovirus (HCMV) genome as an infectious bacterial artificial chromosome in Escherichia coli: a new approach for construction of HCMV mutants. J Virol 73:8320–8329
8. Hahn G, Rose D, Wagner M, Rhiel S, McVoy MA (2003) Cloning of the genomes of human cytomegalovirus strains Toledo, TownevarRIT3, and Towne long as BACs and site-directed mutagenesis using a PCR-based technique. Virology 307:164–177
9. Marchini A, Liu H, Zhu H (2001) Human cytomegalovirus with IE-2 (UL122) deleted fails to express early lytic genes. J Virol 75: 1870–1878
10. Hahn G, Khan H, Baldanti F, Koszinowski UH, Revello MG, Gerna G (2002) The human cytomegalovirus ribonucleotide reductase homolog UL45 is dispensable for growth in endothelial cells, as determined by a BAC-cloned clinical isolate of human cytomegalovirus with preserved wild-type characteristics. J Virol 76:9551–9555
11. Sinzger C, Hahn G, Digel M, Katona R, Sampaio KL, Messerle M, Hengel H, Koszinowski U, Brune W, Adler B (2008) Cloning and sequencing of a highly productive, endotheliotropic virus strain derived from human cytomegalovirus TB40/E. J Gen Virol 89:359–368
12. Murphy E, Yu D, Grimwood J, Schmutz J, Dickson M, Jarvis MA, Hahn G, Nelson JA, Myers RM, Shenk TE (2003) Coding potential of laboratory and clinical strains of human cytomegalovirus. Proc Natl Acad Sci U S A 100:14976–14981
13. Stanton RJ, Baluchova K, Dargan DJ, Cunningham C, Sheehy O, Seirafian S, McSharry BP, Neale ML, Davies JA, Tomasec P, Davison AJ, Wilkinson GW (2010) Reconstruction of the complete human cytomegalovirus genome in a BAC reveals RL13 to be a potent inhibitor of replication. J Clin Invest 120:3191–3208
14. Tischer BK, Smith GA, Osterrieder N (2010) En passant mutagenesis: a two step markerless red recombination system. Methods Mol Biol 634:421–430
15. Sambrook JF, Russel DW (2000) Molecular cloning: a laboratory manual. Cold Spring Harbor Laboratory Press, Cold Spring Harbor, N.Y
16. Borst EM, Irena C-M, Messerle M (2004) Cloning of β-herpesvirus genomes as bacterial artificial chromosomes. In: Zhao SS, Stodolsky M (eds) Bacterial artificial chromosomes: volume 2: functional studies. Humana., Totowa, NJ, pp 221–240
17. Tischer BK, von Einem J, Kaufer B, Osterrieder N (2006) Two-step red-mediated recombina-

tion for versatile high-efficiency markerless DNA manipulation in Escherichia coli. Biotechniques 40:191–197

18. Muyrers JP, Zhang Y, Testa G, Stewart AF (1999) Rapid modification of bacterial artificial chromosomes by ET-recombination. Nucleic Acids Res 27:1555–1557
19. Yu D, Ellis HM, Lee EC, Jenkins NA, Copeland NG, Court DL (2000) An efficient recombination system for chromosome engineering in Escherichia coli. Proc Natl Acad Sci U S A 97:5978–5983
20. Wagner M, Gutermann A, Podlech J, Reddehase MJ, Koszinowski UH (2002) Major histocompatibility complex class I allele-specific cooperative and competitive interactions between immune evasion proteins of cytomegalovirus. J Exp Med 196:805–816
21. Yu D, Smith GA, Enquist LW, Shenk T (2002) Construction of a self-excisable bacterial artificial chromosome containing the human cytomegalovirus genome and mutagenesis of the diploid TRL/IRL13 gene. J Virol 76:2316–2328

# Chapter 6

# The Use of Primary Human Cells (Fibroblasts, Monocytes, and Others) to Assess Human Cytomegalovirus Function

Emma Poole, Matthew Reeves, and John H. Sinclair

## Abstract

The extensive tropism of human cytomegalovirus (HCMV) results in the productive infection of multiple cell types within the human host. However, infection of other cell types, such as undifferentiated cells of the myeloid lineage, gives rise to nonpermissive infections. This has been used experimentally to model latent infection which is known to be established in the pluripotent CD34+ hematopoietic progenitor cell population resident in the bone marrow in vivo. The absence of a tractable animal model for studies of HCMV has resulted in a number of laboratories employing experimental infection of cells in vitro to simulate both HCMV lytic and latent infection. Herein, we will focus on the techniques used in our laboratory for the isolation and use of primary cells to study aspects of HCMV latency, reactivation, and lytic infection.

**Key words** Monocytes, CD34+ cells, Latency, Differentiation, Primary cell isolation

## 1 Introduction

HCMV is a species-specific pathogen which has precluded extensive analyses in animal models. Consequently, many studies have employed the use of cell lines to provide valuable insights into various aspects of HCMV biology. However, the usefulness of cell lines can be compromised by their biological relevance to HCMV as a human pathogen, and as such, much current research has focused on the ex vivo use of primary human cells and, more recently, the development of humanized mouse models [1].

In the human host, latency is established in the CD34+ hematopoietic cell compartment of the bone marrow. Despite the pluripotency of this population, the carriage of viral genomes is restricted to the cells of the myeloid lineage. Viral genome carriage occurs in the absence of the normal lytic transcription program, and reactivation is observed upon terminal differentiation of these myeloid progenitors to a macrophage or dendritic cell (DC) phenotype.

Andrew D. Yurochko and William E. Miller (eds.), *Human Cytomegaloviruses: Methods and Protocols*,
Methods in Molecular Biology, vol. 1119, DOI 10.1007/978-1-62703-788-4_6, © Springer Science+Business Media New York 2014

In this chapter we describe methods we routinely use in our attempts to elucidate the mechanisms that regulate HCMV latency and reactivation. In general, we use two primary cell types: (1) CD34+ cells isolated from G-CSF mobilized patients after leukopheresis which can be differentiated into Langerhans DCs and infected with HCMV either pre-differentiation (for latency and reactivation studies) or post-maturation (for analysis of lytic infection) and (2) CD14+ monocytes isolated from peripheral blood of health donors as latency can also be established in CD14+ monocytes [2]. The advantage of the latter cell type is its relative ease of isolation from venous blood and its ubiquitous availability. Monocytes can also be differentiated along the myeloid lineage to both monocyte-derived DCs and macrophages and, similar to CD34+ cells, can be infected pre-differentiation (for latency and reactivation studies) or post-differentiation (for analysis of lytic infection).

In order to allow us to understand the role of the immune system in the context of both lytic and latent infection, we also isolate autologous fibroblasts from HLA-typed donors to preclude issues arising from class restriction as experiments involving different cell types from non-matched donors may lead to nonspecific immune responses. Thus, for example, fibroblasts isolated from a specific donor may be infected with HCMV and then cocultured with cells mediating immune responses from that donor to analyze HCMV-specific immune responses. Latency in CD34+ and CD14+ cells is marked by the carriage of genome and the presence of the latency-associated transcripts such as UL81-82ast and UL138, but the absence of the immediate early (IE) transcripts. This contrasts with infection in differentiated cells and fibroblasts where the IE RNA and protein is detectable. For the study of natural latency, HCMV genome carriage can be detected directly ex vivo from myeloid cells or progenitors from seropositive individuals using more sensitive methods.

The techniques described in this report are used routinely in our laboratory to underpin our research programs analyzing the precise cellular and molecular mechanisms that govern the establishment, maintenance, and, ultimately, reactivation of HCMV.

## 2 Materials

Ensure that all work with blood products is carried out in accordance with local safety guidelines and with ethical considerations. Isolation of blood cell populations is routinely carried out in a Class II microbiological safety cabinet (Class II MSC) for user protection and sterility usually located in a Containment Level 2/3 (CL2/3) laboratory (UK). Endotoxin-free PBS is purchased as a 1× solution (Sigma).

### 2.1 Mononuclear Isolation

1. Magnetic-activated cell sorting (MACS) magnet, presterilization filter, LS columns.
2. MACS buffer (Miltenyi Biotec).
3. Heparin, Lymphoprep.
4. PBS.
5. Centrifuge (50 ml tube capacity).
6. 50 ml polypropylene conical tubes (Falcon).
7. 14 ml snap-cap polypropylene tubes.
8. Freezing medium (Bambanker™).
9. For isolation of CD14+ monocytes: CD14+ microbeads (Miltenyi Biotec), anti-CD14+ FITC conjugated antibody.
10. For isolation of CD34+ cells: CD34+ microbeads (Miltenyi Biotec), anti-CD34+ FITC conjugated antibody that recognizes a different epitope than QBEND/10 (QBEND/10 is the antibody used in the purification step).

### 2.2 Dermal Fibroblast Isolation

1. Tissue culture grade plasticware: 6-well plates, 25 $cm^2$ flask.
2. 14 ml polypropylene centrifuge tubes.
3. Coverslips.
4. Eagle's Minimum Essential Medium (EMEM) supplemented with 10 % FCS and 5 % penicillin/streptomycin (pen/strep).

### 2.3 Infection and Differentiation

1. Cells as prepared (*see* Subheading 3).
2. Iscove's medium (500 ml Iscove's Modified Eagle's Medium (IMEM) supplemented with 50 ml FCS and 75 ml horse serum as well as 5 ml pen/strep. Medium will look cloudy). Phorbol 12-myristate 13-acetate (PMA).
3. Hydrocortisone.
4. X-vivo 15 medium (BioWhittaker).
5. Cytokines: TGF-beta, TNF-alpha, SCF, Flt-3 L, GM-CSF.
6. L-Glutamine.
7. LPS (derived from *Escherichia coli* serotype 0111:B4).
8. Tissue culture grade plasticware.
9. 14 ml snap-cap polypropylene tubes.
10. 24-well plates.
11. Endotoxin-free PBS.
12. Appropriate isolate of HCMV, e.g., for studies in myeloid or endothelial cells. It is essential to use a myeloid/endothelial tropic HCMV isolate (*see* Chapter 3) which is confirmed to be able to productively infect DCs and endothelial or retinal pigment epithelial cells (RPEs). This can be checked by indirect

immunofluorescent staining for IE72/86 expression, and relative MOIs on these cells can be calculated on the basis of IE-forming foci of infection. The endotheliotropism of the virus can be tested by a side-by-side comparative infection with fibroblasts and RPEs.

### 2.4 Validation of Infection and Reactivation

1. Primers:

   *GAPDH sense*

   5′-GAGTCAACGGATTTGGTCGT

   *GAPDH antisense*

   5′-TTGATTTTGGAGGGATCTCG

   *UL138 sense*

   5′-TGCGCATGTTTCTGAGCTAC

   *UL138 antisense*

   5′-ACGGGTTTCAACAGATCGAC

   *IE sense*

   5′-CGT CCT TGA CAC GAT GGA GT

   *IE antisense*

   5′-ATT CTT CGG CCA ACT CTG GA

   *LUNA*

   *LUNA sense*

   5′-TGACCTCTCCTCCACACC

   *LUNA antisense*

   5′-GGAAAAACACGCGGGGGA
2. 96-well flat-bottomed tissue culture plate.
3. PCR solution A: 100 mM KCl, 10 mM Tris–HCl pH 8.3, 2.5 mM $MgCl_2$.
4. PCR solution B: 10 mM Tris–HCl pH 8.3, 2.5 mM $MgCl_2$, 1 % Tween 20, 1 % Nonidet P-40, 0.4 mg/ml proteinase K.
5. TRIzol (Invitrogen).
6. DEPC-treated water.
7. Primers for GAPDH, UL138, LUNA and IE (see above for sequences).
8. DNA isolation solutions:

   Solution 1: 100 mM NaCl, 5 mM EDTA pH 8.0.

   Solution 2: 10 % SDS.

   Solution 3: 5 M sodium perchlorate, Tris/EDTA (TE) buffer, 10 ml Tris–HCl pH 8.0, 1 mM EDTA.

9. Phenol: chloroform: isoamyl alcohol (25:24:1).
10. Chloroform.
11. Isopropanol.
12. 70 % ethanol.
13. 2 ml Eppendorf tubes.
14. 5× TBE buffer (1 l): 54 g Tris base, 27.25 g orthoboric acid, 4.6 g EDTA. Use at 0.5×.
15. TAE buffer: 0.04 M Tris acetate, 0.001 M EDTA, 17.5 M glacial acetic acid.
16. DNA loading buffers:
    Stop/load buffer: to make 20 ml, first mix 0.2 ml 10 % SDS, 10 ml 0.25 M EDTA.
    Then add 9.8 ml sterile glycerol.
17. DNA gel: 2 % agarose in TAE, 0.85–1 cm thick, containing 1 μg/ml ethidium bromide.
18. 0.25 M HCl.
19. 0.4 M NaOH.
20. Positively charged nylon membrane (N+ Hybond)
21. 20× SSC (1 l): 175.3 g NaCl, 88.2 g Na citrate, pH 7.0 with NaOH.
22. Prehybridization solution (95 ml):10 ml of 50× Denhardt's solution, 30 ml 20× SSC, 500 μl 20 % SDS, 54.5 ml water (note: for GC-rich sequences, 20 ml water may be replaced with formamide).
23. Salmon sperm DNA (Sigma).
24. 50× Denhardt's solution (500 ml): 5 g polyvinyl pyridine, 5 g bovine serum albumin (BSA), 5 g Ficoll, in 500 ml water.
25. Rediprime™ II (Amersham).
26. dCTP α-$^{32P}$.
27. Autoradiograph film.
28. Nuclear Hoechst stain.
29. Mouse anti-IE antibody, Chemicon mAB810.
30. Alexa Fluor 594 or 488 nm conjugated goat anti-mouse antibody.
31. PBS.
32. 0.1 % Triton X-100 in PBS.
33. 4 % formaldehyde in PBS.

## 3 Methods

Ensure that all work with blood products is carried out in accordance with local safety guidelines. Isolation of blood cell populations are routinely carried out in a Class II microbiological safety cabinet (Class II MSC) for user protection and sterility usually located in a Containment Level 2/3 (CL2/3) laboratory (UK).

When isolating cells for natural latency studies, it is essential to use separate HCMV-free facilities to avoid contamination.

### *3.1 CD34+ Isolation*

Blood packs will usually be supplied in two forms: frozen blood packs or freshly isolated samples. CD34+ cells can be mobilized by the administration of G-CSF in healthy donors. The blood from these donors is then isolated directly or frozen in autologous serum and 10 % DMSO. The G-CSF treatment causes these samples to be enriched for white blood cells (WBC) and usually have significant red blood cell depletion.

Frozen packs can be stored short term at −70 °C (for weeks) and long term in liquid nitrogen (for some months) if they cannot be processed the same day they are obtained.

With fresh samples, it is better to process these the same day they are received. However, if this is not possible, they can be stored at 4 °C for up to 24 h.

The procedure for isolation of pure CD34+ cells is essentially the same for both frozen and fresh samples, and yields are dependent on cell number and also the quality of the preparation. Frozen blood packs tend to have more coagulated protein in them because of the serum used to freeze the cells (*see* **Note 1**).

Protocol (For Fresh Blood)

1. Cool MACS magnet adaptor and LS columns at 4 °C. LS columns have a total loading capacity of $2 \times 10^9$ unpurified cells (which would allow the isolation of $10^8$ CD34+ cells). So in this example, for $3 \times 10^9$ total cells use 3 columns.
2. Cool MACS buffer (PBS supplemented with 0.5 % serum and 2 mM EDTA) and PBS on ice.
3. Dilute the blood 1:5 to 1:10 in PBS (*see* **Note 2**).
4. Add 15 ml Lymphoprep to a 50 ml conical tube.
5. Carefully layer 30–35 ml of blood/PBS on top.
6. Spin at $800 \times g$ for 20 min. NO BRAKE on centrifuge.
7. Take WBC at interphase. Avoid aspirating Lymphoprep as it can prevent pelleting of cells.
8. Pellet cells at $300 \times g$ for 10 min. If there is a problem pelleting cells possibly due to contaminating Lymphoprep, then dilute 1:4 in PBS and repeat centrifugation step.
9. Wash in cold PBS $300 \times g$ for 10 min. Add 50 ml of PBS to pelleted cells.

10. OPTIONAL: If you are finding that you have a lot of collagenous material in your prep, then pass cells through pre-separation filter and transfer to a fresh tube. This tends to be more common with the frozen blood packs, but the a pre-separation filter step may be beneficial for the fresh blood as well.
11. Count cells. May need to dilute (1:100).
12. Centrifuge at $300 \times g$ for 10 min.
13. Resuspend pellet in 300 µl of MACS buffer per $10^8$ total cells. So for $3 \times 10^9$ the cells would be resuspended in 9 ml. Add 100 µl of FcR blocking reagent per $10^8$ (so 3 ml in this example) and 100 µl of microbead conjugated anti-CD34 antibodies (again 3 ml in this example).
14. Incubate in a 14 ml polypropylene snap-cap tube with aerated lid for 20 min at 4 °C.
15. Transfer to a 50 ml Falcon and wash in PBS. Spin $300 \times g$ for 10 min.
16. As spin ends, charge the column.
17. To charge the column insert LS column into magnet. Then run 3 ml of MACS buffer through the column. Discard flow through.
18. Resuspend the centrifuged cells in 500 µl of MACS buffer per $10^8$ cells (in this example, this would be 15 ml).
19. The column has a max loading capacity of $2 \times 10^9$ cells, so since we have $3 \times 10^9$ cells, we only add 5 ml per column and in this example 3 columns will be required. As $10^9$ cells are taken up in 5 ml, there should never be more than 5 ml added to 1 column.
20. Collect flow through. Reapply this flow through to the column. Collect again. This represents the CD34+ depleted fraction. This can be frozen as depleted CD34+ PBMC for use later.
21. Wash the column with a 3 ml wash of MACS buffer. Repeat two more times, each time allowing the first wash to run through completely.
22. Finally, remove the column from the magnet and place over collection tube (14 ml snap-cap tube). Elute in 5 ml of PBS OR X-vivo 15 medium. Use plunger to push cells off the column. This represents the purified fraction. Repeat with other 2 columns (*see* **Note 3**).
23. The next stage is dependent on the downstream application.
24. Pool and count the cells. First take a sample (i.e., $10^5$) for flow cytometry analysis to check the purity of the purified CD34+ cells (*see* **Note 4**).
25. Spin at $300 \times g$ for 10 min and remove medium. For freezing, resuspend cells at $5 \times 10^6$ cells per ml in freezing medium and freeze in 1 ml aliquots (*see* **Note 5**).

26. If cells are to be used for downstream experiments within 24 h, then they can be resuspended in X-vivo 15 medium at the appropriate volume and either plated immediately or left overnight in a 14 ml snap-cap tube in a 37 °C 5 % $CO_2$ incubator.

### 3.2 CD34+ Differentiation into CD34+-Derived Langerhans DCs

For studies on naturally latently infected cells, it is critical that all procedures prior to PCR amplification are carried out in HCMV-free facilities. We define HCMV-free facilities as an area that is used exclusively for the preparation, culture, and processing of tissue from seropositive or seronegative donors. All reagents are purchased presterilized, and all solutions are prepared using tissue culture grade sterile water. Under no circumstances are infected cells (or DNA/RNA prepared from them), plasmids, amplified HCMV-PCR products, or autoclaved solutions ever brought into this facility.

CD34+ cells can be differentiated into Langerhans DCs using the following method described by Strobl et al. [3] and modified by MacAry et al. [4]:

1. Following purification or resuscitation, CD34+ cells are incubated in a 14 ml snap-cap tube overnight at 37 °C, 5 % $CO_2$ in X vivo-15 medium.
2. The following morning, count and check the number of viable cells by trypan blue staining to ensure that not all the cells are dead before proceeding (up to 50 % of cells can be lost during the resuscitation procedure); then fill the tube with X-vivo 15 medium and centrifuge at 200 × *g* for 10 min to pellet cells.
3. Resuspend pellet in X-vivo 15 medium supplemented with TGF-β (0.5 ng/ml), TNF-α (2.5 ng/ml), SCF (20 ng/ml), Flt-3 ligand (100 ng/ml), GM-CSF (100 ng/ml), and 2.25 mM L-glutamine.
4. Plate the cells out in 24-well plates at $2 \times 10^4$ cells per well in 1 ml of differentiation medium and left to differentiate for 7 days at 37 °C 5 % $CO_2$ (*see* **Note 6**). Following differentiation, maturation is induced by the addition of 500 ng/ml LPS for 48 h

### 3.3 Infection of CD34+ Cells and CD34+-Derived Langerhans DCs

The following protocol describes the methodology for experimental infection of CD34+ cells for latency or reactivation studies. Purified or resuscitated cells are incubated overnight in X-vivo 15 medium at 37 °C, 5 % $CO_2$ prior to infection. CD34+ cells are suspension cells and therefore can be grown in 14 ml snap-cap tubes. The following example describes infection of $2 \times 10^4$ cells per well in a 24-well format:

1. Pellet cells to be infected in 14 ml snap-cap tubes at 200 × *g* for 15 min.
2. Dilute virus in X-vivo 15 medium (supplemented with 2.25 mM L-glutamine) to a predicted MOI of 5 in approximately 300 μl X-vivo 15 for each $1 \times 10^5$ cell.

3. Resuspend $1 \times 10^5$ CD34+ cells per 300 μl of virus:medium mix. This will give an approximate MOI of 5.
4. Incubate cells for 3 h at 37 °C, 5 % $CO_2$ with gentle agitation every 30 min.
5. Fill tube with X-vivo 15 medium and centrifuge cells at $200 \times g$ for 15 min.
6. Decant medium and discard and resuspend cells in residual medium.
7. Repeat **steps 5** and **6**.
8. Resuspend cells in a final concentration of $2 \times 10^4$ CD34+ cells per 1 ml.
9. Plate out in a 24-well plate at 1 ml per well and incubate at 37 °C and 5 % $CO_2$.
10. At 3 days postinfection (a time at which HCMV latency should be established), replace the medium with fresh X-vivo 15. At this point cells can be harvested or cultured for a further 7 days before harvesting for long-term latency analysis.

Following infection, CD34+ cells may then be differentiated to allow for reactivation analyses. Alternatively, following differentiation of uninfected CD34+ cells into Langerhans DCs, they may then be infected. The infection procedure differs to that for CD34+ cells as they are adherent.

The following protocol describes methodology for experimental infection of CD34+-derived Langerhans DCs:

1. Count one well of a 24-well plate containing adherent Langerhans DCs to determine the amount of virus necessary to achieve an MOI of 5 in 300 μl medium.
2. Dilute the appropriate amount of virus in X-vivo 15 medium and add to adhered Langerhans DCs in 24-well plates.
3. Incubate at RT for 3 h with gentle rocking.
4. Wash cells with X-vivo 15 medium and feed cells with 1 ml of fresh medium.
5. Cells can be harvested at different time points depending on the downstream analysis. Immediate early genes will be detectable as soon as 6 h postinfection. Late genes will be detectable at 72 h postinfection.

### 3.4 CD14+ Monocyte Isolation

An alternative cell type to CD34+ cells for the study of HCMV latency and reactivation is CD14+ monocytes. The advantage of this system is the availability of material as the cells can be isolated directly from small volumes of venous blood.

Protocol (for fresh blood):

Typically 50 ml of venous blood is drawn into tubes containing the anticoagulant heparin to prevent clotting. Generally between

$7 \times 10^7$ and $1 \times 10^8$ total WBCs will be isolated of which CD14+ monocytes represent around 10 %.

Once blood had been drawn:

1. Put MACS magnet adaptor and LS columns at 4 °C (*see* **Note 7**).
2. Cool MACS buffer (PBS supplemented with 0.5 % serum and 2 mM EDTA) and PBS on ice.
3. Dilute the blood 1:1 in sterile PBS.
4. Add 15 ml Lymphoprep to a 50 ml polypropylene conical tube.
5. Layer 35 ml of blood: PBS on top.
6. Spin at $800 \times g$ for 20 min. NO BRAKE on centrifuge.
7. Take of WBC at interphase (*see* **Note 8**).
8. Pellet cells at $300 \times g$ for 10 min.
9. Wash in cold PBS $300 \times g$ for 10 min. Add 50 ml of PBS.
10. Count cells.
11. Pellet cells at $300 \times g$ for 10 min.
12. Resuspend pellet in 80 µl of MACS buffer per $10^7$ total cells. So for $1 \times 10^8$ cells, the cells would be resuspended in 800 µl. Then add 20 µl of microbead conjugated CD14+ antibodies per $10^7$ cells (so 200 µl in this example).
13. Incubate in a 14 ml snap-cap polypropylene tube with aerated lid for 20 min at 4 °C.
14. Transfer cells to a 50 ml polypropylene tube and wash in PBS. Spin $300 \times g$ for 10 min.
15. Charge the MACS column. To charge the column insert LS column into the magnet. Then run 3 ml of MACS buffer/PBS through the column. Discard flow through.
16. Resuspend the centrifuged cells in 500 µl of MACS buffer per $10^8$ cells (in this example, this would be 500 µl).
17. The column has a max loading capacity of $2 \times 10^9$ leading to the potential purification of up to $10^8$ cells so we can load the whole column with sample.
18. Collect flow through. Reapply this flow through to the column. Collect again. The flow through represents the CD14+-depleted fraction. This can be frozen as CD14+-depleted PBMC.
19. Wash the column with 3 ml of MACS buffer/PBS. Repeat two more times, each time allowing the first wash to run through completely.
20. Finally, remove the column from the magnet and place over collection tube. Use the plunger to elute cells off the column. This represents the purified fraction.

21. The next stage is dependent on the downstream application.
22. Pool cells and pellet at 300 × *g* for 10 min.
23. If cells are to be frozen, resuspend at $5 \times 10^6$ cells per ml in freezing medium and freeze in 500 μl aliquots.
24. If cells are to be used immediately for downstream experimentation (which is highly preferable), then they can be resuspended at a volume appropriate for the experimental setup. Typically, we resuspend CD14+ cells at a concentration of $5 \times 10^5$ cells per ml.

### *3.5 Monocyte Differentiation into Monocyte-Derived DCs and Macrophages*

Once monocytes have been isolated, they are plated out at a density of $5 \times 10^5$ per well of a 24-well plate (this can be scaled up or down depending on the downstream application). Cells are plated out in PBS and allowed to adhere for 3 h at 37 °C (*see* **Note 9**), 5 % $CO_2$, before replacing the PBS with medium (X-vivo 15 supplemented with 2.25 mM L-glutamine) or differentiation medium as described below. They can either be infected as monocytes or they can be differentiated along the myeloid lineage into macrophages or DCs prior to infection (*see* **Note 10**).

Monocytes can be differentiated into macrophages essentially as described by Lathey and Spector [5]. The following example uses a 24-well plate format:

1. Remove PBS and incubate the cells with 1 ml medium supplemented with 50 μM hydrocortisone. After 3 days replace with fresh medium supplemented with 50 μM hydrocortisone.
2. Grow cells for an additional 3 days and then replace medium with 1 ml of fresh medium plus 10 ng/ml PMA.
3. After 24–48 h, cellular differentiation will be evident. The cells become more adherent and granular and show the formation of limited processes.

The differentiation of monocytes to DCs is performed essentially as described in Sallusto and Lanzavecchia [6] except that serum was omitted from the growth medium. This example assumes a 24-well plate format:

1. For the generation of immature DCs, replace PBS with 1 ml of X-vivo 15 +2.5 mM L-glutamine supplemented with 1,000 U/ml of IL-4 and 1,000 U/ml of GM-CSF.
2. Leave cells for 6 days at 37 °C, 5 % $CO_2$, to differentiate.
3. To trigger maturation overnight, replace medium with 1 ml of X-vivo 15 +2.5 mM L-glutamine supplemented with 500 ng/ml of LPS.

### 3.6 Infection of Monocytes and Monocyte-Derived DCs and Macrophages

Infecting Adherent Cells with HCMV (96-Well Plate)

1. Prepare stock of virus at desired MOI in 150 μl of X-vivo 15 medium.
2. Aspirate medium from DC cultures using a pipette on the edge of the well.
3. Add virus/medium, pipetting against the edge of the well, and leave for 1 h at 37 °C, 5 % $CO_2$.
4. OPTIONAL: Seal 6-well plates using parafilm and spin down the virus onto the cells. In a standard plate centrifuge, spin at 300×*g* for 15 min before leaving 1 h at 37 °C, 5 % $CO_2$.
5. After 1 h, remove virus medium and replace with fresh X-vivo 15 medium.
6. Cells can be harvested at different time points depending on the downstream analysis.

### 3.7 Dermal Fibroblast Isolation

There are many commercial fibroblast cell lines available. However, if the downstream application requires autologous fibroblasts, then it may be necessary to isolate them specifically from donors of interest. Both commercial and biopsy-derived fibroblasts are fully permissive for HCMV infection. The cell lines take 6–8 weeks to establish from the initial biopsy material and can be readily frozen down and stored in liquid nitrogen. The skin biopsies are performed by qualified medical personnel and obtained according to institutional IRB and ethical guidelines:

1. Obtain skin biopsy material.
2. In a Class II MSC, cut the skin biopsy into fine strips using a rounded scalpel with a guillotine motion to expose straight layers of fibroblasts—these are the only edges where cells will grow out from the tissue.
3. Place a single strip into each well of a 6-well plate.
4. Place a sterile coverslip over the strip. This ensures the fibroblast tissue is immobilized onto the plastic since the outgrowth of fibroblasts is dependent on adherence.
5. Add a small drop of medium (EMEM-10) to the edge of the coverslip and allow capillary action to take the medium up under the coverslip.
6. Add 2 ml of EMEM-10 to each well.
7. Incubate at 37 °C in 5 % $CO_2$.
8. When fibroblasts are confluent under the coverslip, lift coverslip and remove growth medium.
9. Wash with PBS and aspirate.
10. Add 0.5 ml/well of trypsin/EDTA.
11. Incubate at 37 °C and check every couple of minutes to see whether the fibroblasts have detached.

12. Harvest trypsin/fibroblast mix and pool all wells into a centrifuge tube, and top up with EMEM-10 to neutralize trypsin.
13. Centrifuge cells (200×*g*, 5 min) and resuspend in 10 ml EMEM-10.
14. Seed cells into a 25 cm$^2$ flask and incubate at 37 °C, 5 % $CO_2$.

Expand primary fibroblasts when tissue culture flasks become confluent; the cells should not be split higher than 1 in 2.

### 3.8 Infection of Dermal Fibroblasts

Fibroblasts obtained commercially or prepared as described above can be infected with HCMV using the following protocol:

1. Seed fibroblasts in a 24-well plate format in EMEM-10 medium and allow to adhere overnight.
2. Infect at an MOI appropriate for the percentage infection required—this can be worked out empirically using Poisson distribution.
3. Infect in a volume of 300 μl in EMEM-2 and leave on a rocker at RT for 1 h.
4. Wash cells in medium and replace with 1.5 ml EMEM-10.
5. Cells can be harvested at different time points depending on the downstream analysis (*see* **Note 11**).

### 3.9 Validation of Experimental HCMV Latency and Reactivation from CD34+ and Monocyte Models

To analyze latency in CD34+ cells or monocytes, there are currently no commercially available antibodies for the detection of latent products. Therefore, experimental latency is defined by the carriage of viral DNA, the presence of the characterized latency-associated transcripts, e.g., LUNA and/or UL138 [7, 8], and the absence of IE gene transcription or infectious virus production. Reactivation/productive infection is defined as the presence of IE gene transcription and production of infectious virus.

DNA can be isolated from cells latently carrying genome using the following protocol adapted from Roback et al. [9]:

1. Pellet 1×10$^6$ cells.
2. Wash 1× in PBS.
3. Lyse directly in 200 μl solution A.
4. Add 200 μl solution B and transfer to an Eppendorf tube.
5. Heat at 60 °C for 1 h.
6. Place at 95 °C for 10 min (DNA can be used directly in PCR reactions using Bioline Red mix at this point).
7. If necessary (i.e., failure to detect target sequences by PCR), the DNA isolated above can be concentrated using phenol–chloroform extraction and ethanol precipitate.

RNA should also be analyzed for the transcription of latency-associated products in the absence of IE gene transcripts or in the presence of IE depending on whether the cells are latent or reactivating:

1. Productively infected cells, latently infected cells, or cells reactivating virus can be harvested in Trizol following the manufacturer's protocol, solubilizing the RNA at 65 °C for 10 min after resuspension of the pellet in DEPC-treated water.
2. RT is carried out using the Promega reverse transcription kit with the modification of 15 min at RT followed by 1 h incubation at 42 °C.
3. PCR can be carried out using Bioline Red mix using previously published parameters [10].
4. Samples should be run on a 2 % agarose gel in 0.5× TBE (*see* **Note 12**).

For reactivation studies it is advisable to check for the presence of IE by immunofluorescence. This can be carried out in 96 well flat-bottomed plates.

1. Wash cells in PBS.
2. Fix cells in 4 % formaldehyde in PBS for 20 min.
3. Permeabilize in 0.1 % Triton X-100 in PBS for 2 min.
4. Wash 3× 5 min in PBS.
5. To detect IE gene expression, a mouse anti-IE72/86 antibody (Chemicon mAB810) was used. Specifically, antibody was diluted 1:1,000 in PBS and then 100 μl of the diluted antibody added to the well.
6. Incubate 1 h RT.
7. Wash 3× 5 min in PBS.
8. Add 100 μl secondary antibody diluted in PBS with Hoechst stain.
9. Incubate in the dark 1 h RT.
10. Wash 3× 5 min in PBS.
11. Cells can be viewed directly in wells with an immunofluorescence microscope.

### 3.10 Validation of Experimental HCMV Fibroblast Infection

Fibroblast infection can be validated using the same techniques used for cells reactivating virus. However, due to abundance of viral gene expression during lytic fibroblast infection, less sensitive techniques such as western blotting may also be used.

### 3.11 Validation of Natural HCMV Latency and Reactivation

Studies performed on cells derived from healthy seropositive donors (natural latency) are rendered inherently difficult due to the extremely low frequency of genome positive cells within the myeloid compartment—which is estimated to be around 1 genome-positive cell for every $1 \times 10^4$ to $1 \times 10^5$ mononuclear cells [11]. Therefore, extremely sensitive methods need to be used to detect viral genome carriage during natural latency. DNA is isolated from the CD34+ cells (isolated as described) of seropositive individuals (determined by HCMV ELISA). The DNA is then amplified by PCR, Southern blotted, and probed for the presence of HCMV-specific amplification products using radiolabeled probes:

1. Pellet $1 \times 10^6$ mononuclear cells in a 2 ml Eppendorf tube.
2. Resuspend pelleted cells in 600 μl of solution 1.
3. Add 125 μl solution 2 to lyse the cells, and mix by inversion.
4. Add 150 μl solution 3 to aggregate protein from lysed cells, and mix by inversion.
5. Add 1 ml phenol–chloroform: isoamyl alcohol (25:24:1).
6. Centrifuge at $10{,}000 \times g$ for 15 min, and transfer aqueous layer to a fresh tube.
7. Add 1 ml chloroform to supernatant, and mix by inversion.
8. Centrifuge at $10{,}000 \times g$ for 10 min, and transfer aqueous layer to a fresh tube.
9. Add equal volume of isopropanol and leave for 30 min at −20 °C.
10. Centrifuge at $10{,}000 \times g$ for 30 min and wash pellet twice with 70 % ethanol.
11. Resuspend pellet in 100 μl TE buffer.
12. To remove any traces of salts or potential contaminants, dialyze with any appropriate DNA dialysis equipment at 4 °C overnight in TE buffer (*see* **Note 13**). Typically, the isolated DNA is loaded into 3 cm of dialysis tubing (Gibco, ¼ in. by 25 ft, cat#15961-014), sealed at either end with crocodile clips, and then immersed in a 1 l beaker containing ice cold TE buffer. Using a magnetic flea, the TE buffer is mixed constantly at 4 °C on a magnetic stirrer overnight. For studies on natural latency, the removal of salts improves the detection of low copy number sequences by PCR.
13. DNA is then ready for PCR. The PCR reaction needs to be very sensitive, and therefore we have routinely used Amplitaq Gold with Betaine Enhancer (Perkins Elmer) [12]. To detect HCMV DNA, the IE primers above were used at a final concentration of 1 μM. 15 μl of Betaine Enhancer was substituted for water in a 50 μl reaction.
14. Typically, 1 μg of total DNA is amplified by PCR. Controls must include samples of DNA isolated from seronegative donors performed concurrently. To confirm the sensitivity of the PCR, a dilution curve of a positive control plasmid containing the IE target amplimer is diluted from a stock (typically 1 mg/ml) to between $10^{-4}$ and $10^{-7}$ for the PCR. Typically, we use a plasmid DNA sequence of which we have calculated the copy number (*see* **Note 14**).

15. Carry out PCR reactions. 65 cycles will be required to detect very low levels (1–10 copies) of DNA followed by Southern blotting.
16. Run PCR reactions on 2 % agarose gel and photograph gel with ruler alongside so the molecular size markers can be transposed from the gel to the nitrocellulose filter to ensure products detected by Southern blotting are the predicted size.
17. Place gel in 0.25 M HCl with gentle agitation for 20 min (*see* **Note 14**).
18. Place gel in 0.4 M NaOH with gentle agitation for 20 min (caution—gel softens).
19. Assemble semidry apparatus and transfer onto the filter overnight for Southern blotting. In principle, four sheets of 3MM paper are cut concentric to the gel to be blotted but with an extra 6 cm on all sides. The 3MM is soaked in 0.4 M NaOH and then placed on an excess of cling film. The gel is placed on the 3MM topside down and the cling film wrapped over the 3MM to ensure no 3MM is exposed to the air. Then a piece of nitrocellulose filter paper is cut to the size of the gel, wetted with water, and placed on gel. To this two dry pieces of 3MM paper cut to the size of the gel are added followed by a stack of tissues (10 cm depth) on top. Using a glass plate add a 500 g to 1 kg weight. Typically we use 2 × 500 ml glass bottles filled with water. Leave apparatus overnight. Successful transfer is usually implied by a brown discoloration of the tissues due to flow of NaOH through the apparatus.
20. Next day disassemble the Southern blot and rinse the filter in 5× SSC. Prehybridize with prehybridization solution supplemented with sheared and boiled (10 min) salmon sperm DNA at 10 mg/ml.
21. Prehybridize at 65 °C.
22. To generate the probe, Rediprime™ II can be used (Amersham).

    For detection of the amplified product generated in the PCR above (*see* **Step 13**), we use internal primers to amplify a 201 bp product from HCMV DNA which is then gel purified. The following primers are used: 5′-CCCTGATAATCCTGA CGAGG and 5′-CATAGTCTGCAGGAACGTCGT. To generate a radiolabeled probe, 25 ng of this purified DNA product is labeled with dCTP $\alpha$-$^{32P}$ using the Rediprime™ II kit as described by manufacturer. Typically, we use 1.85 MBq of dCTP $\alpha$-$^{32P}$ per Rediprime reaction. In principle, the reaction involves the denaturation of the probe DNA (boiling step) and subsequent incubation with random hexonucleotides and the Klenow fragment of DNA polymerase I which replicates the probe and, simultaneously, incorporates radiolabeled dCTP $\alpha$-$^{32P}$ into the new products.

After the addition of dCTP α-$^{32P}$ to the Rediprime mix, incubate at RT for 1–2 h or for 20 min at 37 °C.

23. Probe is then column purified using Quick Spin G-50 Sephadex spin columns which will bind to nucleotides and sequences below 50 bp in length and incorporation calculated by measuring bound (unincorporated $^{32}$P-αdCTP ) vs. eluted fraction ($^{32}$P-αdCTP incorporated into probe). Typically, we see around 70 % efficiency of incorporation, but if we see lower than 30 %, we will repeat the labeling procedure. Using the known input (1.85 MBq) and % incorporation, you can then aliquot the eluted probe into 0.2 Mbq aliquots which are used per Southern blot. Labeled probes can typically be stored for 2–3 weeks at −20 °C and remain usable for probing Southern blots.
24. Incubate filter with one aliquot of probe and hybridization buffer at 65 °C for 24–48 h.
25. Then wash filter in 2× SSC/0.1 % SDS (low stringency) or 0.2 % SSC/0.1 % SDS (high stringency) for 3×20 min at 65 °C—longer washes will improve signal to noise ratio. Prior to exposure to autoradiographic film, a sweep of the nitrocellulose with a Geiger counter is informative. A uniform signal is indicative of high background, and it is recommended to perform further washes. Usually, counts of less than ten should be detected on areas of the nitrocellulose that have no amplified product, whereas "hot spots" should be readily detectable where positive controls have been loaded.
26. Wrap filter in plastic wrap and expose to autoradiograph film. For the detection of low abundance products, the exposure cassette can be left for up to a week and this is performed in a designated −70 °C freezer.

## 4 Notes

1. Cell counts up to $1 \times 10^{10}$ can be processed directly by CD34+ cell isolation, depending on the availability of magnets, columns, and beads. This is also true of frozen blood packs although the initial recovery of total cells is far less efficient from these packs.
2. For venous blood we use 1:1 but these are highly concentrated samples so a higher dilution factor in PBS is used.
3. A second column purification can be performed on the CD34+-enriched cells for further purity.
4. The antibody used for flow cytometry should recognize a different epitope to the one used for purification.
5. With frozen blood packs it is ideal if the cells are used immediately. Cells from freshly purified blood can be frozen.

6. Typically, single-cell suspensions of CD34+ cells at day 0 grow to loosely adherent clusters by day 4. By Day 6 large clusters of immature Langerhans cells are visible and represent a 50–100-fold increase in cell number following differentiation.
7. LS columns have a total capacity of $10^8$ bound cells.
8. Try not to take Lymphoprep as it can interfere with downstream applications.
9. Cells may adhere within an hour depending on the plastic and health of the cells, and PBS can be replaced with media as soon as cells are adherent.
10. Differentiation to a DC phenotype is determined by a phenotype of $CD14^{neg}$, $HLA\text{-}DR^{+}$, $CD83^{low}$, and $CD86^{low}$ which, upon maturation, is $HLA\text{-}DR^{++}$, $CD83^{+}$, and $CD86^{++}$. Differentiation to a macrophage phenotype is determined by a phenotype of $CD14^{+}$, $CD68^{+}$, and $CD83^{neg}$.
11. Immediate early genes will be detectable as soon as 6 h postinfection. Late genes will be detectable at 72 h postinfection.
12. A 2 % TAE agarose gel is used in preference to lower percentages to aid resolution of small PCR products. DNA can be left up to 48 h dialyzing if necessary. At least one buffer change should be included during dialysis.
13. Using this control we have shown the sensitivity of the PCR is as low as four copies of HCMV target DNA.
14. Bromophenol blue will turn yellow/green.

## References


1. Smith MS, Goldman DC, Bailey AS, Pfaffle DL, Kreklywich CN, Spencer DB, Othieno FA, Streblow DN, Garcia JV, Fleming WH, Nelson JA (2010) Cell Host Microbe 8:284–291
2. Taylor-Wiedeman J, Sissons JG, Borysiewicz LK, Sinclair JH (1991) J Gen Virol 72(Pt 9): 2059–2064
3. Strobl H, Bello-Fernandez C, Riedl E, Pickl WF, Majdic O, Lyman SD, Knapp W (1997) Blood 90:1425–1434
4. MacAry PA, Lindsay M, Scott MA, Craig JI, Luzio JP, Lehner PJ (2001) Proc Natl Acad Sci U S A 98:3982–3987
5. Lathey JL, Spector SA (1991) J Virol 65:6371–6375
6. Sallusto F, Lanzavecchia A (1994) J Exp Med 179:1109–1118
7. Bego M, Maciejewski J, Khaiboullina S, Pari G, St Jeor S (2005) J Virol 79:11022–11034
8. Goodrum F, Reeves M, Sinclair J, High K, Shenk T (2007) Blood 110:937–945
9. Roback JD, Hillyer CD, Drew WL, Laycock ME, Luka J, Mocarski ES, Slobedman B, Smith JW, Soderberg-Naucler C, Todd DS, Woxenius S, Busch MP (2001) Transfusion 41:1249–1257
10. Poole E, McGregor Dallas SR, Colston J, Joseph RS, Sinclair J (2011) J Gen Virol 92:1539–1549
11. Slobedman B, Mocarski ES (1999) J Virol 73:4806–4812
12. Reeves MB, MacAry PA, Lehner PJ, Sissons JG, Sinclair JH (2005) Proc Natl Acad Sci U S A 102:4140–4145

# Chapter 7

# Hematopoietic Long-Term Culture (hLTC) for Human Cytomegalovirus Latency and Reactivation

Mahadevaiah Umashankar and Felicia Goodrum


## Abstract

Of the many research challenges posed by human cytomegalovirus latency, perhaps the most notable is the requirement for primary hematopoietic cell culture. Culturing hematopoietic subpopulations while maintaining physiological relevance must be given utmost consideration. We describe a long-standing primary CD34+ hematopoietic progenitor cell (HPCs) system as an experimental model to study human cytomegalovirus (HCMV) latency and reactivation. Key aspects of our model include infection of primary human CD34+ HPCs prior to ex vivo expansion, maintenance of undifferentiated cells in a long-term culture with a stromal cell support, and an assay to quantitate infectious centers produced prior to and following a reactivation stimulus. Our method offers a unique way to quantitatively assess HCMV latency and reactivation to study the contribution of viral and host genes in latency and reactivation.


**Key words** Human cytomegalovirus, CD34+ hematopoietic cells, Latency, Reactivation, Extreme limiting dilution analysis

## 1 Introduction

Human cytomegalovirus, HCMV, like all herpesviruses, possesses an extraordinary ability to persist in the immunocompetent host by way of a latent infection. HCMV infects a variety of cell types in the host, including hematopoietic progenitor cells, stromal cells, monocytes, monocyte derived macrophages, endothelial cells, epithelial cells, fibroblasts, neuronal cells, and smooth muscle cells [1–4]. While latent reservoirs have not been definitivel defined for HCMV, hematopoietic cells are broadly accepted as an important reservoir of latent virus. HCMV latency has been associated with myeloid lineage cells ranging from CD34+ [5–8] and CD33+ [9–11] progenitor cells to CD14+ monocytes [12–15] both in vivo and in vitro. Studying HCMV latency in any of these cell types has been impeded by the difficulty of culturing primary hematopoietic cell populations while maintaining physiological relevance and the lack of quantitative assays for understanding the molecular mechanisms of latency in these cells.

Andrew D. Yurochko and William E. Miller (eds.), *Human Cytomegaloviruses: Methods and Protocols*, Methods in Molecular Biology, vol. 1119, DOI 10.1007/978-1-62703-788-4_7, © Springer Science+Business Media New York 2014

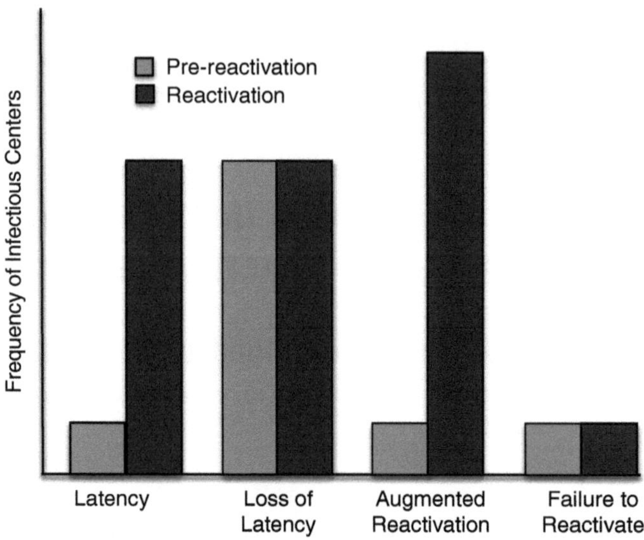


**Fig. 1** The potential latency phenotypes in CD34⁺ hematopoietic progenitor cell model. Pre-reactivation (*light grey bars*) is the virus produced prior to the reactivation as determined by measuring infectious centers in the lysate controls. The frequency of reactivation (*dark grey bars*) is determined by infectious centers obtained following the coculture of viable, latently infected CD34⁺ cells with fibroblast monolayers

Here, we describe methods to efficiently infect and culture highly purified CD34⁺ HPCs in a system designed to support hematopoietic progenitor populations. The methods described are designed for pure populations of infected CD34⁺ HPCs cells, which are achieved by isolating cells infected with recombinant virus expressing a fluorescent marker for infection (i.e., GFP, mCherry) by fluorescent activated cell sorting. The infected HPCs are cultured above a mixed support comprised of two immortalized stromal cell lines, M2-10B4 and Sl/Sl. These murine stromal cell clones have been engineered to express key human cytokines and have been shown to support human hematopoietic progenitor cell proliferation, differentiation and stem cell self-renewal [16]. For these reasons, stromal cell coculture represents the most physiologically relevant means to maintaining hematopoietic progenitor cells in culture. During the 10–14 day culture period, CD34⁺ cells maintain viral genomes, but the genome becomes quiescent, expressing few, if any, genes [17]. At a frequency of 1 in approximately 10,000 cells, virus replication is reactivated by transferring infected CD34⁺ cells into coculture with primary permissive fibroblasts in a variety of cytokine mixtures. Latency and reactivation from latency is determined by calculating the infectious centers produced prior to and following treatment with a reactivation stimulus. Reactivation from latency is defined by a fivefold to tenfold increase in the frequency of infectious centers formation following a reactivation stimulus. The quantitative nature of this assay is able to discern a number of possible phenotypes (Fig. 1). This robust method has allowed for the identification of subpopulations of CD34⁺ cells supporting a latent infection [5] and viral determinants of latency [18–20].

## 2 Materials

All solutions should be prepared using ultrapure water and cell-culture grade reagents. Fetal bovine serum for the culture of hematopoietic progenitor cells must be tested for its ability to support hematopoietic progenitor self-renewal and differentiation.

### 2.1 Human Hematopoietic Long-Term Culture Components

1. M2-10B4 murine (MG3) and Sl/Sl stromal cells: Thaw two vials of each cell type using standard thawing techniques (*see* **Note 1**). Seed the cells into separate 10 cm dishes with appropriate growth media. Once confluent, pass the cells weekly at 1:50 or as appropriate for experimental needs. Do not allow cells to become overgrown. Every second to third passage, seed MG3 and S1/S1 into medium containing hygromycin and G418 to select for stably transfected cells (*see* **Note 2**).
2. MG3 growth medium: RPMI-1640 containing 10 % fetal bovine serum, 100 U/ml penicillin, 100 μg/ml streptomycin, 60 μg/ml Hygromycin, and 400 μg/ml geneticin (G418).
3. Sl/Sl growth medium: Iscove's modified Dulbecco's medium (IMDM) containing 10 % fetal bovine serum, 1 mM sodium pyruvate, 100 U/ml penicillin, 100 μg/ml streptomycin, 125 μg/ml hygromycin, and 800 μg/ml geneticin (G418).
4. Purified CD34$^+$ HPCs: The cells may be isolated fresh from cord blood (*see* **Note 3**) or bone marrow or purchased (Lonza, Walkersville, MD; National Disease Research Interchange, NDRI, Philadelphia, PA). For isolation of CD34$^+$ cells, use the CD34 MicroBead kit (catalogue # 130-046-703) and MACS separation columns (catalogue # 130-042-401) from Miltenyi Biotec according to manufacturer's instructions. Cells should be used or cryopreserved for storage in liquid nitrogen immediately upon isolation (*see* **Note 4**). Purity should be >75 % CD34$^+$ cells as determined by flow cytometry using a fluorescently conjugated CD34 antibody (BD Biosciences or BD Pharmingen).
5. CD34$^+$ hematopoietic cells infection medium: IMDM supplemented with 10 % BIT9500 serum substitute, 2 mM L-Glutamine, 20 ng/ml low-density lipoproteins, and 50 μM 2-mercaptoethanol.
6. Human CD34$^+$ long-term culture medium (hLTCM): MyeloCult H5100 (Catalogue # 05150, Stem Cell Technologies, Vancouver, British Columbia) supplemented with 1 μM hydrocortisone (HC), 100 U/ml penicillin, and 100 μg/ml streptomycin for complete hLTCM.
7. BIT 9500 Serum substitute (Catalogue # 09500, Stem Cell Technologies, Vancouver, British Columbia).
8. High grade, cell culture-tested dimethyl sulfoxide (DMSO; Catalogue # D2650-Hybrimax, Sigma-Aldrich, St. Louis, MO).

9. DNase I (Catalogue # D-4513, cell culture-tested, Sigma-Aldrich, St. Louis, MO) reconstituted to 1 mg/ml in water.
10. Gamma irradiator ($^{135}$Cs source preferable).
11. Collagen (Catalogue # 04902, Stem Cell Technologies, Vancouver, British Columbia).
12. FACS Buffer: PBS + 0.5 % FBS.
13. SlickSeal Microfuge tubes (Catalogue # CN170S-GTS, National Scientific Supply Co, Inc., Claramont, CA).
14. FACSAria, BD Biosciences Immunocytometry Systems or similar cell sorting capability.
15. Transwell-COL (Catalogue # 3491, 24 mm diameter, 0.4 μm pore size, Costar, Corning, NY).
16. hemacytometer and light microscope.

### 2.2 Reactivation Components

1. Human fibroblast growth medium: Dulbecco's modified Eagle's medium (DMEM) supplemented with 10 % FBS, 10 mM HEPES, 1 mM sodium pyruvate, 2 mM L-GLUTAMINE, 0.1 mM nonessential amino acids, 100 U/ml penicillin, and 100 μg/ml streptomycin.
2. Human primary fibroblasts are isolated from human foreskins or human embryonic lungs. The human embryonic lung fibroblast cell line, MRC5, purchased from ATCC (CCL-171™, Manassas, VA) can also be used.
3. Reactivation medium: RPMI supplemented with 20 % fetal bovine serum (FBS), 10 mM HEPES, 1 mM sodium pyruvate, 2 mM L-glutamine, 0.1 mM nonessential amino acids, 100 U/ml penicillin, and 100 μg/ml streptomycin. *Optional*: Reactivation medium can be supplemented with cytokines to promote differentiation and reactivation, including 15 ng/ml IL6, 15 ng/ml G-CSF, 15 ng/ml GM-CSF, 15 ng/ml IL3 (all cytokines purchased from R&D Systems, Minneapolis, MN).
4. Infected CD34+ cells from long-term culture.
5. 96-well deep-well dishes for dilutions (catalogue # 40002-014, VWR, Radnor, PA).
6. 96-well cell culture dishes.
7. Multichannel Pipetman (12-channel).

## 3 Methods

### 3.1 Purification of CD34+ Cells from Cord Blood or Bone Marrow

1. Use fresh cord blood or bone marrow for the isolation of CD34+ Cells (*see* **Note 3**). Dilute cord blood or bone marrow 1:1 with 1× PBS.
2. Layer 35 ml of blood or bone marrow on to 15 ml lymphocyte separation medium (i.e., Ficoll; Catalogue # 25-072-CV,

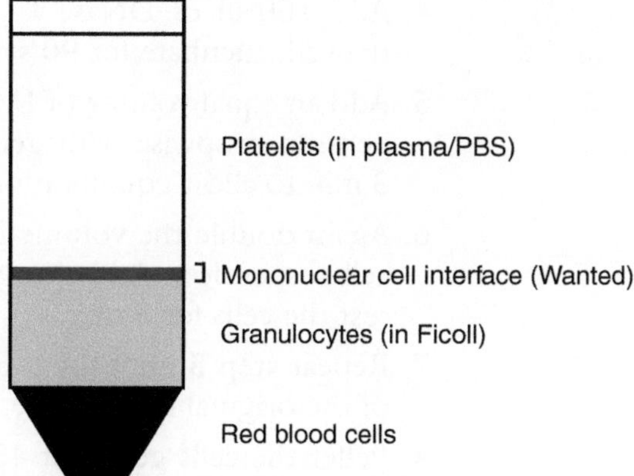


**Fig. 2** Purification of CD34+ cells from cord blood. The mononuclear cells from any blood or bone marrow source are separated by Ficoll gradients, and the CD34+ cells are subsequently purified from the mononuclear fraction using the CD34 MicroBead kit

Mediatech, Inc., Manassas, VA) in 50 ml conical tubes. Centrifuge for 45 min at 450×*g* at 18 °C with NO BRAKE.

3. Carefully aspirate the top layer (plasma) to within 5 ml of the interface.
4. Using a 5 ml pipette (and pipette-aid on slow speed), carefully collect mononuclear cells from the interface into a fresh 50 ml conical tube, being careful to leave as much of the Ficoll layer behind (Fig. 2).
5. Fill conical tube containing collected mononuclear cells with PBS+0.5 % tested-lot FBS and pellet the cells at 450×*g* for 12 min at 4 °C.
6. Resuspend the cells in cold PBS+0.5 % lot-tested FBS. Filter the cells through a 70 μm strainer and pool cells into a single 50 ml conical tube, if necessary.
7. Count the cells using hemocytometer and pellet the cells as in **step 5**.
8. Purify CD34+ cells using the CD34 MicroBead kit from Miltenyi Biotec as per the manufacturer's protocol. Infect the cells immediately or cryopreserve for storage in liquid nitrogen (*see* **Note 4**).

### 3.2 Thaw CD34+ Cells for Infection

1. Prepare to thaw at least $2 \times 10^7$ cells per virus infection assuming 40 % loss of cells in thawing process (*see* **Note 5**).
2. Remove vial(s) from liquid nitrogen storage and immediately transfer vial to 37 °C water bath. Agitate continuously until completely thawed.
3. Submerge vial in 70 % ethanol to disinfect and gently transfer cells from cryovial to a conical tube.

4. Add 100 μl of DNase I (1 mg/ml stock) for each cryovial thawed. Incubate for 90 s (*see* **Note 5**).
5. Add an equal volume of IMDM + 2 % FBS to the total cell suspension dropwise with gentle agitation and rest the cells for 3 min to allow equilibration.
6. Again double the volume by adding IMDM + 2 % FBS to the cell suspension. Add slowly with gentle agitation to mix and rest the cells for 3 min.
7. Repeat **step 5** until the total volume has reached 10–20× that of the original cell volume.
8. Pellet the cells gently at 450 × *g* for 10 min. Aspirate most of the medium without disturbing cell pellet.
9. Resuspend cell pellet in 2/3 the amount of DNase I used in **step 3** by gently flicking the tube.
10. Add 5 ml IMDM + 2 % FBS for every vial originally thawed and pellet the cells at 450 × *g* for 10 min.
11. Resuspend cell pellet in 1/3 the amount of DNase I used in **step 3** by gently flicking the tube.
12. Add 5 ml IMDM + 2 % FBS for every vial originally thawed and pellet the cells at 450 × *g* for 10 min.
13. Resuspend cell pellet in 5 ml of PBS for every vial originally thawed.
14. Count cells using a hemocytometer and trypan blue. Pellet the cells and aspirate the medium.
15. Resuspend cells in $CD34^+$ hematopoietic cells infection medium and incubate cells overnight in the 37 °C + 5 % $CO_2$ incubator (*see* **Note 5**).

### 3.3 Infection of $CD34^+$ Cells

To sufficiently infect $CD34^+$ cells and analyze latency, all viruses should be clinical strains and should be engineered to express a fluorescent protein (i.e., GFP, RFP, mCherry) as a marker for infection. Retain approximately 50 thousand cells for a mock-infected control and for single-color sorting controls.

1. Count the $CD34^+$ cells from the overnight culture and pellet the cells 450 × *g* for 12 min.
2. Resuspend the cells in infection medium at a density of $2 \times 10^7$ cells/ml.
3. Seed the cells into multi-well dishes appropriate for the cell numbers and volumes. The cells will not adhere but they will settle. To keep the settled cells in close proximity with the virus in solution, the medium only needs to cover the bottom of the well. For example, use a 12-well plate for 0.5 ml infection volume or a 6-well plate for 1.0 ml infection volume.

4. Thaw virus stock quickly in a 37 °C bath. Sonicate (three 1-s pulses at 60 % duty; empirically determined) to break aggregates and centrifuge at 10,000 × *g* for 30 s to pellet debris. Add volume of virus equivalent to 2–5 PFU/cell to infection medium (*see* **Note 6**).
5. Place cells in the 37 °C + 5 % $CO_2$ incubator for the infection. Plates should be rocked periodically (2–3 times per hour) during infection to maximize cell contact with virus.
6. After 3–6 h of infection, centrifugally enhance the infection. Place the multi-well dish in the centrifuge and spin cells in dish at 450 × *g* for 20 min at room temperature. After centrifugation, gently resuspend the cells in the medium by pipetting the medium over the surface of the dish. *Do not remove the virus-containing medium.*
7. Virus infection will continue overnight in the 37 °C + 5 % $CO_2$ incubator. Incubate for at least 20–24 h to achieve 50–70 % infection.

### *3.4 Irradiation of Stromal Cells (see Note 7)*

1. Collagen-coat an appropriate amount of 6-well dishes that will be used for human long-term culture (hLTC). You will want 2–3 wells per infection. Distribute 1 ml of 1 mg/ml collagen solution per well and let sit for 2 min (*see* **Note 8**). Collect excess collagen and save for reuse. Allow the plates to dry, uncovered in the hood, for at least 1 h. Rinse wells with PBS before seeding cells.
2. Prepare a 1:1 solution of MG3 and S1/S1 cells for irradiation.

   $3.0 \times 10^5$ cells ($1.5 \times 10^5$ of each cell type) are required per well of a 6-well plate ($0.9 \times 10^6$ of each cell type per 6-well plate). A confluent MG3 or S1/S1 10 cm plate has approximately $4.0 \times 10^6$ cells.

   (a) Trypsinize and count the cells of each cell type. Allocate an appropriate number of cells to a round bottomed15 ml snap cap tube. See Example:

   (b) Pellet the cells 350 × *g* for 7 min and resuspend in 0.5 ml of hLTCM. The cells should be at $10^7$–$10^8$ cells/ml. Keep cells on ice before irradiation.

   (c) Example: For three 6-well dishes.

   $(1.8 \times 10^6) \times (3) = 5.4 \times 10^6$ cells total.

   $(5.4 \times 10^6)/2 = 2.7 \times 10^6$ cells of each type.
3. Irradiate cells using a Gamma cell-40 ($^{135}$Cesium) irradiator at 15 Gy (*see* **Note** 7).
4. After irradiation, bring up the volume of the cells in the snap cap tube to 15 ml using PBS + 0.5 % FBS. Centrifuge 350 × *g* for 7 min to pellet. Resuspend the cells in 2 ml of complete

hLTCM (w/hydrocortisone) per well to be plated. Seed 2 ml of cell suspension per well into 6-well dishes.

5. Incubate the cells at 37 °C+5 % $CO_2$. After the cells have attached, change the medium to remove dead cells and the medium containing free radicals as a by-product of irradiation.

### 3.5 Purify Infected CD34+ Cells

Pure populations of infected (GFP+) cells that are CD34+ or CD34+CD38- are required for quantitative latency/reactivation assays; uninfected cells contaminating the population will lead to inaccurate quantitation.

1. Transfer the cells from the infection well into a 50 ml conical tube through a 70 μm cell strainer, being sure to wash the well with PBS+0.5 % FBS to recover all cells. Pellet the cells for 12 min at 450×*g*. Aspirate the medium thoroughly from cell pellet. Resuspend the cells in 50–100 μl of DNAse I (1 mg/ml) by gently flicking the tube.
2. Add 200–500 μl of Citric Acid Wash Solution (depending on the size of the cell pellet) to the cells to inactivate any remaining extracellular virus. Incubate the cells at room temperature for exactly *1 min*. During the incubation, transfer the cells to a new 50 ml conical tube to leave any residual active virus behind that may aberrantly contribute to the measurement of virus produced following reactivation. At the end of the 1 min incubation period, quickly bring up the volume with αMEM+2 % FBS to 50 ml. The medium color should change from yellow back to a light red, indicating that acid was neutralized. Pellet the cells for 12 min at 450×*g*. Resuspend the cells in 10 ml PBS+0.5 % FBS to wash once more. Filter the cells once or twice through a 70 μm cell strainer and pellet.
3. To stain the cells for the CD34 cell surface marker, aspirate the medium from the pellet and distribute the cells for controls. All samples (controls and sorting tubes) are prepared in polystyrene 4 ml snap cap Falcon tubes for the flow cytometer. In addition to an unstained cell control, a single color control is needed for each fluorescent channel used. Typically, this includes PI, GFP, PE, and sometimes APC.
   (a) *For the unstained and propidium iodide (PI) controls*, transfer ~$10^5$ uninfected cells to each of two tubes in ~200 μl PBS+0.5 % FBS. The unstained control is complete. Add PI to the PI single color control tube from a 100× stock.
   (b) *For the GFP alone control*, transfer ~$10^5$ cells from an infected sample to a 4 ml polystyrene snap cap Falcon tube in 200 μl PBS+0.5 % FBS.
   (c) *For each fluorescent single color control*, transfer ~$10^5$ uninfected cells to a polystyrene 4 ml snap cap Falcon tube in ~200 μl PBS+0.5 % FBS. Add 2 μl of the appropriate antibody to each tube.

4. Prepare experimental samples for sorting: Resuspend cells in PBS + 0.5 % FBS to a concentration of $2 \times 10^6$ cells/ml. Keep the cells in polypropylene tubes (fewer cells will be lost than in polystyrene tubes) that can accommodate large volumes. Add 20 μl of the desired conjugated antibody (i.e., phycoerythrin-conjugated anti-CD34) per $10^6$ cells unless otherwise specified by the manufacturer.
5. Stain cells (controls and experimental samples) for 15 min at 4 °C (unless otherwise specified by the manufacturer); agitating occasionally to keep cells in suspension.
6. Add 10–20 volumes of PBS + 0.5 % FBS to wash and pellet for 12 min at $450 \times g$ at 4 °C.
7. Resuspend the single color controls in 250 μl PBS + 0.5 % FBS (no PI). Resuspend the experimental samples to sort in FACS Buffer at a concentration of $1 \times 10^7$ cells/ml. All cells should be in polystyrene tubes appropriate for the flow cytometer. Keep all tubes on ice and in the dark.
8. Prepare two collection tubes per infection: 1 ml of hLTCM in siliconized microcentrifuge tubes (SlickSeal microfuge tubes) to collect cells from flow cytometer.
9. Using FACSAria or similar cell sorting instrument, isolate $GFP^+/CD34^+$ cells. Yields will depend on the percentage of cells infected (typically 30 % with FIX strain, 60 % with TB40E, and 5 % with AD169) and the gating.

### 3.6 Human CD34+ Cells Long-Term Culture

1. After the sort, pellet the cells at $450 \times g$ for 12 min at 4 °C in a microcentrifuge. Aspirate medium carefully and resuspend in 1 ml of complete hLTCM (with hydrocortisone).
2. Add 1 ml of complete hLTCM into each well that will have a transwell (do not remove the existing 2 ml in the well; total volume will be 3 ml). Place transwells into the top of each well. Transfer the cells resuspended in 2 ml into transwell. Cells densities should range between $10^4$ and $10^5$ cells per transwell; samples may have to be split between multiple transwells. The total volume of complete hLTCM in the well is 5 ml (3 ml in well + 2 ml in transwell). Incubate at 37 °C + 5 % $CO_2$.
3. Feed the cells every 5 days by aspirating approximately half of the medium from below the transwell and replacing it with an equal volume of fresh complete hLTCM. If the stromal cell monolayer begins to lift or appear unhealthy, new stromal cells should be irradiated and the transwells transferred to fresh stromal cell monolayers.

### 3.7 Reactivation

1. Plate human primary fibroblasts in 96-well dishes at a cell density of $5 \times 10^3$ cells/well in reactivation medium (*see* **Note 9**) and incubate overnight at 37 °C + 5 % $CO_2$.

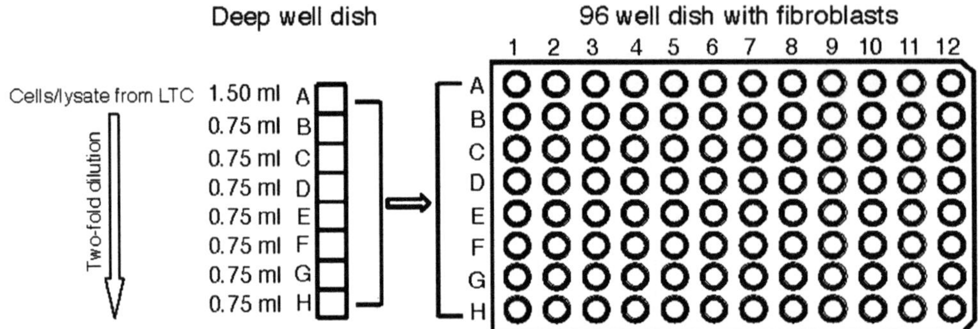


**Fig. 3** Schematic of infectious centers assay. The infected CD34+ cells or the equivalent cell lysate from the long-term culture is diluted twofold in deep-well dishes and transferred to monolayers of MRC5 cell in 96-well dishes for the infectious centers assay

2. Collect the infected CD34+ cells from the trans-wells of the long-term culture into a 15 ml conical tube and wash the trans-wells 3× with PBS + 0.5 % FBS. Pool the cells and the washes. Count the total number of cells.
3. Take $1.2 \times 10^6$ cells for reactivation and $1.2 \times 10^6$ cells for lysate control in two separate siliconized tubes (*see* **Note 10**). Spin down the cells at 450 × g for 12 min. Resuspend the cells in 1.5 ml of reactivation medium.
4. For lysate control, lyse cells mechanically (*see* **Note 11**). To mechanically disrupt cells, resuspend cell pellet in 200 microliters of hypotonic media (one-fourth media to 3-fourths water) and let sit at room temperature for 10 minutes. Then dounce 20 times using an eppendorf pestle. Mix 10 μl of lysate with 10 μl trypan blue and observe under light microscope to ensure the complete lysis of cells. Adjust final volume of lysate to 150 microliters with reactivation media.
5. For each reactivation and lysate control sample, add 1.5 ml of cell slurry or cell lysate (both equivalent to $1.2 \times 10^6$ cells) into each well in row A of the deep-well dish. Fill all other wells with 0.75 ml reactivation medium.
6. Transfer 0.75 ml of each well in Row A to the corresponding well in row B to perform a 1:2 dilution. Repeat for all rows to serially dilute cell slurry and lysate (Fig. 3).
7. Once all dilutions have been made (row A–H), using a multichannel Pipetman, transfer 50 μl of each dilution series (one column of the 96-well dish, eight dilutions) to each well in a row (12 total) of the 96-well dish containing human primary fibroblasts (*see* **Note 12**). Final volume will be 150 μl (50 μl of cell slurry + 100 μl of medium from plating the fibroblasts).
8. Incubate the cells at 37 °C + 5 % $CO_2$ for 15–20 days.

9. Score each well of 96-well dish for GFP expression using a fluorescence microscope. For each of the eight dilutions, determine a fraction of wells that scored positively for GFP expression.
10. Enter the number of input infected CD34+ HPCs per well for each dilution and the fraction of GFP + wells into the online shareware, Extreme limiting dilution analysis (ELDA, http://bioinf.wehi.edu.au/foftware/elda), following the direction of the Web site.
11. Run the program to obtain an estimate of cells required to form an infectious center (*see* **Note 13**) based on the number of GFP+ cells at each dilution. See an example below.

| **Confidence intervals** | | | |
|---|---|---|---|
| **Group** | **Lower** | **Estimate** | **Upper** |
| WT | X1 | X2 | X3 |
| MUTANT 1 | Y1 | Y2 | Y3 |
| MUTANT 2 | Z1 | Z2 | Z3 |

12. Calculate the frequency of infectious centers formed as 1 divided by the number calculated in the Estimate column of the table (*see* **Note 14**). The frequency of infectious centers serves as measure of virus formed before and after reactivation (Fig. 3).

## 4 Notes

1. The M2-10B4 murine stromal cell line (MG3 cells) is engineered to express human interleukin-3 (IL-3) and granulocyte-colony stimulating factor (G-CSF). The Sl/Sl cell line is engineered to express human IL-3 and stem cell factor (SCF). These cells are provided by Stem Cell Technologies Ltd. on behalf of D. Hogge, University of British Columbia (Vancouver, British Columbia). Other stromal cells systems have been used to effectively maintain latently infected CD34+ cells [5, 17]; however, the ability of stromal cells to support HCMV latency must be empirically determined.
2. Hygromycin/G418 selection should be performed every second to third passage. Complete selection requires ~1 week, so it is preferable to initially pass the cells sparsely (1:50) for selection as specified by the protocol from Stem Cell Technologies.
3. Isolate CD34+ cells within 36 h of harvesting the cord blood or bone marrow. Older samples yield fewer CD34+ cells. Typically, cord blood contains 0.5–1.0 % CD34+ cells, and bone marrow contains 2 % CD34+ cells.

4. For freezing, resuspend CD34[+] cells in cold IMDM containing 20 % high grade FBS, 100 U/ml penicillin and 100 μg/ml streptomycin at 1–2 × $10^7$ cells/ml. Add an equal volume of ice-cold solution of 80 % high grade FBS and 20 % cell culture grade DMSO drop wise, mix gently and aliquot into 1.8 ml cryogenic vials. Freeze the vials overnight in a rate-controlled cryopreservation cooler and on the next day, transfer the vials into liquid nitrogen for long-term storage.
5. Plan for 30–45 min for the entire process. DMSO is toxic and makes cells somewhat fragile, and hence work quickly and gently while handling CD34[+] Cells. Use of DNase I will prevent clumping of cells due to DNA released from broken cells. Multiple DNase I treatments and washes with large volumes are necessary to remove cells from DNA and cell debris. Usually, 50–60 % of the cells should be viable at the end of thawing process. The overnight incubation period in infection medium (low serum, low cytokine) enhances infectivity of thawed cells.
6. Virus stocks are concentrated stocks at $10^7$–$10^8$ PFU/ml stored at −80 °C. Viruses stocks are concentrated to high titer by sedimenting virus from infected culture supernatant through a 20 % D-Sorbitol cushion. High titer virus is required for high titer infection in low volumes. Depending on virus strains, 2–5 PFU per cell as determined from titering on fibroblasts will infect 10–80 % of CD34[+] cells. Viruses used for infection should be marked with a fluorescent tag (i.e., GFP driven by the SV40 promoter) so that infected cells can be purified from uninfected cells. This is essential for accurate quantitation of latency/reactivation.
7. Stromal cells should be irradiated ~24 h prior to setting up long-term cultures. This period of time allows the stromal cells to condition the medium. Irradiation dose may need to be determined empirically based on the calibration of irradiator. With a well-calibrated irradiator, a dose of 15–20 Gy is adequate to arrest cell division.
8. Collagen solution is provided at a concentration of 3 mg/ml in acetic acid. Dilute in sterile water. Collagen can be saved for reuse up to six times as long as sterility is maintained.
9. Seed 100 μl cells per well of a 96-well dish. Plan for 10 ml per dish. 1 dish per reactivation and 1 dish per lysate control are required for each virus or condition to be tested (i.e., a minimum of two 96-well dishes per transwell of long-term culture).
10. 1.2 × $10^6$ cells will result in 40,000 cells (or lysate equivalent) per well for the first dilution (row A). The cells numbers are for single replicates of reactivation and lysate controls. Many replicates of reactivation and lysate control can be done if the number of infected cells available from the long-term culture is not limiting. This protocol uses 40,000 cells per well of the

96-well dish to achieve optimal numbers to calculate infectious centers from latency and reactivation; however, the first dilution for reactivation can vary between 15,000 and 40,000 cells per well depending on the cell numbers available.

11. Lysate control provides an estimate of preformed virus prior to reactivation and hence it is important to achieve complete cell lysis without damaging the virus to accurately quantify the virus preexisting in the $CD34^+$ cell cultures. Complete lysis of the cells can be analyzed using trypan blue. To estimate the loss of virus due to the lysis, resuspend an aliquot of virus of known concentration in hypotonic buffer and dounce homogenize as done to prepare the lysate control. Then titer this virus (by plaque assay or $TCID_{50}$) in parallel with an equivalent aliquot that was not disrupted. From the titers, calculate the percentage of infectious virus remaining after the treatment.
12. Mix the cells intermittently in the dilution series to avoid settling of the cells at the bottom of the deep-well dish and to obtain even distribution of cells across the 96-well dish.
13. The program performs a statistical analysis to show a goodness of fit accompanying graphs for the groups of data analyzed and a table showing confidence Intervals [21].
14. The number corresponding to the estimate is the number of cells required for one reactivation event. The inverse number of the estimate will give the frequency of infectious centers.

## Acknowledgments

This work was supported by Public Health Service Grants CA11343 and AI079059 to F.G. from the National Cancer Institute (NCI) and the National Institute of Allergy and Infectious Disease (NIAID), respectively.

### References

1. Sinzger C, Grefte A, Plachter B, Gouw AS, The TH, Jahn G (1995) Fibroblasts, epithelial cells, endothelial cells and smooth muscle cells are major targets of human cytomegalovirus infection in lung and gastrointestinal tissues. J Gen Virol 76:741–750
2. Soderberg C, Larsson S, Bergstedt-Lindqvist S, Moller E (1993) Definition of a subset of human peripheral blood mononuclear cells that are permissive to human cytomegalovirus infection. J Virol 67:3166–3175
3. Schrier RD, Nelson JA, Oldstone MB (1985) Detection of human cytomegalovirus in peripheral blood lymphocytes in a natural infection. Science 230:1048–1051
4. Boeckh M, Hoy C, Torok-Storb B (1998) Occult cytomegalovirus infection of marrow stroma. Clin Infect Dis 26:209–210
5. Goodrum F, Jordan CT, Terhune SS, High KP, Shenk T (2004) Differential outcomes of human cytomegalovirus infection in primitive hematopoietic subpopulations. Blood 104:687–695
6. Reeves MB, MacAry PA, Lehner PJ, Sissons JG, Sinclair JH (2005) Latency, chromatin remodeling, and reactivation of human cytomegalovirus in the dendritic cells of healthy carriers. Proc Natl Acad Sci U S A 102: 4140–4145
7. Sindre H, Tjoonnfjord GE, Rollag H, Ranneberg-Nilsen T, Veiby OP, Beck S, Degre

M, Hestdal K (1996) Human cytomegalovirus suppression of and latency in early hematopoietic progenitor cells. Blood 88:4526–4533

8. von Laer D, Meyer-Koenig U, Serr A, Finke J, Kanz L, Fauser AA, Neumann-Haefelin D, Brugger W, Hufert FT (1995) Detection of cytomegalovirus DNA in CD34+ cells from blood and bone marrow. Blood 86: 4086–4090
9. Hahn G, Jores R, Mocarski ES (1998) Cytomegalovirus remains latent in a common precursor of dendritic and myeloid cells. Proc Natl Acad Sci U S A 95:3937–3942
10. Kondo K, Kaneshima H, Mocarski ES (1994) Human cytomegalovirus latent infection of granulocyte-macrophage progenitors. Proc Natl Acad Sci U S A 91:11879–11883
11. Kondo K, Xu J, Mocarski ES (1996) Human cytomegalovirus latent gene expression in granulocyte-macrophage progenitors in culture and in seropositive individuals. Proc Natl Acad Sci U S A 93:11137–11142
12. Hargett D, Shenk TE (2010) Experimental human cytomegalovirus latency in CD14+ monocytes. Proc Natl Acad Sci U S A 107: 20039–20044
13. Smith MS, Bentz GL, Alexander JS, Yurochko AD (2004) Human cytomegalovirus induces monocyte differentiation and migration as a strategy for dissemination and persistence. J Virol 78:4444–4453
14. Soderberg-Naucler C, Fish KN, Nelson JA (1997) Reactivation of latent human cytomegalovirus by allogeneic stimulation of blood cells from healthy donors. Cell 91:119–126
15. Soderberg-Naucler C, Streblow DN, Fish KN, Allan-Yorke J, Smith PP, Nelson JA (2001) Reactivation of latent human cytomegalovirus in CD14(+) monocytes is differentiation dependent. J Virol 75:7543–7554
16. Miller CL, Eaves CJ (2002) Long-term culture-initiating cell assays for human and murine cells. In: Klug CA, Jordan CT (eds) Hematopoietic stem cell protocols. Humana, Totowa, pp 123–141
17. Goodrum FD, Jordan CT, High K, Shenk T (2002) Human cytomegalovirus gene expression during infection of primary hematopoietic progenitor cells: a model for latency. Proc Natl Acad Sci U S A 99:16255–16260
18. Goodrum F, Reeves M, Sinclair J, High K, Shenk T (2007) Human cytomegalovirus sequences expressed in latently infected individuals promote a latent infection in vitro. Blood 110:937–945
19. Petrucelli A, Rak M, Grainger L, Goodrum F (2009) Characterization of a novel golgi-localized latency determinant encoded by human cytomegalovirus. J Virol 83:5615–5629
20. Umashankar M, Petrucelli A, Cicchini L, Caposio P, Kreklywich CN, Rak M, Bughio F, Goldman DC, Hamlin KL, Nelson JA, Fleming WH, Streblow DN, Goodrum F (2011) A novel human cytomegalovirus locus modulates cell type-specific outcomes of infection. PLoS Pathog 7(12):e1002444
21. Hu Y, Smyth GK (2009) ELDA: extreme limiting dilution analysis for comparing depleted and enriched populations in stem cell and other assays. J Immunol Methods 347:70–78

# Chapter 8

# Analysis of Cytomegalovirus Binding/Entry-Mediated Events

Gary C.T. Chan and Andrew D. Yurochko

## Abstract

The broad cellular tropism of human cytomegalovirus (HCMV) is a direct consequence of the multifaceted viral entry process involving a combination of viral glycoprotein and cellular receptor interactions that carefully orchestrate viral binding and penetration events. Although recent strides have been made in elucidating the molecular mechanisms of HCMV entry, it has become increasingly clear that the first step of the viral life cycle is exquisitely complex and dependent on several factors including virus strain and cell type. The lack of a full understanding about HCMV entry emphasizes the need for molecular techniques that can help to identify the specific roles of viral glycoproteins and cellular receptors during the viral entry process. Here, we describe various methodologies used in our laboratory and others to examine the different steps required for HCMV entry into target cells.

**Key words** Gradient purification, Immunofluorescence, Membrane fusion, Real-time polymerase chain reaction (PCR), Western blot

## 1 Introduction

The cellular plasma membrane represents the initial barrier that viruses must traverse in order to deliver viral genetic information into the host cell. Viruses have evolved a multitude of strategies to navigate the plasma membrane ranging from simple approaches, involving single viral glycoprotein and cellular receptor interactions, to complicated tactics involving a myriad of different glycoprotein–cellular receptor associations. In general, herpesviruses, including HCMV, are structurally complex viruses that are characterized by the expression of several viral membrane glycoproteins, which can engage and activate a multitude of different cellular receptors to cooperatively initiate viral entry into target cells [1]. In the case of HCMV, the sophistication of the viral entry mechanism allows for the virus to enter the cell either by direct fusion of viral and cellular lipid membranes or by fusion of viral and endosomal membranes following internalization of the virus into endosomes [2]. Consequently, HCMV has a broad cellular tropism

Andrew D. Yurochko and William E. Miller (eds.), *Human Cytomegaloviruses: Methods and Protocols*,
Methods in Molecular Biology, vol. 1119, DOI 10.1007/978-1-62703-788-4_8, © Springer Science+Business Media New York 2014

enabling for the infection of a variety of biologically distinct cell types [3], which can lead to a diverse set of organ pathologies [4]. Despite this complexity, the process of HCMV entry can be separated into two distinct phases: (1) *attachment* of the viral particle to the cell surface and (2) *penetration* of the virus across cellular membranes into the cytoplasmic space. Below, we will briefly summarize the roles of the major viral glycoproteins and cellular receptors that control binding and penetration of HCMV.

Attachment of HCMV is mediated by binding of viral gB to ubiquitously expressed heparin sulfated proteoglycans (HSPGs) on the cell surface [5, 6]. Although an essential step, because HSPGs cannot initiate cellular signaling pathways that are required for viral entry [7], engagement of HSPGs by gB alone does not stimulate penetration of the virion into the cytoplasmic space. This initial tethering of HCMV virions to HSPGs is reversible and believed to function to stabilize the virus at the cell surface until engagement of secondary signaling receptors can occur, at which time penetration takes place [8–11].

Penetration of HCMV is mediated by fusion between the viral envelope with either the plasma membrane or with the endosomal membrane [2]. Viral gB and gH engagement and activation of cellular platelet derived growth factor receptor (PDGFR)-α, epidermal growth factor receptor (EGFR), and integrin heterodimers have been shown to be required for penetration of surface-bound HCMV into several different cells types [8–14]. Although many of the cellular receptors targeted by HCMV during viral entry are widely expressed, none is expressed on all HCMV-infectable cells, suggesting that other cellular receptors on different cell types may mediate compensatory or redundant signaling events needed for viral entry. It should also be pointed out that gH can form different complexes with various viral glycoproteins, and that the composition of these complexes is a key determining factor for regulating viral tropism via the fusion process. The heterotrimeric complex composed of gH, gL, and gO mediates HCMV entry into fibroblasts, while the gH/gL/UL128-131 complex is essential for viral entry into epithelial cells, endothelial cells, and monocytes [3, 15]. However, the specific cellular receptor(s) that these complexes target remains unknown. Moreover, expression of these glycoprotein complexes is dependent on virus strain and passage number, thus further adding to the complexity of HCMV entry [2].

Despite recent advances in unraveling the mechanism of HCMV entry, overall this complex process, which is dependent on virus strain and cell type, still remains a poorly understood event. Indeed, nucleotide sequence analysis of the HCMV genome revealed 54 open reading frames (ORFs) that potentially encode for glycoproteins [16], yet only a handful of these ORFs have been functionally explored, emphasizing the importance of techniques that can delineate key steps required in the viral entry process. This chapter details a variety of methodologies used for examining attachment, internalization, and/or penetration into host cells.

## 2 Materials

### 2.1 Purification of Virions

1. Cell lines: Human fibroblasts (HF).
2. Cell culture medium: Dulbecco's minimal essential medium (DMEM) supplemented with 4 % fetal bovine serum (FBS).
3. Virus stain: Towne or other clinical/laboratory isolates of HCMV (*see* **Note 1**).
4. Tris-buffered saline; TBS (10 mM Tris–HCl, 150 mM NaCl, pH 7.5).
5. Sorbitol gradient solutions: 20, 30, 40, and 60 % sorbitol solutions (w/v) in TBS.

### 2.2 Cellular Binding Assay

1. DMEM + 4 % FBS.
2. Phosphate-buffered saline; PBS (10 mM phosphate buffer, 140 mM NaCl, pH 7.4).
3. Purified virus (described in Subheading 2.1).
4. QIAamp DNA Mini kit (Qiagen).
5. SYBR green (Applied Biosystems).
6. Forward and reverse primers: HCMV genomic immediate early (IE) region, 5′- ACACGATGGAGTCCTCTGCC-3′ (forward) and 5′-TTCTATGCCGCACCATGTCC -3′ (reverse); glyceraldehydephosphate dehydrogenase (GAPDH), 5′-GAAG GTGAAGGTCGGAGTC-3′ (forward) and 5′-GAAGATGG TGATGG GATTTC-3′ (reverse).

### 2.3 Virus Entry Assay

1. DMEM + 4 % FBS.
2. Phosphate-buffered saline; PBS (10 mM phosphate buffer, 140 mM NaCl, pH 7.4).
3. Purified virus (described in Subheading 2.1).

#### 2.3.1 Detection of the Major IE Protein by Immunofluorescence

1. Methanol.
2. Blocking solution: 10 % bovine serum albumin (BSA) (or serum from species of animal in which secondary antibody was prepared) + human Fc block (Miltenyi Biotec) in PBS.
3. Anti-IE1 antibody (Commercially available from several biotechnology companies).
4. Appropriate fluorescently conjugated secondary antibody.
5. 4′,6-Diamidino-2-phenylindole (DAPI).

#### 2.3.2 Detection of the IE Protein by Western Blot

1. Sodium dodecyl sulfate–polyacrylamide gel electrophoresis (SDS-PAGE) reagents.
2. Polyvinylidene fluoride (PVDF) membrane.
3. TBS containing 0.1 % Tween-20.
4. 5 % skim milk (Carnation nonfat dry milk) and 0.1 % Tween-20 diluted in TBS.

5. Anti-IE1 antibody (Commercially available from several biotechnology companies).
6. Horseradish peroxidase (HRP)-conjugated anti-species antibodies.
7. Chemiluminescence detection reagents [i.e., ECL plus (Amersham)].

*2.3.3 Examination of pp65 Localization*

1. 2 % paraformaldehyde in PBS.
2. 0.1 % Triton X-100 in blocking solution.
3. Blocking solution: 10 % BSA (or serum from species of animal in which secondary antibody was prepared) + human Fc block (Miltenyi Biotec) in PBS.
4. Anti-pp65 antibody (Commercially available from several biotechnology companies).
5. Fluorescently conjugated secondary antibody.
6. DAPI.

*2.3.4 Quantitative PCR Analysis of HCMV Genome Uptake*

1. Materials are same as Subheading 2.2.
2. Proteinase K; PK (1 mg/mL).

### 2.4 Purification of Labeled Virions

1. Materials are same as Subheading 2.1.
2. Fluoro-Link antibody Cy3 labeling kit (Amersham).

### 2.5 Viral and Plasma Membrane Fusion Assay

1. DMEM + 4 % FBS.
2. Blocking solution: 10 % BSA (or serum from species of animal in which secondary antibody was prepared) + human Fc block (Miltenyi Biotec) in PBS.
3. FITC-conjugated anti-cell surface receptor antibody (i.e., FITC-EGFR).
4. Purified Cy3-HCMV as described in Subheading 2.4.
5. 2 % paraformaldehyde in PBS.
6. DAPI.

## 3 Methods

### 3.1 Purification of Virions

1. Infect HF at 100 % confluency with HCMV (MOI 0.01–0.1) in DMEM + 4 % FBS.
2. Incubate infected cells and change media every 2–3 days until a cytopathic effect (CPE) of 90 % is achieved (approximately 10–14 days).
3. Change medium and continue to incubate infected HF for an additional 3–5 days.

4. Collect the supernatant (*see* **Note 2**) and spin at 400×*g* for 10 min to remove cells and cellular debris.
5. Concentrate virions by centrifugation through a 20 % sorbitol cushion at 20,000×*g* for 1 h at room temperature (RT).
6. Resuspend pellet in TBS and layer onto 20–70 % sorbitol step gradients.
7. Band virus by centrifugation at 100,000×*g* for 1 h at RT.
8. Collect band at the 50–60 % density interface, which represents intact enveloped virus.

### 3.2 Cellular Binding Assay

1. Adhere HFs on tissue culture plates and incubate overnight (*see* **Note 3**).
2. Wash the cells with PBS and equilibrate at 4 °C for 1 h in DMEM+4 % FBS.
3. Add virus (MOI 1) for 90 min at 4 °C (*see* **Note 4**).
4. Wash infected HFs extensively with cold PBS to remove unbound virus.
5. Isolate total DNA using a QIAamp DNA Mini kit (Qiagen).
6. Run real-time RT-PCR analysis with the following PCR mix: total DNA (50 ng), SYBR green (Applied Biosystems), and primers for HCMV genomic IE region or GAPDH.

### 3.3 Virus Entry Assay

1. Perform **steps 1–4** as described in Subheading 3.2.
2. Temperature shift HCMV-infected HFs to 37 °C to allow for penetration.
3. Perform viral entry analysis as described below in Subheadings 3.3.1–3.3.4.

#### *3.3.1 Detection of the Major IE Protein by Immunofluorescence (See* ***Note 5****)*

1. Perform **steps 1–4** as described in Subheading 3.2.
2. Incubate HCMV-infected HFs at 37 °C overnight.
3. Fix the cells with cold methanol for 10 min at −20 °C.
4. Wash the cells with PBS and block with 10 % BSA (or serum from species of animal in which secondary antibody was prepared)+human Fc block in PBS solution.
5. Add primary anti-IE antibody for 60 min at RT.
6. Wash with PBS and add appropriate fluorescently conjugated (i.e., Alexafluor-488) secondary antibody for 60 min at RT.
7. Counterstain with DAPI (1.4 μg/mL).
8. Use flow cytometry or fluorescent microscope to examine IE protein expression.
9. Determine the frequency of IE-positive cells (*see* **Note 6**).

#### 3.3.2 Detection of the IE Protein by Western Blot (See **Note 5**)

1. Perform **steps 1–4** as described in Subheading 3.2.
2. Incubate HCMV-infected HFs at 37 °C overnight.
3. Harvest infected cells in SDS-PAGE sample buffer.
4. Run gel electrophoresis.
5. Transfer onto PVDF membrane.
6. Block membrane blot in 5 % skim milk and 0.1 % Tween-20 diluted in TBS.
7. Add primary anti-IE1 antibody for 1 h at RT.
8. Wash extensively with TBS containing 0.1 % Tween-20.
9. Add HRP-conjugated secondary antibody for 1 h at RT.
10. Wash with TBS containing 0.1 % Tween-20 and detect band with chemiluminescence detection reagent [i.e., ECL plus (Amersham Life Sciences)].
11. Quantify with Molecular Analyst Software (*see* **Note 6**).

#### 3.3.3 Examination of pp65 Localization

1. Perform **steps 1–4** as described in Subheading 3.2.
2. Incubate HCMV-infected HFs at 37 °C for 1 h to allow for internalization of surface-bound virion.
3. Fix the cells with 2 % paraformaldehyde and permeabilize in block solution [10 % BSA (or serum from species of animal in which secondary antibody was prepared) + human Fc block in PBS] containing 0.1 Triton X-100 for 10 min at RT.
4. Block with blocking solution for 60 min at RT.
5. Perform immunofluorescence protocol as in Subheading 3.3.1 (**steps 4–6**), but use an anti-pp65 antibody instead of an anti-IE antibody (*see* **Note 6**).

#### 3.3.4 Quantitative PCR Analysis of HCMV Genome Uptake (See **Note 7**)

1. Perform **steps 1–4** as described in Subheading 3.2.
2. Incubate infected cells for 1 h at 37 °C to allow for internalization of surface-bound virion.
3. Treat the cells with 1 mg/mL of PK for 60 min at 4 °C to detach surface-bound virions.
4. Wash PK treated cells with PBS to remove detached virus.
5. Perform **steps 5–6** as described in Subheading 3.2.

### 3.4 Purification of Labeled Virions

1. Perform **steps 1–8** as described in Subheading 3.1.
2. Label purified virions with Cy3 using the Fluoro-Link antibody Cy3 labeling kit as described by manufacturer (Amersham).

### 3.5 Viral and Plasma Membrane Fusion Assay (See Note 8)

1. Adhere HFs on tissue culture plates and incubate overnight.
2. Wash the cells with PBS and block with 10 % BSA + human Fc block in PBS solution.

3. Label cell membrane with a FITC-conjugated cell surface antibody (i.e., FITC-EGFR) on ice.
4. Wash the cells with cold PBS and maintain culture at 4 °C in DMEM.
5. Infect with Cy3-virus (MOI 10) for 90 min at 4 °C.
6. Wash infected HFs extensively with cold PBS to remove unbound virus.
7. Temperature shift cultures to 37 °C for 30 min to allow for fusion of viral and cellular membranes.
8. Fix the cells with 2 % paraformaldehyde.
9. Counterstain with DAPI (1.4 μg/mL).
10. Use fluorescent microscope to detect colocalization of membranes (*see* **Note 9**).

### 3.6 Conclusions

HCMV entry is a dynamic and complex process dependent on several cellular and viral factors. We, and others, have only begun to touch the surface of deciphering the mechanisms by which HCMV can enter distinct and divergent cell lineages. Because many aspects of the entry process remain to be elucidated, having well-defined methodologies to clearly delineate each viral entry step is essential to addressing fundamental questions about HCMV entry. The procedures outlined in this chapter provide a foundation to begin examining the molecular mechanisms of HCMV entry; however, each protocol has its own advantages and limitations. The choice of which methodology to utilize depends not only on the specific viral entry steps being examined but also on the unique characteristic of each model system [i.e., the delayed kinetics of viral replication in monocytes (IE gene expression is not observed until infected monocytes have differentiated into macrophages) forgo the use of methods involving IE gene expression]. It is therefore important to identify the techniques (or combination thereof) that can address the cell-type specific entry question being asked. Overall, the methods described in this chapter are effective in examining HCMV entry into multiple different cell types.

## 4 Notes

1. The protocols described in this chapter use the Towne/E strain; however, other HMV strains can be used.
2. Certain strains of HCMV are mostly cell-associated and thus not present at high levels in the culture media. Under those circumstances, HCMV stocks can be generated by mechanical lysis of the infected HEL fibroblasts.
3. The protocol can also be modified for other cell types. For non-adherent cells, washing steps are done with cold PBS by

centrifugation to 400 × $g$ for 10 min to collect cells and remove unbound virus.

4. Higher MOIs can be used in the binding and entry assays; however, we have found that high MOIs can mask the blocking effects of known viral entry inhibitors, such as EGFR and integrin neutralizing antibodies.
5. It is important to be aware that HCMV can enter a variety of cell types including those that are not permissive for viral replication or where viral replication is significantly delayed. Consequently, the use of IE gene expression to examine internalization and/or penetration is not always appropriate.
6. Because IE gene expression is also dependent on the cellular transcription environment, to confirm that inhibitory effects are specifically due to blocks in the viral entry process, the uptake and nuclear localization of pp65 (Subheading 3.3.3) should also be followed.
7. Quantitative PCR analysis of HCMV genome uptake is the most sensitive of the described assays for detecting subtle changes in HCMV entry. However, a limitation of this assay is the inability to separate internalization from penetration in cell types where HCMV entry occurs through endocytosis. Thus, viral and plasma membrane fusion (Subheading 3.5) and/or nuclear localization of pp65 (Subheading 3.3.3) should be determined if internalization vs. penetration needs to be distinguished.
8. The protocol was adapted from [10].
9. Membrane fusion accessed by the co-localization (yellow) of FITC (green; cellular membrane) and Cy3 (red; viral membrane).

## References


1. Compton T (2004) Receptors and immune sensors: the complex entry path of human cytomegalovirus. Trends Cell Biol 14:5–8
2. Adler B, Sinzger C (2009) Endothelial cells in human cytomegalovirus infection: one host cell out of many or a crucial target for virus spread? Thromb Haemost 102:1057–1063
3. Sinzger C, Digel M, Jahn G (2008) Cytomegalovirus cell tropism. In: Shenk T, Stinski M (eds) Current topics in microbiology and immunology: human cytomegalovirus. Springer, Berlin, pp 63–83
4. Britt W (2008) Manifestations of human cytomegalovirus infection: proposed mechanisms of acute and chronic disease. In: Shenk T, Stinski M (eds) Current topics in microbiology and immunology: human cytomegalovirus. Springer, Berlin, pp 417–470
5. Compton T, Nowlin DM, Cooper NR (1993) Initiation of human cytomegalovirus infection requires initial interaction with cell surface heparan sulfate. Virology 193:834–841
6. Kari B, Gehrz R (1992) A human cytomegalovirus glycoprotein complex designated gC-II is a major heparin-binding component of the envelope. J Virol 66:1761–1764
7. Dreyfuss JL, Regatieri CV, Jarrouge TR, Cavalheiro RP, Sampaio LO, Nader HB (2009) Heparan sulfate proteoglycans: structure, protein interactions and cell signaling. An Acad Bras Cienc 81:409–429
8. Chan G, Nogalski MT, Yurochko AD (2009) Activation of EGFR on monocytes is required for human cytomegalovirus entry and mediates cellular motility. Proc Natl Acad Sci U S A 106: 22369–22374

9. Wang X, Huang DY, Huong SM, Huang ES (2005) Integrin αβ3 is a coreceptor for human cytomegalovirus. Nat Med 11:515–521

10. Wang X, Huong SM, Chiu ML, Raab-Traub N, Huang ES (2003) Epidermal growth factor receptor is a cellular receptor for human cytomegalovirus. Nature 424:456–461

11. Feire AL, Koss H, Compton T (2004) Cellular integrins function as entry receptors for human cytomegalovirus via a highly conserved disintegrin-like domain. Proc Natl Acad Sci U S A 101:15470–15475

12. Soroceanu L, Akhavan A, Cobbs CS (2008) Platelet-derived growth factor-alpha receptor activation is required for human cytomegalovirus infection. Nature 455(7211):391–395

13. Nogalski MT, Chan G, Stevenson EV, Gray S, Yurochko AD (2011) Human cytomegalovirus-regulated paxillin in monocytes links cellular pathogenic motility to the process of viral entry. J Virol 85:1360–1369

14. Maidji E, Genbacev O, Chang HT, Pereira L (2007) Developmental regulation of human cytomegalovirus receptors in cytotrophoblasts correlates with distinct replication sites in the placenta. J Virol 81:4701–4712

15. Scrivano L, Sinzger C, Nitschko H, Koszinowski UH, Adler B (2011) HCMV spread and cell tropism are determined by distinct virus populations. PLoS Pathog 7:e1001256

16. Chee MS, Bankier AT, Beck S, Bohni R, Brown CM, Cerny R, Horsnell T, Hutchison CA 3rd, Kouzarides T, Martignetti JA et al (1990) Analysis of the protein-coding content of the sequence of human cytomegalovirus strain AD169. Curr Top Microbiol Immunol 154: 125–169

# Chapter 9

# Use of 5-Ethynyl-2′-Deoxyuridine Labelling and Flow Cytometry to Study Cell Cycle-Dependent Regulation of Human Cytomegalovirus Gene Expression

Lüder Wiebusch and Christian Hagemeier

## Abstract

The cell cycle position at the time of infection has a profound influence on human cytomegalovirus (HCMV) gene expression and therefore needs consideration in the design and control of HCMV experiments. While G0/G1 cells support the immediate onset of viral transcription, cells progressing through the S and G2 cell cycle phases prevent HCMV from entering the lytic replication cycle. Here, we provide two fast and reliable protocols that allow one to determine the cell cycle distribution of the designated host cells and monitor viral protein expression as a function of the cell cycle state. Both protocols make use of the thymidine analogue 5-ethynyl-2′-deoxyuridine and "click" chemistry to label HCMV-non-permissive S phase cells in a gentle and sensitive way.

**Key words** Cell cycle, Cytomegalovirus, EdU labelling, Flow cytometry, HCMV antibodies, Immediate early genes, Propidium iodide staining

## 1 Introduction

High Cyclin A2-CDK1/2 activity in the S/G2 phase of the cell division cycle interferes with the onset of viral immediate early (IE) gene expression, leading to a significant delay of HCMV lytic replication [1–3]. This effect is independent of the host cell type and applies to HCMV laboratory strains as well as clinical isolates [2]. Still, the role of the cell cycle is often neglected in daily laboratory practice and cells not sufficiently synchronized in G0/G1 are frequently used for infection experiments. The subsequent heterogeneity of infected cell populations can cause distorted or ambiguous results. This chapter provides easy-to-perform protocols that enable researchers who have access to a flow cytometer (equipped with an argon ion laser or similar) to quickly check the cell cycle distribution before virus inoculation (Subheading 3.1), analyze representative IE, early and late gene expression, and correlate this with cell cycle position at the time of infection (Subheading 3.2).

Andrew D. Yurochko and William E. Miller (eds.), *Human Cytomegaloviruses: Methods and Protocols*, Methods in Molecular Biology, vol. 1119, DOI 10.1007/978-1-62703-788-4_9, © Springer Science+Business Media New York 2014

These protocols rely on S phase labelling achieved by incorporation of the nucleoside analogue 5-ethynyl-2′-deoxyuridine (EdU) into nascent DNA. After cell permeabilization, the incorporated EdU can be detected by copper-catalyzed cycloaddition of a fluorescent azide dye to its terminal alkyne group [4]. This reaction is quick and efficient and occurs at very mild conditions, following the principles of "click" chemistry [5]. A further advantage over the conventional antibody-based 5-bromo-2′-deoxy (BrdU) labelling method is that EdU labelling can be more flexibly combined with immunostaining techniques (*see* **Note 1**).

## 2 Materials

### 2.1 EdU Labelling

The following materials are required for both protocols (Subheadings 3.1 and 3.2).

1. EdU (Life Technologies). Prepare 10 mM stock solution in dimethylsulfoxide (DMSO). Store aliquots at −20 °C.
2. Phosphate-buffered saline (PBS), free of $Ca^{2+}$ and $Mg^{2+}$.
3. 0.25 % trypsin–0.02 % EDTA solution.
4. 15 ml screw-cap conical centrifuge tubes.
5. 1 M Tris–HCl in deionized $H_2O$, adjust to pH 8.5 with HCl.
6. $CuSO_4$. Prepare 100 mM stock solution in deionized $H_2O$.
7. Alexa Fluor 488-Azide. Prepare 1 mM stock solution in DMSO. Store this solution light-protected at −20 °C.
8. Ascorbic acid. Prepare 1 M stock in deionized $H_2O$. Storage at −20 °C.
9. 1 % (w/v) bovine serum albumin (BSA), fraction V in PBS. Prepare freshly before use.

### 2.2 Propidium Iodide (PI) Staining

In addition to the above listed items the following materials are needed for protocol in Subheading 3.1.

1. Citrate buffer: 40 mM sodium citrate, 0.25 M sucrose in $H_2O$, adjust to pH 7.4 with HCl. Store at 4 °C.
2. Permeabilization buffer: 0.1 % Igepal CA-630 and 0.5 mM EDTA in PBS. Store at 4 °C.
3. Ribonuclease A (RNase A). Prepare 10 mg/ml stock solution in 10 mM Tris–HCl, pH 7.5, 15 mM NaCl. If the RNase A is not DNase free, inactivate DNase enzymes by boiling the stock solution for 10 min. Precipitation of protein impurities will occur. The RNase solution is then clarified by centrifugation and stored in aliquots at −20 °C.
4. PI. Prepare 1 mg/ml stock solution in deionized $H_2O$ and store it protected from light at 4 °C.

### 2.3 Immuno-Detection of Viral Proteins

The following additional reagents are required for protocol in Subheading 3.2.

1. Absolute ethanol. Store at −20 °C.
2. Mouse monoclonal antibodies against major IE proteins pUL123/IE1 and pUL122/IE2 (clone E13, Argene, France, or clone 8B1.2, Millipore, Temecula, CA, USA), early viral protein pUL84 (clone Mab84, Santa Cruz Biotechnology, Santa Cruz, CA, USA), early-late viral protein gB (clone 1-M-12, Santa Cruz), or late viral protein pUL99/pp28 (clone CH19, Santa Cruz). Other HCMV-specific antibodies may also be suitable (*see* **Note 2**). Dilute antibody solutions with an equal volume of glycerol and store at −20 °C. The final concentration of 50 % glycerol prevents freezing and therefore protects the antibody from denaturation.
3. Goat anti-mouse immunoglobulin G (IgG) antibody coupled to Alexa Fluor 546.

### 2.4 Flow Cytometry

1. Flow cytometer equipped with a laser for 488 nm excitation and with suitable band pass filters to collect emission of green (Alexa Fluor 488) and orange-red (PI, Alexa Fluor 546) fluorescence. The data shown in Figs. 1 and 2 were obtained with a FACS-Canto II system (BD Biosciences, San Jose, CA, USA) using 530 ± 15 and 585 ± 21 nm band pass filters for fluorescence measurements (*see* **Note 3**).
2. Acquisition and analysis software such as FACS-Diva (BD Biosciences) or a similar company's analysis software.

## 3 Methods

### 3.1 Rapid Analysis of Host Cell Cycle Distribution by EdU Labelling and PI Staining

This protocol is a combination of EdU labelling [4] and DNA content analysis according to Vindelov [6]. It is designed and optimized to get high resolution cell cycle data in a short time. With some experience, it takes no more than 90 min from adding the EdU reagent to the culture medium to the recording of quantitative flow cytometry results. These results can help HCMV researchers in their decision about whether a given time point is suitable for starting the infection. The protocol can also be used to study the influence of HCMV on cell cycle progression. However, once viral DNA replication has started (in fibroblasts typically around 24 h post infection) its use is not further recommended, as data interpretation would be complicated by the fact that EdU incorporates into newly synthesized viral DNA [7] and the accumulation of virus DNA leads to broadening and shifting of DNA peaks [8] (*see* **Note 4**).

1. Add EdU to a final concentration of 10 μM (1:1,000 dilution of the EdU stock solution) to the cell culture medium and incubate the cells for 30 min under normal culture conditions (*see* **Note 5**).

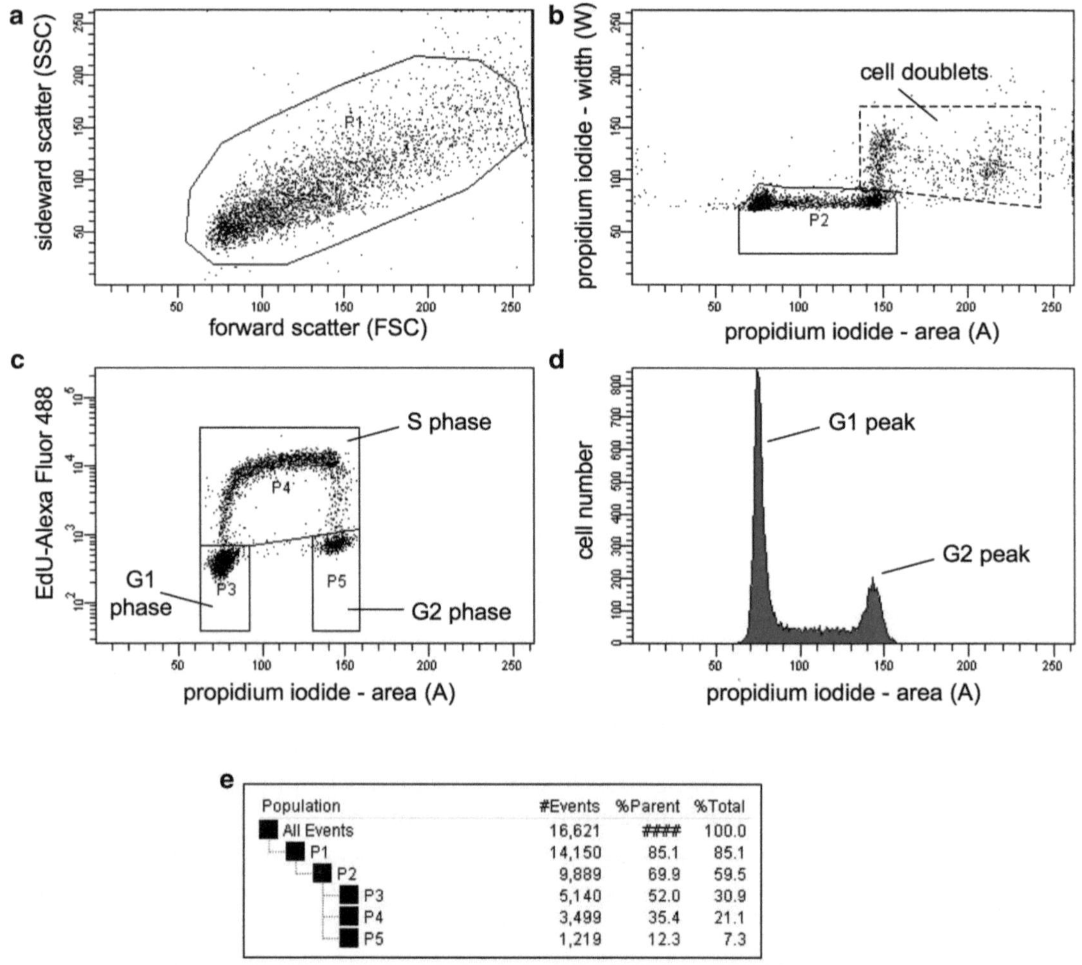


| Population | #Events | %Parent | %Total |
|---|---|---|---|
| All Events | 16,621 | #### | 100.0 |
| P1 | 14,150 | 85.1 | 85.1 |
| P2 | 9,889 | 69.9 | 59.5 |
| P3 | 5,140 | 52.0 | 30.9 |
| P4 | 3,499 | 35.4 | 21.1 |
| P5 | 1,219 | 12.3 | 7.3 |

**Fig. 1** Flow cytometry methodology used to analyze cell cycle distribution of EdU and propidium iodide (PI)-labelled cells. A growing culture of human embryonic lung fibroblasts (at approximately 70 % confluency) was EdU and PI-labelled as described in Subheading 3.1. Thereafter, the cells were analyzed by flow cytometry. (**a**) First, a dot plot acquisition window was created displaying on a linear scale the forward and sideward light scatter (FSC and SSC) properties of events. FSC is a measure of cell size, SSC of cellular granularity. A region (P1) was set that excludes cellular debris and larger cell aggregates from further analysis. (**b**) On a second dot blot, cells from the P1 region were analyzed for area (*A*) and width (*W*) values of their PI fluorescence signal. A region, P2, was set to gate out events with high W-values from further analysis, representing doublets of G1, S, or G2 cells. (**c**) The third and final dot plot was used to analyze cells from the P2 region for PI (linear scale) versus Alexa Fluor-488 fluorescence (logarithmic scale). PI-A fluorescence intensity is directly proportional to the cellular DNA content, whereas Alexa Fluor 488 signal strength indicates the amount of incorporated EdU. Three populations were defined: EdU-positive cells (S-phase cells, P4), EdU-negative cells with a G1 DNA content (G0/G1 cells, P3), EdU-negative cells with a G2 DNA content (G2/M cells, P5). (**d**) Shown is the corresponding DNA histogram of cells from the P2 region where PI-A fluorescence is plotted against cell number. (**e**) The hierarchy and percentages of analytical regions P1 to P5 are given in this table

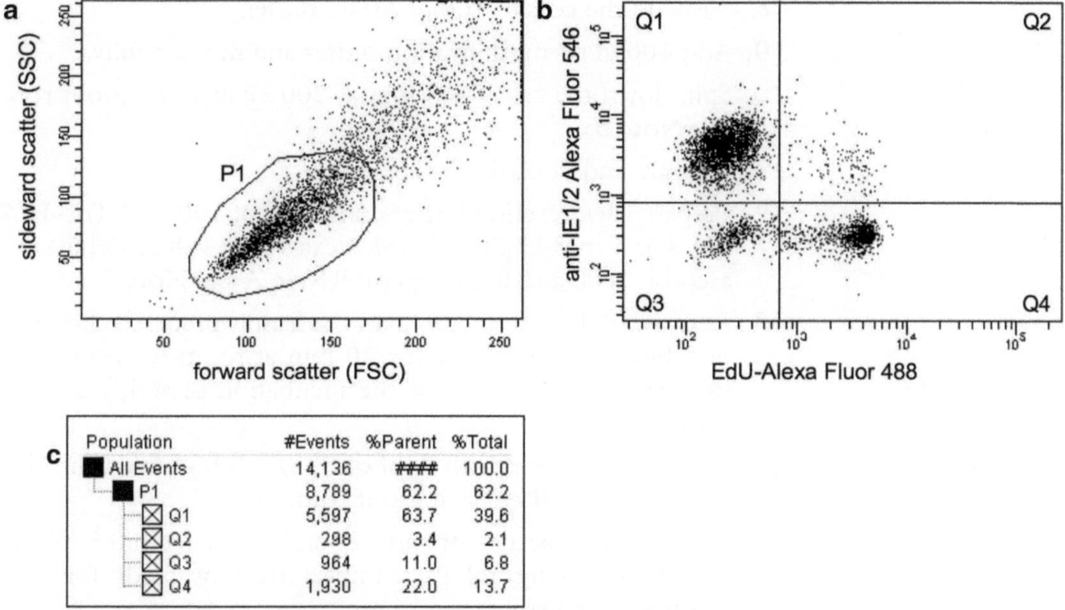


| Population | #Events | %Parent | %Total |
|---|---|---|---|
| All Events | 14,136 | #### | 100.0 |
| P1 | 8,789 | 62.2 | 62.2 |
| Q1 | 5,597 | 63.7 | 39.6 |
| Q2 | 298 | 3.4 | 2.1 |
| Q3 | 964 | 11.0 | 6.8 |
| Q4 | 1,930 | 22.0 | 13.7 |

**Fig. 2** Flow-cytometric analysis of cell cycle-dependent HCMV major immediate protein expression. Human embryonic lung fibroblasts were first incubated for 30 min with 10 μM EdU and then infected with HCMV strain AD169 using an MOI of 10. Five hours post infection, the cells were harvested. After fixation and permeabilization in ethanol, the cells were stained for EdU incorporation and IE1/IE2 protein expression as described in Subheading 3.2. (**a**) Shown is an FSC-SSC-diagram where a region P1 was defined that excludes cellular debris and cell aggregates from further analysis. (**b**) Cells from the P1 region were analyzed for Alexa Fluor 488 (*x*-axis) and Alexa Fluor 546 fluorescence (*y*-axis). Based on the resulting populations, four quadrants were defined. Q1 contains the large fraction of IE1/IE2-positive, EdU negative cells. Q2 contains the few IE1/IE2 and EdU-positive cells. Q3 contains cells that were negative for both IE1/IE2 and EdU. Q4 contains the many IE1/IE2-negative cells that were in S phase (EdU positive) at the beginning of infection. (**c**) Shown are the hierarchy and percentages of analytical regions P1, Q1–4

2. Remove the EdU containing medium and transfer it to a 15 ml conical tube. If suspension cells are used proceed to **step 6**.
3. Wash adherent cells once with PBS.
4. Add 0.25 % trypsin–0.02 % EDTA solution to the cells and wait until cells detach from the surface of the culture dish or can be easily dislodged by pipetting.
5. Pipet cell suspension up and down to resolve cell clumps and combine the suspension with the medium in the 15 ml conical tubes. If working under low-serum conditions add 1/10 volume of fetal or newborn calf serum in order to inactivate the trypsin.
6. Centrifuge the cells for 3 min at 200 × *g* using a swing-out rotor.
7. Aspirate and discard the supernatant. Carefully remove any residual amount of medium.

8. Suspend the cells in 100 μl citrate buffer.
9. Add 400 μl permeabilization buffer and mix carefully.
10. Spin down the cells for 3 min at 200 × *g* in a swing-out rotor (*see* **Note 6**).
11. Aspirate and discard the supernatant.
12. Prepare click reaction buffer consisting of 100 mM Tris–HCl, pH 8.5, 4 mM $CuSO_4$, 1 μM Alexa Fluor 488 azide, 50 mM ascorbic acid, and 0.25 mg/ml RNase A (*see* **Note 7**).
13. Resuspend the cells in 100 μl of the freshly prepared click reaction buffer and incubate for 20 min at room temperature in the dark. After 10 min of the incubation period, resuspend cells.
14. Dilute the reaction with 2 ml of PBS/1 % BSA and centrifuge for 3 min at 200 × *g* (swing-out rotor).
15. Aspirate the supernatant and suspend cell pellet in 0.5 ml PBS containing 25 μg/ml PI. Samples are now ready for flow-cytometric analysis.
16. Analyze the cells by flow cytometry using the gating strategy described in Fig. 1. Include unstained and single-stained control samples to compensate for fluorescence overlap between Alexa Fluor 488 and PI emissions. Refer to the cytometer software manual for a detailed protocol of compensation setup. The compensation settings can be stored and used in future experiments.

### 3.2 Analysis of Cell Cycle-Dependent HCMV Gene Expression

In this protocol, the EdU labelling method is used in combination with immunofluorescence staining of HCMV proteins. EdU labelling takes place just prior to virus inoculation, leaving a stable mark in cells being positioned in non-permissive S-phase at the beginning of infection. This allows one to identify those cells in later phases of the viral replication cycle and track their fate in terms of IE, early and late protein expression [2]. In experiments where a "background" level of non-permissive S-phase cells at the start of infection would disturb the generation of clean and reliable data, the EdU label can also be used to exclude the S-phase population from analysis by an appropriate flow-cytometric gating strategy. Flow cytometry offers several other advantages over the standard immunoblot analysis of HCMV gene expression as it provides information on a single cell level that can easily be quantified and statistically evaluated.

1. Add EdU to a final concentration of 10 μM (1:1,000 dilution of the EdU stock solution) to the cell culture medium and incubate the cells for 30 min at 37 °C (*see* **Note 5**).
2. Remove the EdU containing medium and wash the cells three times with PBS (*see* **Note 8**).

3. Start HCMV infection by incubating the cells with virus-containing medium (*see* **Note 9**).
4. At desired times, harvest infected cells by trypsinization as described in **steps 4–7** of Subheading 3.1.
5. Resuspend the cells in 1 ml ice-cold PBS.
6. Add 3 ml of pre-cooled (−20 °C) absolute ethanol drop-wise over 5–10 s while vortexing cells at medium speed. This procedure reduces cell clumping caused by the increased adhesiveness of permeabilized cells.
7. Incubate the cells for at least 5 min on ice (**Note 10**).
8. Spin down the cells for 3 min at 200 × *g* in a swing-out rotor.
9. Aspirate and discard the supernatant.
10. Carefully resuspend the cells in 1 ml ice-cold PBS to wash away residual ethanol. After prolonged storage in 75 % ethanol, the cells can be very sticky and may require repeated but gentle pipetting for complete resuspension (*see* **Note 11**).
11. Spin down the cells for 3 min at 200 × *g* in a swing-out rotor and aspirate the supernatant.
12. Prepare click reaction buffer consisting of 100 mM Tris–HCl, pH 8.5, 4 mM $CuSO_4$, 1 μM Alexa Fluor 488 azide, and 50 mM ascorbic acid.
13. Resuspend the cells in 100 μl of the freshly prepared click reaction buffer and incubate for 20–30 min at room temperature in the dark. Every 10 min during the incubation period, resuspend sedimented cells.
14. Dilute the reaction with 2 ml of PBS/1 % BSA and centrifuge for 3 min at 200 × *g* using a swing-out rotor.
15. After aspiration of the supernatant, resuspend cell pellet in 100 μl of primary anti-CMV antibody (*see* **Note 12**), diluted to 1 μg/ml in PBS/1 % BSA. Incubate for 30–60 min at room temperature. Resuspend the cells periodically to facilitate antibody binding.
16. After incubation with the primary antibody, wash the cells 1× with 2 ml PBS/1 % BSA and centrifuge for 3 min at 200 × *g*.
17. Discard the supernatant and incubate the cells for 30–60 min in 100 μl secondary antibody (Alexa Fluor 546 conjugated goat anti-mouse IgG), diluted to 1 μg/ml in PBS/1 % BSA.
18. Wash the cells as in **step 16** and suspend the cells in 0.5 ml PBS.
19. Analyze the cells by flow cytometry as described in Fig. 2. Fluorescence overlap between Alexa Fluor 488 and 546 emissions needs to be compensated (*see* **step 16** in Subheading 3.1 and **Note 13**).

## 4 Notes

1. EdU detection by click chemistry is a non-antibody-based method and imposes no restrictions on the choice of the primary antibody host species when combined with immunofluorescence staining of viral or cellular antigens. This is in contrast to traditional BrdU staining where in most cases high affinity BrdU antibody clones of mouse or rat origin are employed. Considering that the majority of HCMV-specific antibodies in use are also of mouse origin, the EdU labelling technique represents a versatile approach for HCMV research. A further benefit over BrdU immunostaining is the small size of the fluorescently conjugated azide dye that is used for covalent EdU labelling. This ensures quick penetration of cellular and nuclear structures and binding to incorporated EdU under non-denaturing conditions. The only disadvantage of EdU lies in its higher toxicity. This problem, however, applies mainly to long-term labelling approaches and may be overcome in the near future by less toxic EdU derivatives [9].
2. The indicated antibody clones have been tested and are routinely used for this purpose in our laboratory. Many HCMV proteins are expressed at high levels during the course of the lytic replication cycle. This facilitates their detection by flow cytometry and makes it likely that other existing virus-specific antibodies also qualify for this method. In our experience, most antibodies that give a strong and specific signal in fluorescence microscopy work as well for flow cytometry.
3. If the flow cytometer is equipped with different filter sets it is recommended to use one of the following interactive online tools to find out the most efficient combinations of filters and fluorophores:
   - http://www.bdbiosciences.com/research/multicolor/spectrum_viewer/index.jsp.
   - http://probes.invitrogen.com/resources/spectraviewer/.
   - http://www.mcb.arizona.edu/ipc/fret/.
4. This caveat applies particularly to high multiplicity infections of fully permissive cells like primary human fibroblasts. Due to the large size of the HCMV genome, its high rate of replication and the relatively long time span between onset of viral DNA synthesis and lytic release of viral progeny, the virus has a sizeable impact on the total DNA content of the host cell. This is illustrated by the fact that viral replication compartments occupy large areas of the cell nucleus at late times of infection. In the early days of HCMV cell cycle research, the progressive increase of total DNA content has been misinterpreted as arising from cellular DNA synthesis [10, 11].

5. A minimum of $1–2 \times 10^5$ cells is required for both methods. It is recommended to start with cell numbers above $5 \times 10^5$ cells as these higher cell numbers increase the visibility and sometimes also the stability of the cell pellets and therefore lower the risk of cell loss during the repeated centrifugation/aspiration cycles.
6. This treatment leads to disruption of the cellular membrane. For some cell types the isolated nuclei may require higher centrifugal forces to sediment. If you do not see a visible pellet, increase the centrifugal force up to $500 \times g$.
7. RNase A does not participate in the click reaction itself. Instead, removal of cellular RNA from samples is needed because PI intercalation into RNA would preclude a proper measurement of cellular DNA content. As buffer conditions, incubation time and temperature of the click reaction will not affect the action of RNase A, this enzyme is included in the reaction mix to streamline the protocol.
8. Minimizing the amount of residual EdU by thorough washing is important as EdU can inhibit HCMV replication (L.W., unpublished data).
9. To calculate the multiplicity of infection (MOI) use a virus with known titer and determine the total cell number of a non-infected control culture.
10. If necessary, cells can be stored at this stage for days or weeks before staining. This is convenient, for example, when studying infection kinetics. Samples collected at different time points can thus be analyzed in parallel at the end of the experiment. It should be noted that cell swelling during overnight storage in 75 % ethanol leads to significantly enlarged and a snow-white appearance of cell pellets after centrifugation. These changes are reversed after the first wash in PBS (**step 8** in Subheading 3.1).
11. To minimize shearing forces, do not vortex and do not use Pasteur pipettes. This may result in significant cell loss. Micropipettes fitted with disposable 1 ml polypropylene tips are recommended.
12. If using whole IgG rabbit antibodies as a primary immunoreagent, care has to be taken because rabbit IgG binds nonspecifically to the HCMV-encoded Fc-receptor [12]. This binding can be reduced by including 10 % horse serum in the PBS/1 % BSA buffer which is used for blocking, antibody incubation and washing steps. Furthermore, the amount of nonspecific antibody binding needs to be controlled by replacing the primary antibody with nonspecific rabbit IgG.
13. If the flow cytometer is equipped with more than one laser, the choice of fluorophores can be adapted to lower the spectral overlap. In the case of 488 and 633 nm lasers, for instance, it is convenient to use a combination of Alexa Fluor 488 and Alexa Fluor 647 conjugated reagents.

## Acknowledgement

This work has been supported by the DFG grant WI2043/3-1 to L.W. and C.H.

### References

1. Fortunato EA, Sanchez V, Yen JY, Spector DH (2002) Infection of cells with human cytomegalovirus during S phase results in a blockade to immediate-early gene expression that can be overcome by inhibition of the proteasome. J Virol 76(11):5369–5379
2. Zydek M, Hagemeier C, Wiebusch L (2010) Cyclin-dependent kinase activity controls the onset of the HCMV lytic cycle. PLoS Pathog 6(9):e1001096. doi:10.1371/journal.ppat.1001096
3. Oduro JD, Uecker R, Hagemeier C, Wiebusch L (2012) Inhibition of human cytomegalovirus immediate-early gene expression by cyclin A2-dependent kinase activity. J Virol 86(17):9369–9383.doi:10.1128/JVI.07181-11,JVI.07181-11 [pii]
4. Salic A, Mitchison TJ (2008) A chemical method for fast and sensitive detection of DNA synthesis in vivo. Proc Natl Acad Sci U S A 105(7):2415–2420. doi:10.1073/pnas.0712168105, 0712168105 [pii]
5. Kolb HC, Finn MG, Sharpless KB (2001) Click chemistry: diverse chemical function from a few good reactions. Angew Chem Int Ed Engl 40(11):2004–2021, doi:10.1002/1521-3773(20010601)40:11<2004::AID-ANIE2004>3.0.CO;2-5 [pii]
6. Vindelov LL, Christensen IJ (1994) Detergent and proteolytic enzyme-based techniques for nuclear isolation and DNA content analysis. Methods Cell Biol 41:219–229
7. Strang BL, Boulant S, Chang L, Knipe DM, Kirchhausen T, Coen DM (2012) Human cytomegalovirus UL44 concentrates at the periphery of replication compartments, the site of viral DNA synthesis. J Virol 86(4):2089–2095.doi:10.1128/JVI.06720-11,JVI.06720-11 [pii]
8. Lu M, Shenk T (1996) Human cytomegalovirus infection inhibits cell cycle progression at multiple points, including the transition from G1 to S. J Virol 70(12):8850–8857
9. Neef AB, Luedtke NW (2011) Dynamic metabolic labeling of DNA in vivo with arabinosyl nucleosides. Proc Natl Acad Sci U S A 108(51):20404–20409. doi:10.1073/pnas.1101126108, 1101126108 [pii]
10. Bain M, Sinclair J (2007) The S phase of the cell cycle and its perturbation by human cytomegalovirus. Rev Med Virol 17(6):423–434. doi:10.1002/rmv.551
11. Jault FM, Jault JM, Ruchti F, Fortunato EA, Clark C, Corbeil J, Richman DD, Spector DH (1995) Cytomegalovirus infection induces high levels of cyclins, phosphorylated Rb, and p53, leading to cell cycle arrest. J Virol 69(11):6697–6704
12. Antonsson A, Johansson PJ (2001) Binding of human and animal immunoglobulins to the IgG Fc receptor induced by human cytomegalovirus. J Gen Virol 82(Pt 5):1137–1145

# Chapter 10

# Methods for Studying the Function of Cytomegalovirus GPCRs

Christine M. O'Connor and William E. Miller

## Abstract

All of the cytomegaloviruses discovered to date encode two or more genes with significant homology to G-protein coupled receptors (GPCRs). The functions of these cytomegalovirus GPCRs are just beginning to be elucidated; however, it is clear that they exhibit numerous interesting activities in both in vitro and in vivo systems. In this chapter, we review the various methodologies that can be used to examine biochemical aspects of viral GPCR signaling in vitro as well as examine the biological activity of these viral GPCRs in vitro and in vivo in virus infected cells using recombinant cytomegaloviruses.

**Key words** G-protein coupled receptors, Human cytomegalovirus, Murine cytomegalovirus, Virus genetics, Signal transduction, Virological methods

## 1 Introduction

Human cytomegalovirus (HCMV) is a betaherpesvirus that infects a large majority of the world's population. Infection with HCMV in utero is the leading cause of infectious congenital birth defects in developed countries, resulting in developmental disabilities. Although infection with the virus remains, for the most part, asymptomatic in healthy individuals, immunocompromised individuals who undergo viral reactivation or receive a primary infection suffer from severe morbidity and often mortality as a direct consequence of HCMV-associated disease [1].

Sequence analysis of the HCMV genome reveals that this virus encodes at least four G protein-coupled receptors (GPCRs), including UL33, UL78, US27, and US28 [2, 3]. GPCRs are cell surface molecules that contain seven transmembrane domains and function in signal transduction [4]. The binding of an appropriate ligand to a given GPCR activates the receptor while dissociation of the ligand converts the GPCR to an inactive state. Some GPCRs exhibit significant activity in the absence of a bound ligand and in this case the signaling is termed constitutive.

Andrew D. Yurochko and William E. Miller (eds.), *Human Cytomegaloviruses: Methods and Protocols*, Methods in Molecular Biology, vol. 1119, DOI 10.1007/978-1-62703-788-4_10, © Springer Science+Business Media New York 2014

When in their active state, GPCRs induce a variety of signal transduction pathways that alter the cellular environment by activating molecules involved in adhesion, migration, proliferation, differentiation, cytoskeletal dynamics, contractility, etc. [5]. Both the primate (i.e., human, rhesus) and non-primate (i.e., murine, rat) CMVs encode members of the UL33 and UL78 family, whereas only the primate CMVs additionally encode US27 and US28 [6].

Cytomegalovirus encoded GPCRs have been demonstrated to respond to external ligands and/or signal constitutively (some examples are provided with references—HCMV US28 [7–11], HCMV UL33 [12], Rat CMV (RCMV) R33 [12, 13], and Murine CMV (MCMV) M33 [9, 14]). The murine UL33 orthologue M33 contributes in vivo to pathogenesis, as assessed by a requirement of the GPCR and its constitutive signaling for viral replication within the host's salivary glands [15–18]. Similar results were demonstrated with the RCMV UL33 orthologue, R33 [19]. Although members of the UL33 gene family are required for pathogenesis in vivo; UL33, R33, and M33 are dispensable for replication in fibroblasts [20, 21]. Similar to UL33, HCMV UL78 has orthologues across the betaherpesvirus family. MCMV M78 is a virion constituent, and upon infection of host cells, promotes immediate early (IE) viral mRNA accumulation [22]. Infection of the respective host with a virus harboring a deletion of M78 or R78 decreases viral titers in the spleen, liver, and salivary glands, while increasing the survival rates in these animals [22, 23], suggesting a role for these GPCRs in viral pathogenesis in both the mouse and rat CMV models. HCMV UL78 is assembled into the mature viral particle [24], and although not essential for efficient viral replication in fibroblasts [24, 25] or in a renal artery tissue culture model [25], it is critical for replication in epithelial and endothelial cells [24]. In epithelial cells, UL78 is necessary for appropriate delivery of the viral particle to the nucleus [24]. Recent investigations into the role of US27 during viral infection have revealed that this protein is important for efficient spread of HCMV via the extracellular route in both endothelial cells and fibroblasts [26]. Currently UL78 and US27 signaling (constitutive and/or ligand-induced), chemokine interaction(s), and natural ligand(s) remain unknown.

Arguably the most-studied HCMV GPCR is encoded by US28. US28 exhibits both constitutive and agonist-dependent signaling and has been demonstrated to bind numerous C–C chemokines (RANTES, MIP-1α, MIP-1β, and MCP-1) and the CX3C-chemokine fractalkine [7, 8, 10, 27–29]. US28 constitutive signaling is exemplified by its ability to activate phospholipase C–β in the absence of ligand, while its agonist-dependent signaling is exemplified by its ability to modulate $Ca^{2+}$ flux and direct vascular smooth muscle cell migration [7, 10, 11, 30–32]. HCMV US28 appears to be "promiscuous" in its G-protein coupling as it is able to activate either $G_{q/11}$or $G_{12/13}$ G proteins, or both [7, 32–34]. As a consequence of this G-protein signaling, US28 is able to influence the activity of a variety of downstream effectors such as NFκB, NFAT, CREB, MAPK, Rho, and STAT/IL6 [9, 33, 35–38]. In addition to its involvement in cell migration, HCMV US28 has been shown to have oncogenic potential. Expression of US28 enhances cell growth and cell cycle progression, and induces a pro-angiogenic and transformed phenotype in vitro [39, 40]. In vivo, injection of NIH3T3 fibroblasts expressing US28 into nude mice does indeed promote tumorigenesis, possibly via COX-2 up-regulation [40]. More recently, investigators have shown that the HCMV US28 RNA is found in both glioblastomas [36] and medulloblastomas [41], and in the former, promotes an invasive and angiogenic phenotype [36]. Taken together, these results argue for a role of US28 as a viral oncogene.

This chapter focuses on recent advancements in methodologies used for studying the function of CMV GPCRs, emphasizing protocols that can be performed in the context of infection. In particular, we will address methods for examining viral GPCR signaling by transient assays and focus on the recent breakthroughs in generating viral mutants making possible the investigation of GPCR function in the context of infection. Finally, we will discuss the importance of animal models to our understanding of the CMV GPCRs in viral pathogenesis.

## 2 Materials

### 2.1 Cell Culture

1. Primary human fibroblasts, such as HS68 cells (ATCC, CRL-1635) used between passages 10 and 20 (*see* **Note 1**).
2. Human embryonic kidney cell line HEK-293 (ATCC, CRL-1573).
3. Primary human retinal pigment epithelial cells (ARPE19 [ATCC, CRL-2302]) used between passages 22 and 35.
4. Human umbilical vascular endothelial cells (HUVECs) isolated from umbilical cords by collagenase digestion (or purchased from Lonza) maintained on either Primaria tissue culture plates (BD Falcon) or plates pre-coated with 0.1 % pig gelatin (Sigma) in 1× PBS, and used between passages 2 and 8.
5. Dulbecco's modified eagle medium (DMEM) containing 10 % Fetal Clone III serum (Hyclone), and supplemented with 100 U/ml each of penicillin and streptomycin is used to culture primary human fibroblasts.
6. Minimal essential medium MEM containing 10 % fetal bovine serum (Hyclone), and supplemented with 100 U/ml each of penicillin and streptomycin is used to culture HEK-293 cells.

7. DMEM-HAM's F12 containing 10 % FBS, 2.5 mML-GLUTAMINE, 0.5 mM sodium pyruvate, 15 mM HEPES, 1.2 g/l $NaHCO_3$, and 100 U/ml each of penicillin and streptomycin is used to culture ARPE19 cells.
8. EGM-2 medium supplemented with the EGM-2 additives (Lonza) is used to culture HUVECs.
9. Trypsin–EDTA: 0.05 % Trypsin, 0.53 mM EDTA.
10. Transfection reagent (e.g., Mirus TransIT® LT1 or Invitrogen Lipofectamine 2000).

### 2.2 Assessing PLC-β Activity by Measuring $IP_3$ Accumulation

1. Wash dowex in formate phase (AG1-X8, Bio-Rad 140-1444) with 20 l $dH_2O$. Store as 50 % slurry in $dH_2O$ at 4 °C. Add 1 ml of slurry to column prior to use. A variety of reusable columns can be used. Bio-Rad Poly-Prep® chromatography columns (part no. 731-1550) work well.
2. [2-$^3$H(N)]-myo-inositol (PerkinElmer, NET-114A).

### 2.3 Cell Lysis

1. Standard lysis buffer: 50 mM Hepes, pH 7.4, 0.5 % NP-40, 250 mM NaCl, 10 % Glycerol, 2 mM EDTA, 1 mM PMSF, 2.5 μg/ml aprotinin, 5.0 μg/ml leupeptin, 200 μM activated sodium orthovanadate, 1 mM Sodium Fluoride (*see* **Note 2**).
2. RIPA lysis buffer: 10 mM Tris pH 7.5, 0.1 % SDS, 1.0 % Triton X-100, 1.0 % deoxycholate, 150 mM NaCl, 5 mM EDTA, 1 mM PMSF, 2.5 μg/ml Aprotinin, 5.0 μg/ml Leupeptin, 200 μM activated sodium orthovanadate, 1 mM sodium fluoride.
3. Laemmli Sample Buffer: 14.0 ml 4× Tris Stacking buffer pH 6.8, 14.4 ml 50 % Glycerol, 2.0 g SDS, 240 μl beta-mercaptoethanol, 9.4 ml $dH_2O$.
4. 10× Red Blood Cell (RBC) Lysis Buffer: 40.15 g $NH_4Cl$, 5.0 g $NaHCO_3$, 0.186 g EDTA in 200 ml $dH_2O$. Dilute RBC Lysis buffer to 1× prior to use.

### 2.4 Western Blotting

1. Tris buffered saline containing Tween-20 (TBST): 50 mM Tris–HCl, pH 7.4, 150 mM NaCl, 0.1 % Tween-20.
2. Blocking Buffer: TBST containing 5 % nonfat dried milk.
3. Phospho-p38 MAPK (Thr180/Tyr182) Antibody, Cell Signaling Technology, Cat#9211.
4. Supported nitrocellulose (e.g., Schleicher and Schuell).
5. Chemiluminescence Detection Kit (e.g., Amersham ECL or Pierce SuperSignal western blotting kits).
6. Infrared Detection Kit (e.g., Li-Cor Odyssey western blotting kits).

### 2.5 Luciferase Reporter Assays

1. Dual-Luciferase Reporter Assay System, Promega, Cat#E1910.

### 2.6 BAC Recombineering

1. 0.2 mg/ml D-biotin: sterile filtered, made fresh.
2. 20 % galactose: autoclaved, stored at 4 °C.
3. 20 % 2-deoxy-galactose: autoclaved, made fresh.
4. 20 % glycerol: autoclaved, stored at room temperature.
5. 10 mg/ml L-leucine: Heat to get into solution but do not let boil. Sterile filtered, stored at 4 °C.
6. 12.5 mg/ml chloramphenicol in EtOH, stored at −20 °C.
7. 1 M $MgSO_4 \cdot 7H_2O$ stored at room temperature.
8. 1× M9 medium (1 l, autoclaved, stored at room temperature): 42.3 mM $Na_2HPO_4$ (6 g/l), 22 mM $KH_2PO_4$ (3 g/l), 18.7 mM $NH_4Cl$ (1 g/l), 8.6 mM NaCl (500 mg/l).
9. 5× M63 (1 l, autoclaved, stored at room temperature): 75.5 mM $(NH_4)_2SO_4$ (10 g/l), 0.5 M $KH_2PO_4$ (68 g/l), 9.0 μM $FeSO_4 \cdot 7H_2O$ (2.5 mg/l). Adjust to pH 7 with KOH.
10. M63 minimal plates (500 ml makes 20–25 plates): 7.5 g agar in 400 ml $ddH_2O$ in a 500 ml bottle with a stir bar and autoclaved, cooled to a "touchable" temperature of ~50–55 °C, 100 ml 5× M63 medium, 500 μl 1 M $MgSO_4 \cdot 7H_2O$ (1 μM), 500 μl chloramphenicol (12.5 μg/ml), 2.5 ml biotin (0.5 mg) (*see* **Note 3**), 5 ml galactose (0.2 %), 2.25 ml leucine (45 mg).
11. 2-DOG plates (500 ml makes 20–25 plates)
    7.5 g agar in 400 ml $H_2O$ in a 500 ml bottle with a stir bar and autoclaved, cooled to a "touchable" temperature of 50–55 °C, 100 ml 5× M63 medium, 500 μl 1 M $MgSO_4 \cdot 7H_2O$ (1 μM), 500 μl chloramphenicol (12.5 μg/ml), 2.5 ml biotin (0.5 mg) (*see* **Note 3**), 5 ml 2-deoxy-D-galactose (0.2 %), 2.25 ml leucine (45 mg), 5 ml glycerol (0.2 %).
12. MacConkey indicator plates: Prepare MacConkey agar plus galactose according to manufacturer's instructions (e.g., BD, Cat#281810), 12.5 μg/ml chloramphenicol.
13. PCR cleanup columns (e.g., GE Healthcare GFX columns).

### 2.7 Purification of BAC DNA

1. CMPS1 [Similar to Qiagen P1 buffer, +RNAse]: 50 mM Tris–HCl, pH 8.0, 10 mM EDTA, pH 8.0, 200 μg/ml RNAse A added *just prior to use* (20λ of 10 mg/ml stock, per 1.0 ml CMPS1).
2. Alkaline SDS Solution [Similar to Qiagen P2 buffer]: Final concentrations: 0.2 N NaOH, 1 % SDS. Make stock solutions at 2× concentrations, so mix equal parts *just prior to use.*
3. TEN Solution: 10 mM Tris–HCl, pH 7.4 or pH 8.0, 1 mM EDTA, pH 8.0, 150 mM NaCl.

4. 10.1 TE Solution: 10 mM Tris–HCl, pH 7.4 or pH 8.0, 0.1 mM EDTA, pH 8.0.
5. Endotoxin Removal Kit: Sigma, Cat#E4274.
6. Column BAC purification kit (e.g., Machery-Nagel Nucleobond BAC purification kit).

### 2.8 Plasmids

1. pcDNA3 (or similar) vector for viral GPCR of choice (HCMV-US27, HCMV-US28, MCMV-M33, etc.).
2. pcDNA3 (or similar) vector for pp71.
3. pGL3 3× MHC-Luc (or similar) to assess NFκB activity.
4. pGL3 9× NFAT-LUC (or similar) to assess NF-AT activity.
5. pFRLUC and pFA2CREB to assess CREB activity (Stratagene).
6. pHRG-TK to control for transfection variation and generalized effects of viral GPCRs on basal transcriptional activity (Promega).

### 2.9 Miscellaneous

1. Anti-fade mounting medium (e.g., Vector Labs VECTASHIELD or Molecular Probes SlowFade).
2. Biotinylated anti-FLAG antibody (e.g., Sigma M2 biotinylated anti-FLAG).
3. Cell Surface Protein Isolation kit (Thermo Scientific).

## 3 Methods

While it is not possible to generalize the signaling activities of the cytomegalovirus GPCRs into a single pathway, it is clear that at least several of these receptors (US28/M33/R33) signal via G-proteins such as $G_{q/11}$ and drive a number of downstream signals including accumulation of the second messenger $IP_3$, activation of protein kinases, and stimulation of transcription factor activity. In this section, we will describe basic methodology that can be used to assess these particular signaling activities.

### 3.1 Measuring Viral GPCR Stimulated Inositol Triphosphate ($IP_3$) Accumulation

The following protocol was designed for the study of HCMV US28 stimulated $IP_3$ accumulation (a.k.a. $PIP_2$ hydrolysis, PLC activity, inositol triphosphate accumulation) in HCMV infected fibroblasts or in transiently transfected HEK-293 cells, but can easily be adapted for use in a number of different cell types and can be modified to study other cytomegalovirus GPCRs in conditions of virus infection or transient transfection [7, 9, 10, 32, 42–44]. In the case of virus infection, the methodology described uses the primary human fibroblast cell line, HS68 (ATCC, CRL-1635) and in the case of transient transfection, the methodology described uses the embryonic kidney cell line HEK-293 (ATCC, CRL-1573). $G_{q/11}$ stimulated PLC-β activity cleaves phosphatidylinositol 4,5-bisphosphate ($PIP_2$) into inositol 1,4,5-trisphosphate ($IP_3$)

and diacylglycerol (DAG). Accumulation of $IP_3$ is easily measured in the lab with standard equipment and reagents.

1. Plate the cells into 12-well culture plates so that they will be ~75 % confluent at time of plating. Incubate in humidified incubators at 37 °C and 5 % $CO_2$.
   (a) For infection of HS68 fibroblasts with HCMV, the suggested cell number is ~100,000 cells per well in a total volume of 1 ml of medium.
   (b) For transient transfection of HEK-293 cells, the suggested cell number is ~250,000 cells per well in a total volume of 1 ml medium. HEK-293 cells and their derivatives are not tightly adherent, and care should be taken to facilitate adherence, such as coating the culture wells with 5 mg/ml collagen prior to plating cells.
2. Let the cells adhere overnight.
3. Infect with virus or transfect with appropriate viral GPCR expression construct. The length of time to let the infection or transfection proceed prior to harvesting should be determined empirically depending on timing of viral GPCR expression, etc. In the case of HCMV infection experiments, US28 reaches maximal expression at approximately 48 h post-infection (hpi), and therefore, 48 h would be an appropriate time to analyze US28 dependent $IP_3$ accumulation. Similarly, 48 hpi is a typical time at which to analyze transfection experiments as this is the time at which most transient gene expression peaks.
   (a) Infection of HS68 fibroblasts with HCMV. Adsorb virus to cells at appropriate multiplicity of infection (MOI) for 3–6 h. Both wild-type and ΔUS28 strains should be used to ascertain the specific effects of US28 on driving the activation of this signaling pathway. To achieve roughly 95–99 % infection, an MOI of 3–5 should be chosen. At the end of the adsorption period, remove the medium containing virus and feed with fresh medium.
   (b) Transient transfection of HEK-293 cells. The following describes the amount of DNA and lipid required for each well of a 12-well plate, although the amounts can be scaled up or down depending on the scale of transfection required. 250 ng of plasmid DNA is diluted in 50 μl serum free medium, supplemented with 1 μl of Mirus TransIT® LT1 transfection reagent and incubated for 15 min at room temperature. The 50 μl transfection reaction is then transferred to the appropriate wells of a 12-well plate and the transfection is allowed to proceed for 6 h. At the end of the 6 h incubation, remove the medium containing DNA/transfection reagent and feed with fresh medium. The TransIT® LT1 transfection reagent is highly efficient and

exhibits low toxicity and therefore can be left on the cells overnight if desired. The viral GPCRs themselves are somewhat toxic in nature and thus should be tested at various concentrations (i.e., 10, 50, and 250 ng of DNA per well). All transfections should contain a total of 250 ng plasmid DNA, so in cases where less than 250 ng of viral GPCR DNA is used, the transfection cocktail should be supplemented with an empty vector. It is recommended that the experiments be performed in duplicate or triplicate.

4. The next day (18–24 h post-infection/transfection), aspirate the medium and add 1 ml/well of fresh medium containing 1.0 μCi/ml [2-$^{3}$H(N)]-myo-inositol. The concentration of myo-inositol can be increased if necessary and the cells can be labeled in either serum free medium or serum containing medium (*see* **Note 4**).
5. The following day (approximately 40–48 h post-infection/transfection, wash the cells 1× with 1 ml serum free medium.
6. Feed the cells with 1 ml serum free medium containing 20 mM LiCl. If using chemokines or other potential agonists, add simultaneously with serum-free medium containing LiCl. The LiCl inhibits endogenous inositol phosphatase activity and enables newly produced, receptor stimulated $IP_3$ to accumulate.
7. Let inositol phosphates accumulate for 2–3 h.
8. Stop the reaction by aspirating medium, adding 1 ml of 0.4 M perchloric acid per well, and incubating for 15 min in the cold room. The perchloric acid will not cause the cells to lift off, but the perchloric acid at this point will contain the accumulated IPs.
9. Transfer 800 μl of perchloric acid from each well to a microfuge tube containing 400 μl of 0.72 M KOH/0.6 M $KHCO_3$. This will form a white fluffy precipitate.
10. Vortex and centrifuge for 1 min at 15,000 × *g*.
11. Transfer 50 μl of supernatant to scintillation vials, add 10 ml scintillation fluid, and count. (This step is optional and may be used to internally control for the relative labeling and cell number used.)
12. Transfer 1 ml of the supernatant to Fisherbrand 12 × 75 mm glass tubes containing 3 ml $dH_2O$.
13. Prepare dowex columns by adding 1 ml of dowex slurry (described in Subheading 2) and let settle. A variety of reusable columns can be used for this step.
14. Pour sample from **step 12** over column. Let sample flow through.
15. Wash the columns with 2 bed volumes (~25 ml) of $dH_2O$.

16. Wash the columns with 1 bed volume of 60 mM sodium formate/5 mM disodium tetraborate.
17. Elute bound IPs. Each of the following elution steps should be performed by transferring the chromatography column into a fresh scintillation vial prior to elution (*see* **Note 5**).
    (a) Elute $IP_1$ with 4 ml of 0.2 M ammonium formate/0.1 M formic acid. Wash the column with 1 bed volume of the same buffer.
    (b) Elute $IP_2$ with 4 ml of 0.4 M ammonium formate/0.1 M formic acid. Wash the column with 1 bed volume of the same buffer.
    (c) Elute $IP_3$ with 4 ml of 0.8 M ammonium formate/0.1 M formic acid.
18. Add 10 ml scintillation fluid to each eluted sample. Count in scintillation counter.

### 3.2 Measuring Viral GPCR Stimulated Protein Kinase Activation

The following protocol is specifically designed for the detection of US28 or M33 stimulated p38-MAPK kinase activation in transiently transfected HEK-293 cells, but can easily be adapted for use in a number of different cell types or for different protein kinases [38, 44, 45]. Moreover, the protocol can be modified to study cytomegalovirus GPCRs in conditions of virus infection or transient transfection. The protocol takes advantage of phospho-specific antibodies (which recognize activated forms of protein kinases) to assess viral GPCR mediated activation of the protein kinase in question.

1. Plate HEK-293 cells into 12-well culture plate so that they will be ~75 % confluent at time of plating. Incubate in humidified incubators at 37 °C and 5 % $CO_2$. The suggested cell number is ~250,000 cells per well in a total volume of 1 ml medium. HEK-293 cells and their derivatives are not tightly adherent, and care should be taken to facilitate adherence, such as coating the culture wells with 5 mg/ml collagen prior to plating cells.
2. Let the cells adhere overnight.
3. Transfect with appropriate viral GPCR expression construct as described above in Subheading 3.1 **step 3b**.
4. Forty-eight hour post-transfection, lyse the cells and prepare protein extracts for gel electrophoresis. Protein extracts can be prepared using several different lysis buffers, depending on the preference of the investigator (see Subheading 2). Extracts prepared directly in Laemmli sample buffer (**step 4a**) maintains the phosphorylation status of most kinases. However, the use of this buffer eliminates the possibility of quantifying protein concentrations and therefore requires accurate cell counts prior to preparation of the extracts. When extracts are prepared

in standard lysis buffer (**step 4b**), one must ensure that phosphatase activity does not affect the results of the experiments. In particular, it is important to use NaF and activated $Na_3VO_4$ in lysis buffers to inhibit serine/threonine and tyrosine phosphatases respectively (*see* **Note 2**).

(a) To prepare whole cell extracts directly in Laemmli sample buffer, the medium is aspirated from the 12-well plates, 250 μl of Laemmli sample is added directly to the wells, the wells are scraped briefly with a cell scraper, and the extracts are transferred to microcentrifuge tubes. The extracts are sonicated briefly to disrupt chromosomal DNA.

(b) To prepare whole cell extracts in standard lysis buffer, the medium is aspirated from the 12-well plates, the wells are washed 1× with 1× PBS, and 250 μl of standard lysis buffer is added directly to the wells. The wells are scraped briefly with a cell scraper, the extracts are transferred to microcentrifuge tubes, and incubated on ice for 15–30 min. The extracts are clarified by centrifugation at 12,000 × *g* and supernatant is transferred to a fresh tube. Protein concentration is then quantified by standard protein assays (Bradford, Bio-Rad Protein Assay, etc.).

5. The extracts prepared by either procedure in **step 4** are then subjected to SDS-PAGE using standard protocols for gel electrophoresis.
6. Transfer resolved proteins to supported nitrocellulose membranes and block nonspecific reactivity with Tris-buffered saline containing 0.1 % Tween 20 (TBST) and 5 % nonfat dried milk. In some cases, blocking with 5 % nonfat dried milk can increase nonspecific reactivity due to the presence of phosphoproteins present in the milk. In this case, 1 % bovine serum albumin (BSA) can be used as a substitute for the milk.
7. Antibody directed against the phosphorylated/activated form of the protein kinase of interest is then used to probe western blots. In the case of p38-MAPK, the anti-phosphospecific p38 antibody is diluted 1:1,500 in TBST. Bound primary antibody is then detected using the appropriate secondary antibodies using enhanced chemiluminescence or infrared fluorescence systems.

### 3.3 Measuring Viral GPCR Stimulated Transcription Factor Activity

The cytomegalovirus GPCRs have been reported to activate a number of transcription factors including NFκB, CREB, and NFAT [7, 12, 16, 18, 38]. The following protocol is specifically designed for the detection of US28 or M33 stimulated transcriptional reporter activity in transiently transfected HEK-293 cells, but can easily be adapted for use in a number of different cell types or transcription factors and can be modified to study other cytomegalovirus GPCRs in conditions of virus infection or transient transfection. Two important considerations should be taken into account when assessing viral GPCR stimulated transcription factor activity in infected cells. First, it is important to use wild-type and viral GPCR null mutants (i.e., ΔUS28 mutants) to differentiate between viral GPCR effects and those due to either the virion itself or other cytomegalovirus proteins. Many of the transcription factors stimulated by the viral GPCRs are in fact activated during cytomegalovirus infection, but it is clear that the viruses use multiple mechanisms to activate transcription factors at different stages of infection. Such is the case for NFκB, which is activated within minutes after virion binding, presumably due to virus engagement of NFκB linked cell surface receptors and also during the IE and E phases of infection. Second it is also important to use internal controls such as the pHRG-TK Renilla luciferase control reporter. This will allow the investigator to control for generalized effects of cytomegalovirus infection on the basal transcription machinery itself, which can lead to artifactual conclusions regarding specific changes in transcription factor activity.

1. Plate HEK-293 cells into 12-well culture plate so that they will be ~75 % confluent at the time of plating. Incubate in humidified incubators at 37 °C and 5 % $CO_2$. The suggested cell number is ~250,000 cells per well in a total volume of 1 ml medium. HEK-293 cells and their derivatives are not tightly adherent, and care should be taken to facilitate adherence, such as coating the culture wells with 5 mg/ml collagen prior to plating cells.
2. Let the cells adhere overnight.
3. Transfect with appropriate viral GPCR expression constructs and reporter genes. For transient transfection of HEK-293 cells in one well of a 12-well culture plates, 250 ng of plasmid DNA is diluted in 50 μl serum free medium, supplemented with 1 μl of Mirus TransIT® LT1 transfection reagent and incubated for 15 min at room temperature. The 50 μl transfection reaction is then transferred to cells plated as described in **step 1** above and the transfection is allowed to proceed for 6 h. At the end of the 6 h incubation, remove the medium containing DNA/transfection reagent and feed with fresh medium. The TransIT® LT1 transfection reagent is highly efficient and exhibits low toxicity and therefore can be left on the cells overnight if desired. The viral GPCRs themselves are somewhat toxic in nature and thus should be tested at various concentrations (i.e., 10, 50, and 250 ng of DNA per well). The concentration of the reporter gene DNA per well is as follows: for assessing NFκB activity (15 ng of pGL3-3X MHC-Luc), for assessing NF-AT activity (15 ng of pGL3-9X NFAT-Luc) and for assessing CREB activity (30 ng pFR-LUC/10 ng pFA2-CREB) (*see* **Note 6**). The control pHRG-TK renilla luciferase plasmid should be included in all transfections at a concentration of 15 ng per well. All transfections should contain a total of 250 ng plasmid DNA, so in cases where less than 250 ng of

viral GPCR DNA is used, the transfection cocktail should be supplemented with an empty vector. To control for transfection variability, it is recommended that the experiments be performed in duplicate or triplicate.

4. Forty-eight hour post-transfection, aspirate medium and wash wells with PBS.
5. Add 200 μl of 1× Passive Lysis Buffer (PLB) per well (*see* **Note 7**). Incubate for 15–30 min at room temperature. Transfer lysate to microcentrifuge tubes. The samples can be stored at −80 °C at this point.
6. Experimental luciferase (firefly) and control luciferase (renilla) can be examined on a luminometer using Luciferase Assay Reagent II (LAR II) and Stop&Glo Reagent according to the manufacturer's instructions (*see* **Note 7**).

These techniques provide the basis from which to examine proximal ($IP_3$ accumulation), intermediate (p38-MAP kinase), and distal (transcription factor) signaling activity emanating from cytomegalovirus encoded GPCRs. The number of antibodies that recognize phosphorylated and thus activated protein kinases is increasing at a rapid pace, thus enabling researchers to continue to explore a variety of signaling pathways that lie downstream of the viral GPCRs. There are numerous other methodologies that have been used to examine viral GPCR signaling; however, space constraints simply prevent us from covering each of these techniques in detail. For example, Smit and colleagues have used limited microarray analyses to uncover genes upregulated in response to US28 expression, and data mining approaches could easily be combined with large scale gene expression studies to identify networks of signaling pathways downstream of the viral GPCRs [40]. Finally, several investigators have used pharmacological inhibitors of signaling proteins such as PLC-β, PKC, PI3-K, etc. to identify additional signaling proteins downstream of the viral GPCRs [9, 18, 38–40, 46]. It is important to note that most of the signaling information that we currently have regarding the viral GPCRs has been generated in in vitro transfection/overexpression systems and it will be essential to use this current knowledge and extend these studies to identify what signals are truly generated in cytomegalovirus infected cells.

### 3.4 Generating and Analyzing Recombinant Cytomegaloviruses with Mutant GPCRs

Although significant advancements in our understanding of cytomegalovirus GPCRs have been made using the transient systems described above, recombineering methodologies have made possible the ability to investigate the CMV GPCRs in the context of viral infection.

*3.4.1 BAC Recombineering of GPCR Genes*

Early work aimed at generating recombinant viruses with mutations in viral GPCR genes took advantage of homologous recombination in mammalian cells, whereby a selectable marker was introduced by site-directed mutagenesis (Table 1). This method proved laborious and inefficient, and for some genes, impossible. Open reading frames (ORFs) that were essential for growth or those that conferred a severe growth defect when compared to wild-type, could only be mutated by this method if they were generated on complementing cell lines, which expressed the ORF of interest in *trans* [47]. Fortunately RCMV, RhCMV, MCMV, guinea pig CMV (GPCMV), and a variety of laboratory and clinical HCMV strains have been cloned into bacterial artificial chromosomes (BACs). Original protocols for BAC recombineering generated recombinant viral BACs that remained "marked" with either an insertion cassette or partial sequence from the shuttle plasmids used for RecE/T-mediated recombineering techniques [48]. Thus, this procedure does not yield seamless recombinants, which complicates the generation of viral BACs with multiple mutations and/or tags as well as revertants.

The field of bacterial recombineering has advanced greatly over the recent years, and researchers have adapted these methods for generating recombinant CMV BACs to study the function of the CMV GPCRs (Table 1). In particular, two methods including *galK* [49] and I-SceI [50] recombineering have proved extremely useful, as each of these protocols results in recombinant viral BAC DNA that is seamless at the site of recombination. The advantage of seamless recombineering is such that one can generate multiple site-specific mutations, epitope tags, fusion proteins, gene insertions, or whole ORF deletions within a single background. Additionally, these recombineering protocols are more efficient than previous BAC-mediated methods or site-directed mutagenesis in mammalian cells, have lower rates of off-site spontaneous recombination, require less time to generate mutants, permit reversion of the mutation, and unlike homologous recombination in mammalian cells, support the mutagenesis of essential ORFs. *GalK* and I-SceI recombination each take advantage of the Red recombinase system [49, 50]. The I-SceI method has previously been described for the study of HCMV GPCRs, and thus this chapter focuses on the utilization of the *galK* recombineering system [46]. Recombineering by *galK* uses a straightforward methodology, which involves a positive selection of the *galK* insertion cassette, followed by homologous recombination of either a double-stranded oligonucleotide or a purified PCR product by counterselection.

1. PCR amplify the galactokinase (*galK*) gene using no more than 2 ng of pGalK plasmid as the template with primers that contain a minimum of 50 bp of homologous sequence to the intended site of mutagenesis within the BAC. The underlined sequences below are complimentary to the pGalK cassette:

**Table 1**
**Cytomegalovirus GPCR recombinant virus constructs**

| Viral GPCR | Host | Strain | Recombination method |
|---|---|---|---|
| M33 | Murine | K181 | Homologous recombination in mammalian cells [15–17, 65] |
| M33 | | K181 | Kan-*frt* BAC [18] |
| M78 | Murine | K181 | Homologous recombination in mammalian cells [22] |
| R33 | Rat | Maastricht | Homologous recombination in mammalian cells [19, 74] |
| R78 | Rat | Maastricht | Homologous recombination in mammalian cells [23, 74] |
| UL33 | Human | AD169 | Kan-*frt* BAC [12, 21], homologous recombination in mammalian cells [20], *galK* BAC [21] |
| | | FIX | Kan-*frt* BAC, *galK* BAC [21] |
| | | TB40/E | Kan-*frt* BAC, *galK* BAC [21] |
| UL78 | Human | AD169 | Kan-*frt* BAC [21], *galK* BAC [21], pST shuttle vector BAC [25] |
| | | FIX | Kan-*frt* BAC [24], *galK* BAC [24] |
| | | TB40/E | Kan-*frt* BAC [24], *galK* BAC [24] |
| US27 | Human | AD169 | Homologous recombination in mammalian cells [28] |
| | | AD169 | Kan-*frt* BAC [21], *galK* BAC [21] |
| | | FIX | Kan-*frt* BAC [26], *galK* BAC [26] |
| | | TB40/E | Kan-*frt* BAC [26], *galK* BAC [26] |
| US28 | Human | AD169 | Homologous recombination in mammalian cells [28] |
| | | Toledo | Homologous recombination in mammalian cells [31] |
| | | Towne | Homologous recombination in mammalian cells [29] |
| | | AD169 | Kan-*frt* BAC, *galK* BAC [21], pST shuttle vector BAC [10, 42] |
| | | FIX | Kan-*frt* BAC [32, 43], *galK* BAC [21] |
| | | TB40/E | Kan-*frt* BAC [21], *galK* BAC [75] |
| | | Titan | Kan-*frt* BAC [35, 40] |
| US27:US28 multiple deletion mutant | Human | AD169 | Homologous recombination in mammalian cells [28] |
| UL33:UL78:US27:US28 multiple deletion mutant | Human | AD169 | I-SceI BAC [46] |
| | | FIX | *galK* BAC [21] |
| | | TB40/E | *galK* BAC [75] |

Forward Primer: 5′–50′ bp homology-CCTGTTGACAATTAATCATCGGCA-3′

Reverse Primer: 5′–50′ bp complimentary strand homology-TCAGCACTGTCCTGCTCCTT-3′

(a) To remove the pGalK template, digest the PCR product by adding 1 μl of DpnI directly into the PCR reaction and incubate at 37 °C for 1 h.

(b) Use PCR cleanup columns to purify the PCR product and elute the PCR product in 25 μl of $dH_2O$.

2. This cassette is then inserted into the CMV BAC genome by homologous recombination mediated by heat shock induced Red recombinase enzymes. It is critical to maintain the BAC at 32 °C in a recombination competent bacterial strain, such as SW102, SW105, or SW106 all of which are *galK*$^-$.

   (a) Following overnight culture of the BAC-containing bacteria at 32 °C in 5–10 ml of medium containing 12.5 μg/ml chloramphenicol, inoculate 25 ml of medium containing 12.5 μg/ml chloramphenicol with a 1:50 dilution of the overnight culture. Grow the bacteria to an $OD_{600}$ of 0.5–0.6. Heat-shock the cells at 42 °C for 15 min in a shaking water bath.

   (b) Quickly cool the bacteria in an ice bath slurry, with shaking, then transfer 10 ml of the culture to a pre-chilled conical and pellet the bacteria at 4 °C. Gently resuspend the pellet in 1 ml ice-cold $ddH_2O$ and transfer to a microcentrifuge tube. Wash the cell pellet three additional times in ice-cold $ddH_2O$, and resuspend the pellet after the final wash in 100 μl of ice-cold $ddH_2O$.

   (c) Use 2.5 μl of the PCR product to transform 50 μl of the bacteria by electroporation in a pre-chilled 2 mm gap electroporation cuvette. Recover for 1 h in 1 ml of medium without antibiotic at 32 °C, and then wash the pellet three times in M9 salts taking care not to vortex or pipette too harshly. Resuspend the final pellet in 1 ml M9 salts.

3. Plate 100 μl of undiluted cells and 100 μl of a 1:10 dilution onto M63 minimal medium plates and incubate at least 3 days at 32 °C. BACs that recombine to express *galK* are chosen by positive screen for growth on minimal medium containing galactose as the sole carbon source. Although all colonies that grow on these plates should ideally contain an integrated *galK* cassette, these plates screen for the preferential growth of potentially successful recombinants.

4. Successful *galK* recombinants are further selected using MacConkey's indicator plates containing galactose, on which *galK*-positive clones will grow as single red colonies, while *galK*-negative clones will grow as white colonies. This step is

critical to ensuring that the clones used in the counter-selection step do indeed contain the *galK* cassette.

5. Patch single red colonies from **step 4** onto LB/chloramphenicol plates and confirm the insertion of the *galK* cassette at the proper location of interest by PCR using flanking primer sets.
6. Next, counterselect against *galK*, by substituting either a PCR product or a double stranded oligo that contains the mutation or epitope tag of one's choice, all of which also contain flanking arms to the region in the BAC DNA being mutated. In this step the PCR product or double stranded oligo is inserted into the CMV BAC genome by homologous recombination mediated by red recombinase as described in **step 2**, with slight modification.
   (a) Prepare competent bacteria as described above (*see* Subheading 3.4.1, **steps 2a** and **2b**).
   (b) Mix 10 mg of each oligo in a volume of 100 ml 1× PCR buffer. Boil for 5 min and cool slowly to room temperature. EtOH-precipitate the annealed oligos and resuspend the final pellet in 100 μl $ddH_2O$ to yield a final concentration of 200 ng/ml. Use 1 μl per transformation. For reversion using PCR products, PCR amplify the desired insert with at least 50 bp of flanking sequence to the site of recombination, generating a PCR product of ~1,000 bp. If the product is larger, increase the size of the flanking sequence. For example, use 500 bp for products >2 kb.
   (c) Following transformation, recover the bacteria in 10 ml medium without antibiotic in a 100 ml baffled flask for 4.5 h in a 32 °C shaking incubator. Remove 1 ml of the culture and wash with ice-cold $ddH_2O$ in a refrigerated microcentrifuge as above in **step 2c**. The cells are diluted and plated (as described in **step 3**) on 2-deoxy-D-galactose (2-DOG) plates. Incubate the plates for 3–4 days at 32 °C.
7. Selection against *galK* involves resistance to 2-DOG on minimal plates with glycerol as the carbon source. 2-DOG is harmless to bacteria, unless phosphorylated by functional *galK*. As a result, 2-DOG becomes 2-deoxy-galactose-1-phosphate, which bacteria cannot metabolize, and thus it is a toxic intermediate to those clones that still harbor the *galK* cassette. The resulting 2-DOG-resistant colonies are recombinant clones that have no residual foreign DNA sequences as a result of the recombineering protocol.
8. Patch colonies on both 2-DOG and M63 minimal plates to ensure for the absence of *galK*. Confirm clones for the absence of *galK* by PCR and finally sequence the recombinants to ensure genomic integrity at the site of the recombination. One can now use this clone to generate additional recombinants within the same background, or alternatively, reconstitute infectious virus (*see* Subheading 3.4.2).

#### 3.4.2 Reconstitution of Infectious Virus

Following the successful generation of recombinant BACs for the GPCR(s) of interest, one can easily reconstitute infectious virus. The HCMV clinical strain TB40/E [51], for example, yields high titers following reconstitution, providing ample virus with which to perform a multitude of experiments. Additionally, as none of the GPCR mutants, including those recombinants that harbor multiple GPCR deletions, show a particle to PFU defect, each recombinant will indeed yield a stock with a titer that is sufficient [21]. The first step in reconstituting recombinant viruses is purifying the BAC DNA by either alkaline lysis/precipitation or column purification kit. The protocol described here involves purification of BAC DNA by alkaline lysis and we have modified the protocol to also include an additional step to remove endotoxins. Compositions of the buffers used in this protocol are given in Subheading 2 of this chapter.

1. Grow 10 ml overnight (~16–18 h) culture of bacteria containing the BAC of interest.
2. Pellet the bacteria, resuspend in 200 μl CMPS1 w/RNAseA solution, lyse with 400 μl alkaline SDS solution, and neutralize with 300 μl potassium acetate. Pellet debris and treat clarified supernatant with the Endotoxin Removal solution (Sigma) according to the manufacturer's instructions (*see* **Note 8**). Precipitate DNA with 1.0 ml isopropanol.
3. Dissolve the resulting pellet in 500 μl TEN buffer at room temperature. Once the pellet has dissolved (roughly 10 min) centrifuge briefly to remove any remaining cellular debris, and precipitate the DNA from the supernatant with two volumes ethanol. The resulting BAC DNA is resuspended in 10.1 TE buffer, and should be used within 24 h for transfection. Importantly, one should refrain from freezing BAC DNA that is slated for transfection, as this greatly reduces the efficiency. Additionally, when manipulating BAC DNA, one should take care not to shear the DNA by rapid pipetting or using standard pipette tips (wide-bore tips are optimal).
4. To reconstitute virus, transfect low-passage, primary fibroblasts ($1.7 \times 10^6$ cells) in a 4 mm cuvette by electroporation (960 μF, 0.26 V) in 500 μl of Opti-MEM with 1 μg of pCGN-pp71 (or equivalent pp71-expressing plasmid) and the BAC DNA of interest. Plate transfected cells in either T75 flasks or 100 mm dishes.
5. Feed transfected cells every 2–3 days until significant cytopathic effect (CPE) is observed.
6. To generate TB40/E stocks proceed to **step** 7, to generate FIX stocks proceed to **step 10**.
7. When fibroblasts have reached 100 % CPE, scrape cells into the infectious supernatant and collect by low-speed centrifugation.
8. Reserve the supernatant and bath-sonicate the cells to release the cell-associated virus, spin the cell debris as before, and combine the supernatants.

9. This combined supernatant can now be used to generate high titer stocks by further expansion.
   (a) For generating TB40/E stocks, only 1/10th of the infectious supernatant from **step 8** is necessary to infect at least $5 \times 10^7$ cells. The remaining supernatant can be stored at −80 °C in 1–2 ml aliquots and used to generate additional future stocks at another time.
10. For generating FIX stocks, when fibroblasts have reached 100 % CPE, remove the medium and trypsinize the cells. Seed the infected cells onto approximately $4 \times 10^7$ to $5 \times 10^7$ cells. FIX is highly cell-associated, and thus seeding the cells rather than the medium from the transfection plate is critical.
11. For both FIX and TB40/E, generating a high titer usable stock may require one to concentrate the viral stock. To concentrate virus, harvest cells and medium from the expanded stock by scraping cells into the infectious media, as described above in **step 7**. Pipette the cleared medium into ultracentrifugation tubes, and underlay with 20 % D-sorbitol, containing 50 mM Tris–HCl, pH 7.2 and 1 mM $MgCl_2$. Concentrate infectious virus by ultracentrifugation at 72,128 × $g$ for 90 min at 25 °C. Virus stocks can be stored at −80 °C in complete medium containing 1.5 % BSA for long-term storage. Store sterile-filtered 3 % BSA in 1 × PBS at 4 °C and resuspend virus 1:1 in complete media: 3 % BSA.

*3.4.3 Assessment of Viral Growth Properties*

Many of the early assessments of the HCMV ORFs' necessities for viral replication in tissue culture were performed using fibroblasts [52, 53]. Although fibroblasts are invaluable to the study of HCMV lytic replication, they do not afford the ability to uncover functions of HCMV genes that are essential for growth in other clinically relevant cell types and tissues. Such is the case for the viral GPCRs that are not required for HCMV replication in fibroblasts. Thus, many investigators have taken advantage of clinical strains of HCMV that exhibit a broader cell tropism. The use of these clinical strains (i.e., TB40/E, TR, or FIX) allows for studies in an expanded repertoire of cell types including, but not limited to fibroblasts, hematopoietic progenitor cells, monocytes, macrophages, epithelial and endothelial cells. Assessing the growth properties of a mutant virus in a range of cell types is critical, as CMV pathogenesis in vivo is complicated and involves a plethora of different cells and tissues.

Assessing Production and/or Spread of Virus Occurring Via the Extracellular Route

1. Plate cells in 6-well plates (~$5 \times 10^5$ to $1 \times 10^6$ cells, depending on cell type), designating two wells for each virus being tested. These two wells will serve as duplicate infections. Note that infection of ARPE19 cells does not result in efficient extracellular spread of HCMV (e.g., FIX and TB40/E) therefore quantifying viral replication is performed by cell-associated viral assays (see below).

2. For multi-step growth curves in fibroblasts or endothelial cells use a low MOI (e.g., 0.01–0.1 PFU/cell). For single-step growth curves, a high MOI between 1.0 and 3.0 PFU/cell is recommended.
   (a) Dilute the viral stock in the medium specific to the cell type being used. Ensure that enough inoculum is prepared to cover each well in addition to some that is reserved to assess the input titer. This is important when comparing viral growth between wild-type and recombinants, as one needs to ensure an equal amount of virus was used in the initial infection. Thus, reserving some inoculum that was not put onto cells is important when titering the growth curve.
   (b) Remove the medium and wash one time with 1× PBS. Add the inoculum to the cells in low volume (750 μl for a well of a 6-well plate) to ensure sufficient contact of the virus with the cells. Incubate at 37 °C/5 % $CO_2$ for 1 h, rocking the plate every 15 min.
   (c) Remove the inoculum, wash three times with 1× PBS, add fresh medium to the cells, and return to the incubator.
3. For low MOI infections, suggested time points include 0, 4, 8, 12, and 15 days post-infection (dpi). For high MOI infections, suggested time points include 0, 24, 48, 72, 96, and 120 hpi. The investigator should adjust these, as necessary. At each time point, remove a portion of the supernatant. This will vary depending on the assay being used to titer the viral growth curve. One should reserve enough supernatant from the cells such that the titering assay can be performed in triplicate. Replenish the cultures with the same volume of fresh medium that was removed for the time point. Store all of the collected samples at −80 °C until the time course is completed.
4. Once all of the time points have been collected, thaw the samples in a 37 °C water bath, and assess the titers by plaque assay, $TCID_{50}$ analyses, or modified IFA for IE protein expression. Each time point for each virus should be measured in triplicate.

#### Assessing Production and/or Spread of Virus Occurring Via the Cell-Associated Route

1. Plate cells in 6-well plates (~$5\times10^5$ to $1\times10^6$ cells, respectively). Dedicate one well for each time point for every virus being assessed.
2. Both endothelial cells and fibroblasts support infection with low MOIs of 0.01–0.1 PFU/cell or high MOI infections of at least 1.0 PFU/cell. Viral infection of ARPE19 cells spreads exclusively by cell-to-cell contact following either FIX or TB40/E infection. For ARPE19 cells, it is advisable to use a MOI of approximately 0.1 PFU/cell for multi-step growth curves and a MOI of at least 1.0 PFU/cell for single-step growth curves (*see* **Note 9**).

3. Prepare the inoculum as above, diluting virus in the appropriate medium.
   (a) Remove the medium from the cells, and wash the cells with 1× PBS as described in **step 2c** of Assessing Production and/or Spread of Virus Occurring Via the Extracellular Route.
   (b) Add the inoculum to the appropriate wells as above reserving an aliquot of the inoculum, and infect the cells for 1 h at 37 °C/5 % $CO_2$, rocking the plates every 15 min. Although not required, infection of ARPE19 cells is increased by centrifugal enhancement at 1,000 × *g* for 30 min at room temperature. If this step is performed, the cells should next be incubated at 37 °C/5 % $CO_2$ for an additional 1 h with rocking every 15 min.
   (c) Remove the inoculum and wash the cells three times with 1× PBS to remove any residual virus that had not entered the cells. Replenish the cultures with fresh medium and return to the incubator.
4. For either fibroblasts or endothelial cells, collect the cell-associated virus at the times described above in **step 3** of Assessing Production and/or Spread of Virus Occurring Via the Extracellular Route for low and high MOIs. For ARPE19 cells, infection progresses at a slower rate, and thus, low MOI time points include 0, 10, 20, and 30 dpi. Additionally, the medium on ARPE19 cultures should be changed every 5 dpi to ensure cell health over the time course of infection. For single-step growth analyses at high MOI, suggested time points include 0, 4, 8, and 12 dpi.
5. To collect cell-associated virus at each time point, remove the medium from the cultures, and wash 2–3 times with 1× PBS. Add back at least 1 ml of fresh medium, and scrape the cells into the medium. Samples should be stored at −80 °C until the time course is completed.
6. Evaluating the titer of the cell-associated virus requires three freeze-thaw cycles to disrupt the cells thereby releasing the virus. Thaw samples in a 37 °C water bath, ensuring that the samples completely thaw, and then quickly re-freeze in liquid nitrogen. Following the third thaw, spin down cellular debris and transfer the medium to a fresh tube for use in a titering assay as described above.

Undoubtedly, the types of approaches described in this section will generate important information on the roles of viral GPCRs in viral replication and spread in clinically relevant cell types in vitro, and when combined with in vivo viral replication experiments in animal models, will provide clues as to how these proteins function to facilitate replication and pathogenesis during the natural course of cytomegalovirus infection.

### 3.5 Detecting Viral GPCR Proteins in Infected Cells

As mentioned above, bacterial recombineering techniques have afforded investigators the ability to epitope tag viral proteins, in particular the viral GPCRs. Previous studies assessing the expression and subcellular localization of the CMV GPCRs generally included the overexpression of individual GPCRs in cell types in which CMV infection is not supported. Although these studies yield important information about potential function, these were not performed in the context of viral infection. Moreover, the expression level of viral GPCRs in infected cells may be very different than that observed in transient assays and may result in qualitative and quantitative differences in signaling. Antibodies directed against several of the CMV GPCRs including US27 [54], US28 [55], UL33 [20, 56], and MCMV M78 [22] have been generated. However, construction of viral recombinants expressing epitope-tagged GPCRs allows investigators to utilize commercially available validated antibodies that work across a variety of techniques including immunofluorescence assay (IFA), immunoprecipitation/western blot, immuno-electron microscopy, and fluorescence activated cell sorting (FACS).

#### 3.5.1 Detection and Localization of Viral GPCRs by Immunofluorescence Assay

Immunofluorescence assay (IFA) provides a useful platform for determining the cellular localization of a given GPCR and provides a convenient tool for determining the percentage of cells expressing the GPCR in question.

1. Grow cells on gelatin coated glass coverslips and infect at an MOI of at least 0.5 PFU/cell. Using an MOI of 3 PFU/cell will typically guarantee that >95 % of the cells are infected.
2. At the desired time post-infection, wash cells with 1× PBS, and fix with 2 % paraformaldehyde at 37 °C for 15 min. Alternatively, one can use cold 100 % EtOH to fix the infected cells, although it is important to note that this will destroy any color marker (e.g., eGFP or mCherry) that is expressed from the viral genome.
3. Following fixation, wash the cells three times with 1× PBS at room temperature, and then permeabilize with 0.1 % Triton X-100 for 15 min at room temperature.
4. Wash the cells with PBS containing 0.2 % Tween 20, and then block for at least 1 h at room temperature in 2 % BSA/0.2 % Tween 20 in 1× PBS. Alternatively, one can block the coverslips overnight at 4 °C. If multiple time points are necessary, the blocking step is an excellent step at which to stop until the remaining slides are harvested.
5. After blocking all of the coverslips, stain with primary antibody in blocking buffer for at least 1 h at room temperature.
6. After staining with primary antibody, wash the coverslips with 1× PBS containing 0.2 % Tween 20 at least three times at room temperature.

7. Stain the cells/coverslips with secondary antibody containing the appropriate conjugated fluorophore for at least 1 h in the dark at room temperature. A nuclear dye, such as 4′,6-diamidino-2-phenylindole (DAPI) or Hoechst should also be included, as this serves as an excellent control.
8. Wash the cells/coverslips three times in 1× PBS containing 0.2 % Tween 20.
9. Mount and seal the coverslips onto slides using an appropriate anti-fade mounting medium.
10. View the cells using standard fluorescent or confocal microscopy techniques.

We have taken advantage of recombineering techniques to generate FLAG-tagged GPCR recombinants in the AD169, FIX, and TB40/E backgrounds. Using these recombinants coupled with IFA, we have shown cellular localization for each of the GPCRs and have determined the presence of each in the mature HCMV virion [21, 24, 26].

#### 3.5.2 Detection and Localization of Viral GPCRs (and Interacting Partners) by FLAG Immunoprecipitation/Western Blot

Immunoprecipitation followed by western blotting is a very sensitive technique that can be used to detect viral GPCRs. This sensitivity is essential as future investigations designed to analyze vial GPCR expression function are likely to be performed in clinically relevant cell types that may not exhibit lytic expression patterns comparable to that observed in standard HCMV infected fibroblasts.

1. Plate cells in 100 mm dishes at 50–75 % confluent. Infect with viruses or expression constructs expressing FLAG-tagged GPCRs as described elsewhere in this report.
2. At the appropriate times post-infection or post-transfection, remove medium and wash cells 1× with PBS.
3. Add 1.0 ml of RIPA buffer containing protease and phosphatase inhibitors.
4. Transfer the lysate to microcentrifuge tubes and shear the DNA by passing lysates through a 22 G needle and syringe 15–20 times.
5. Clarify the supernatant by centrifugation at 15,000 × *g* for 15 min at 4 °C.
6. Transfer supernatant to a clean microcentrifuge tube and preclear lysates by adding 50 μl Sepharose 4B and rotating for 30–60 min at 4 °C.
7. Pellet Sepharose 4B by centrifugation at 15,000 × *g* for 15 min at 4 °C.
8. Transfer the supernatant to fresh tube and add 20 μl of anti-FLAG M2 beads. Rotate for ≥4 h at 4 °C.
9. Pellet M2 beads by centrifugation at 15,000 × *g* for 15 min at 4 °C. Wash beads four times with 1 ml RIPA buffer.

10. Resuspend washed beads in 50 μl of 3× sample buffer and incubate for 30 min at room temperature or 10 min at 42 °C. It is important to avoid boiling the immunoprecipitated samples as GPCRs have a tendency to aggregate and can form altered species that do not migrate at the predicted molecular weight on SDS-PAGE gels.
11. Separate immunoprecipitates by SDS-PAGE and analyze by western blot as described above in Subheading 3.2.
12. For western blot analyses of immunoprecipitated viral GPCRs and interacting proteins, it is important to use antibody reagents that are derived from a species different from that used in the immunoprecipitation step (in this case, the immunoprecipitating M2 antibody is mouse) to prevent cross reactivity between the immunoprecipitating and primary western antibody.

#### 3.5.3 Detection and Localization of Viral GPCRs by Fluorescence Activated Cell Sorting

Fluorescence Activated Cell Sorting (FACS) analysis is a powerful and rapid tool for assessing the expression of a given viral GPCR. Investigators have successfully used this method to demonstrate cell surface expression of epitope-tagged CMV GPCRs [32, 43, 55, 57–62]. For example, Stropes and Miller generated a variety of FLAG-tagged US28 recombinants to study US28 signaling in infected cells and demonstrated that while wild-type and a N-terminal truncation mutant exhibited similar constitutive signaling activities, the N-terminal truncation mutant exhibited decreased cell surface accumulation in comparison to wild-type US28 [43].

1. Using as few as $1 \times 10^5$ cells, infect the cells with a recombinant virus of choice as described throughout this chapter. The cells can be infected with a wide range of MOIs as FACS can accurately detect a positive cell population as small as 2–3 %.
2. Harvest the infected cells by trypsinization Trypsin–EDTA (0.05 % Trypsin, 0.53 mM EDTA) and neutralize the trypsinized population by resuspending cells in complete medium containing serum.
3. Wash the cells two times with 1× PBS.
4. Stain the cells with a primary antibody directed at the epitope tag for ≥1 h at 4 °C. The antibody should be diluted in 1× PBS containing 0.5 % BSA. When using FLAG-tagged viral GPCRs, it may be beneficial to use biotinylated anti-FLAG antibody as this coupled with fluorophore conjugated streptavidin can enhance the signal significantly.
5. If the primary antibody used is not preconjugated with a fluorophore, wash the cells as above in PBS, and then stain the cells with the appropriate secondary antibody in the aforementioned buffer for ≥1 h at 4 °C.
6. After a final series of washes in PBS, analyze cells by FACS.

#### *3.5.4 Other Potential Methodologies for the Detection and Localization of Viral GPCRs*

Investigators have also utilized a variety of additional techniques, including enzyme-linked immunosorbent assay (ELISA) and immune-electron microscopy (immuno-EM) to assess the intracellular and/or surface expression of viral GPCRs [9, 20, 22, 24, 26, 42, 44, 54, 56, 58, 63, 64]. The finding that US27 also localizes to the membranes of the cells was demonstrated by immunoprecipitation of cell surface proteins following infection with a US27 FLAG-tagged, yet localizes as well to the perinuclear region as shown by IFA [26]. More recently, Tschische et al. described the heteromerization of HCMV UL33, UL78, and US27 each with US28 in transient transfection assays, and provided evidence of their colocalization using a combination of IFA, immunoprecipitation, and bioluminescence resonance energy transfer (BRET) analyses [64]. Additionally, Fraile-Ramos and colleagues utilized immuno-EM to discern the intracellular localization of HCMV UL33 and US27 [56]. Taken together, these methodologies provide useful tools in examining the expression and localization of the CMV GPCRs within infected cells.

### *3.6 Methods for Studying Viral GPCR Function in Animal Models*

The experimental approaches and methodology described thus far enable a thorough examination of the biochemical and molecular signaling activities of the viral GPCRs and can be used to study the in vitro function of these interesting and conserved cytomegalovirus proteins. However, they fall short of addressing perhaps the most important fundamental questions regarding the CMV GPCRs: (1) What are the primary biological functions of these CMV GPCRs in vivo? (2) How do these functions affect pathogenesis? and (3) How does the signaling activity of the CMV GPCRs mediate their roles in pathogenesis? Therefore, it is essential to extend the biochemical and molecular genetic experiments described thus far with pathogenesis experiments performed in animal models. The results obtained from in vivo studies will provide the genesis for the rational design of experiments aimed at exploring the molecular functions of cytomegalovirus GPCRs in biologically relevant cellular models. Of the models available for cytomegalovirus research, the mouse model appears to be the best suited for studies on the role that the GPCRs play in pathogenesis in vivo. Both the M33 and M78 genes exhibit profound growth defects in organs important for viral persistence such as the salivary gland [15–18, 22, 65]. Moreover, the mouse is easily amenable to genetic manipulation such as transgenesis and gene knockout, thus allowing investigators to extend pathogenesis studies and potentially explore detailed mechanisms underlying pathogenic processes. Finally, the mouse is a cost-effective model in which one can functionally and mechanistically examine these cytomegalovirus encoded GPCRs before moving on to more complex primate models, if warranted. In this section, we will describe basic methodology to assess cytomegalovirus replication/dissemination in the mouse using wild-type and M33 null MCMVs as an example. BACs containing the Smith and K181 strains have been generated and can be manipulated to delete entire GPCR ORFs or one can make more subtle mutations in signaling motifs, etc. using recombineering methodologies similar to that discussed above [66, 67].

There are a multitude of different strains of mice that have been used to study cytomegalovirus pathogenesis, many offering unique attributes that can be exploited to gain additional insight into the mechanisms of cytomegalovirus replication and spread in vivo. Briefly, strains such as Balb/C are relatively sensitive to MCMV infection, while other strains, such as C57BL/6 are much more resistant to MCMV infection [68, 69]. The nature of this difference lies in the *cmv1* locus which encodes the activating NK receptor LY49H in the resistant, but not sensitive strains. Severely immunodeficient mouse strains such as CB17$^{SCID}$ and NOD-SCID-gammaCnull (NSG) mice have emerged as useful models to explore CMV replication and trafficking in the absence of adaptive (CB17$^{SCID}$) or adaptive/innate NK (NSG) immune function [70–73]. Particular care must be given to dosage and duration of infection when using the immunodeficient animals, as these animals are particularly sensitive to cytomegalovirus and quickly succumb to the infection.

#### 3.6.1 Examination of MCMV Replication and Spread in the Mouse Using Viral Recombinants with Deletions/Mutations in the GPCR Genes

1. Five- to six-week-old female mice are obtained from the appropriate vendor and housed under pathogen-free conditions in barrier-filtered SMI cages according to Association for Assessment and Accreditation of Laboratory Animal Care (AALAC) approved guidelines. The mice are given water and chow ad libitum for the duration of the experiment.
2. Six- to twelve-week-old mice are infected with $1 \times 10^5$ to $1 \times 10^6$ PFU/animal of tissue culture derived wild-type or M33 null viruses. Salivary gland derived stocks of many MCMV strains can alternatively be used. However, mutants such the M33 null viruses do not exhibit strong salivary gland tropism and thus do not allow for the generation of salivary gland stocks. Thus, in the case of M33 null viruses or other mutants that do not grow in the salivary gland, one is limited to tissue culture derived virus.
3. Virus is injected into animals via one of several routes including intraperitoneal (i.p.), intravenous (i.v.), or subcutaneous (s.c.) into the rear footpad. The i.p. route is the most convenient for routine assessment of MCMV growth in various tissues and in this case a 28 gauge insulin syringe containing up to 300 μl of virus diluted in PBS is used for the infection.
4. At appropriate times post-infection (*see* **Note 10**) animals are sacrificed by $CO_2$ asphyxiation and blood is immediately obtained by cardiac puncture and placed into EDTA treated

blood collection tubes. The blood can be used to assess the number of MCMV infected blood leukocytes as described in **step 6**.

5. Internal organs and/or tissues of interest (such as spleen, liver, and salivary gland) are removed via dissection and placed into 1 ml DMEM, flash-frozen, and stored at −80 °C until use. To assess virus titers in organs and/or tissues proceed to **step 7**.
6. Assess the number of infected leukocytes by infectious center assay.
   (a) Dilute 500 μl of blood in 5 ml RBC lysis buffer. Mix and incubate at room temperature until the RBCs lyse (the solution will change from opaque to clear, but remain deep red). Pellet WBCs for 10 min at 400 × *g*. Wash the WBCs three times with sterile 1× PBS to remove hemoglobin and platelets. Resuspend washed WBCs in 1 ml of 1× PBS.
   (b) Transfer $1 \times 10^4$ to $1 \times 10^5$ WBC to MEF monolayers (*see* **Note 11**), incubate for 3–4 h to allow WBCs to settle to bottom of well and come in contact with MEFs. Carefully remove the medium without disturbing settled WBCs and overlay with DMEM containing 0.75 % carboxymethyl cellulose (CMC). Incubate undisturbed for 6–7 days.
   (c) Remove the medium, fix monolayers with methanol, and stain with Giemsa diluted 1:5. Count plaques. Plaques that develop arise as a consequence of a single infected WBC that is productively shedding virus, and hence the term infectious center.
7. Assess virus titers in organs/tissues by plaque assay.
   (a) Thaw tissue that was suspended in 1 ml DMEM and flash-frozen as described in **step 5**. Transfer to Dounce and homogenize organ using 20–30 passes with the tight fitting glass pestle. Ensure visually that the tissue is completely homogenized—if not proceed with additional passes until the tissue is completely disrupted.
   (b) Centrifuge for 5 min in a microcentrifuge at 5,000 × *g* to pellet cellular and tissue debris. Transfer supernatant to fresh tube.
   (c) Transfer dilutions of tissue supernatant to MEF monolayers (*see* **Note 11**), and incubate for 3–4 h to allow virus adsorption. Carefully remove the medium and overlay with DMEM containing 0.75 % CMC. Incubate undisturbed for 4–5 days. Virus titers in organs vary dramatically depending on initial virus dose and dpi, so care should be taken to ensure that the dilutions of tissue supernatant used in the assay will allow for quantitation of plaques in each well.

(d) Remove the medium, fix monolayers with methanol, and stain with Giemsa diluted 1:5. Count plaques.

Using in vivo assays like the one just described, it is evident that the cytomegalovirus GPCRs confer important activities that facilitate viral replication in the whole organism. It is important to ensure that the observed phenotype is due to deletion/alteration of the targeted gene, and this can be accomplished by using "rescue" viruses in which the mutated region is reverted to wild-type. While it is clear that the GPCRs and their ability to signal through G-proteins are essential for replication in vivo, it is not clear what specific signaling pathways are involved or how activation of these signaling pathways facilitate replication. The power of mouse genetics combined with in vivo growth assessment of viruses with GPCR mutations should provide important answers to these questions. It is the answers to these questions that should be at the forefront of future investigations aimed at exploring molecular and biochemical properties of the viral GPCRs.

### 3.7 Conclusions and Discussion of Current State-of-the-art Techniques Useful for Studying Viral GPCR Signaling/Function

Techniques such as transient transfections and related gene delivery methodologies have proved to be invaluable in providing a basic understanding of the CMV GPCRs and how they function in vitro. However, continued vertical advancement of our understanding of the CMV GPCRs requires us as investigators to distance ourselves from standard in vitro techniques and begin to perform studies in the context of virus-infected cells using clinical strains of HCMV and cell types important for in vivo pathogenesis. Taking advantage of CMV GPCR mutants constructed by recombineering techniques is critical for the successful transition to these more sophisticated types of experiments. The tools and resources, including bacterial recombineering techniques, now exist for the cytomegaloviruses, therefore making such studies possible. BAC recombineering protocols provide efficient means to derive viral recombinants for use in both in vitro and in vivo studies. The benefit to generating mutants in BAC viral DNA as opposed to utilizing expression plasmids is that one can investigate the function of the GPCRs in the milieu of the remaining viral ORFs and at physiologically relevant expression levels. Finally, the tools now available for studying CMV GPCRs affords us as investigators the ability to perform high-throughput screens to search for novel viral GPCR therapeutics that may influence HCMV infection and/or replication. Over one third of marketed drugs target cellular GPCRs, and thus the CMV GPCRs are attractive targets. Both the gammaherpesvirus and betaherpesvirus subfamilies encode GPCRs, and in animal models, these proteins have been shown to aid in viral pathogenesis. Thus, it seems likely that herpesviruses have hijacked cellular GPCRs to promote viral replication and dissemination in the host. Utilizing the current resources and technologies, we will undoubtedly uncover the function of these proteins, and perhaps exploit their activities in an effort to develop novel anti-viral therapies to combat HCMV infections.

## 4 Notes

1. Typical doubling times are 48–72 h for HS68 fibroblasts and 18–24 h for HEK-293 cells.
2. To activate the sodium orthovanadate, prepare a 200 mM stock solution, adjust the pH to 10 using NaOH/HCl and boil until colorless. Cool to room temperature, readjust to pH 10 and repeat until the solution stabilizes at pH 10 and remains colorless. Store the activated sodium orthovanadate in aliquots in the −20 °C freezer. The protease inhibitors aprotinin, leupeptin, and PMSF can be substituted for Complete Mini Protease tabs (Roche).
3. Biotin is degraded by light—make fresh and use plates ≤1 month.
4. The different medium formulations contain various amounts of unlabelled myo-inositol, and therefore, it may be beneficial to use MEM as the inositol concentration is lower and gives better labeling.
5. In many cases, simply eluting the total inositol phosphates will provide an extremely accurate measurement of receptor signaling. In this case after **step 15**, simply transfer the columns to scintillation vials and elute the total inositol phosphates with 4 ml 0.1 M formic acid/1.0 M ammonium formate.
6. The pFR-LUC and pFA2 plasmids are part of the PathDetect In Vivo Signal Transduction *trans*-reporting system available from Agilent Technologies™. More information about these plasmids and the *trans*-reporting system can be found at: http://www.genomics.agilent.com/files/Manual/219000.pdf.
7. Passive Lysis Buffer (PLB), Luciferase Assay reagent II (LAR II), and Stop&Glo are components of the Promega Dual-Luciferase® Reporter Assay System. More information about this system can be found at: http://www.promega.com/products/reporter-assays-and-transfection/reporter-assays.
8. The bacterial strains SW102, SW105, and SW106 produce endotoxins, which are co-purified with the BAC DNA and thus can be introduced into mammalian cells during the transfection process. Endotoxins stimulate components of the innate immune response in mammalian cells, and therefore are toxic to cells in tissue culture. Thus, adding a step within the alkaline lysis protocol to remove endotoxins will greatly improve the health of the mammalian cells post-transfection

of the BAC DNA and greatly improve the overall transfection efficiency.

9. Infections below a MOI of 0.1 PFU/cell in ARPE19 cells result in insufficient viral output for assessment by plaque assay and are therefore not recommended.
10. MCMV replicates in a large number of cell types and tissues. In particular the virus can be found at high levels during the acute phase in organs such as spleen and liver (3–5 dpi) and during the persistent phase in tissues such as the salivary gland (12–21 dpi).
11. 1 day prior to using for plaque or infectious center assays, plate 100,000 primary MEFs (passages 2–8) into each well of a 12-well plate.

## References


1. Khanna R, Diamond DJ (2006) Human cytomegalovirus vaccine: time to look for alternative options. Trends Mol Med 12:26–33
2. Chee MS, Bankier AT, Beck S, Bohni R, Brown CM, Cerny R, Horsnell T, Hutchison CA 3rd, Kouzarides T, Martignetti JA et al (1990) Analysis of the protein-coding content of the sequence of human cytomegalovirus strain AD169. Curr Top Microbiol Immunol 154:125–169
3. Chee MS, Satchwell SC, Preddie E, Weston KM, Barrell BG (1990) Human cytomegalovirus encodes three G protein-coupled receptor homologues. Nature 344:774–777
4. Dorsam RT, Gutkind JS (2007) G-protein-coupled receptors and cancer. Nat Rev Cancer 7:79–94
5. Sodhi A, Montaner S, Gutkind JS (2004) Does dysregulated expression of a deregulated viral GPCR trigger Kaposi's sarcomagenesis? FASEB J 18:422–427
6. Miller-Kittrell M, Sparer TE (2009) Feeling manipulated: cytomegalovirus immune manipulation. Virol J 6:4
7. Casarosa P, Bakker RA, Verzijl D, Navis M, Timmerman H, Leurs R, Smit MJ (2001) Constitutive signaling of the human cytomegalovirus-encoded chemokine receptor US28. J Biol Chem 276:1133–1137
8. Kuhn DE, Beall CJ, Kolattukudy PE (1995) The cytomegalovirus US28 protein binds multiple CC chemokines with high affinity. Biochem Biophys Res Commun 211:325–330
9. Waldhoer M, Kledal TN, Farrell H, Schwartz TW (2002) Murine cytomegalovirus (CMV) M33 and human CMV US28 receptors exhibit similar constitutive signaling activities. J Virol 76:8161–8168
10. Minisini R, Tulone C, Luske A, Michel D, Mertens T, Gierschik P, Moepps B (2003) Constitutive inositol phosphate formation in cytomegalovirus-infected human fibroblasts is due to expression of the chemokine receptor homologue pUS28. J Virol 77:4489–4501
11. Gao JL, Murphy PM (1994) Human cytomegalovirus open reading frame US28 encodes a functional beta chemokine receptor. J Biol Chem 269:28539–28542
12. Casarosa P, Gruijthuijsen YK, Michel D, Beisser PS, Holl J, Fitzsimons CP, Verzijl D, Bruggeman CA, Mertens T, Leurs R, Vink C, Smit MJ (2003) Constitutive signaling of the human cytomegalovirus-encoded receptor UL33 differs from that of its rat cytomegalovirus homolog R33 by promiscuous activation of G proteins of the Gq, Gi, and Gs classes. J Biol Chem 278:50010–50023
13. Gruijthuijsen YK, Casarosa P, Kaptein SJ, Broers JL, Leurs R, Bruggeman CA, Smit MJ, Vink C (2002) The rat cytomegalovirus R33-encoded G protein-coupled receptor signals in a constitutive fashion. J Virol 76:1328–1338
14. Sherrill JD, Miller WE (2006) G protein-coupled receptor (GPCR) kinase 2 regulates agonist-independent Gq/11 signaling from the mouse cytomegalovirus GPCR M33. J Biol Chem 281:39796–39805
15. Davis-Poynter NJ, Lynch DM, Vally H, Shellam GR, Rawlinson WD, Barrell BG, Farrell HE (1997) Identification and characterization of a G protein-coupled receptor homolog encoded by murine cytomegalovirus. J Virol 71:1521–1529

16. Case R, Sharp E, Benned-Jensen T, Rosenkilde MM, Davis-Poynter N, Farrell HE (2008) Functional analysis of the murine cytomegalovirus chemokine receptor homologue M33: ablation of constitutive signaling is associated with an attenuated phenotype in vivo. J Virol 82:1884–1898
17. Cardin RD, Schaefer GC, Allen JR, Davis-Poynter NJ, Farrell HE (2009) The M33 chemokine receptor homolog of murine cytomegalovirus exhibits a differential tissue-specific role during in vivo replication and latency. J Virol 83:7590–7601
18. Sherrill JD, Stropes MP, Schneider OD, Koch DE, Bittencourt FM, Miller JL, Miller WE (2009) Activation of intracellular signaling pathways by the murine cytomegalovirus G protein-coupled receptor M33 occurs via PLC-{beta}/PKC-dependent and -independent mechanisms. J Virol 83:8141–8152
19. Beisser PS, Vink C, Van Dam JG, Grauls G, Vanherle SJ, Bruggeman CA (1998) The R33 G protein-coupled receptor gene of rat cytomegalovirus plays an essential role in the pathogenesis of viral infection. J Virol 72: 2352–2363
20. Margulies BJ, Browne H, Gibson W (1996) Identification of the human cytomegalovirus G protein-coupled receptor homologue encoded by UL33 in infected cells and enveloped virus particles. Virology 225:111–125
21. O'Connor CM, Shenk T. Unpublished observations
22. Oliveira SA, Shenk TE (2001) Murine cytomegalovirus M78 protein, a G protein-coupled receptor homologue, is a constituent of the virion and facilitates accumulation of immediate-early viral mRNA. Proc Natl Acad Sci U S A 98:3237–3242
23. Beisser PS, Grauls G, Bruggeman CA, Vink C (1999) Deletion of the R78 G protein-coupled receptor gene from rat cytomegalovirus results in an attenuated, syncytium-inducing mutant strain. J Virol 73:7218–7230
24. O'Connor CM, Shenk T (2012) Human cytomegalovirus pUL78 G protein-coupled receptor homologue is required for timely cell entry in epithelial cells but not fibroblasts. J Virol 86:11425–11433
25. Michel D, Milotic I, Wagner M, Vaida B, Holl J, Ansorge R, Mertens T (2005) The human cytomegalovirus UL78 gene is highly conserved among clinical isolates, but is dispensable for replication in fibroblasts and a renal artery organ-culture system. J Gen Virol 86:297–306
26. O'Connor CM, Shenk T (2011) Human cytomegalovirus pUS27 G protein-coupled receptor homologue is required for efficient spread by the extracellular route but not for direct cell-to-cell spread. J Virol 85:3700–3707
27. Kledal TN, Rosenkilde MM, Schwartz TW (1998) Selective recognition of the membrane-bound CX3C chemokine, fractalkine, by the human cytomegalovirus-encoded broad-spectrum receptor US28. FEBS Lett 441:209–214
28. Bodaghi B, Jones TR, Zipeto D, Vita C, Sun L, Laurent L, Arenzana-Seisdedos F, Virelizier JL, Michelson S (1998) Chemokine sequestration by viral chemoreceptors as a novel viral escape strategy: withdrawal of chemokines from the environment of cytomegalovirus-infected cells. J Exp Med 188:855–866
29. Vieira J, Schall TJ, Corey L, Geballe AP (1998) Functional analysis of the human cytomegalovirus US28 gene by insertion mutagenesis with the green fluorescent protein gene. J Virol 72:8158–8165
30. Billstrom MA, Johnson GL, Avdi NJ, Worthen GS (1998) Intracellular signaling by the chemokine receptor US28 during human cytomegalovirus infection. J Virol 72:5535–5544
31. Streblow DN, Soderberg-Naucler C, Vieira J, Smith P, Wakabayashi E, Ruchti F, Mattison K, Altschuler Y, Nelson JA (1999) The human cytomegalovirus chemokine receptor US28 mediates vascular smooth muscle cell migration. Cell 99:511–520
32. Stropes MP, Schneider OD, Zagorski WA, Miller JL, Miller WE (2009) The carboxy-terminal tail of human cytomegalovirus (HCMV) US28 regulates both chemokine-independent and chemokine-dependent signaling in HCMV-infected cells. J Virol 83: 10016–10027
33. Melnychuk RM, Streblow DN, Smith PP, Hirsch AJ, Pancheva D, Nelson JA (2004) Human cytomegalovirus-encoded G protein-coupled receptor US28 mediates smooth muscle cell migration through Galpha12. J Virol 78:8382–8391
34. Moepps B, Tulone C, Kern C, Minisini R, Michels G, Vatter P, Wieland T, Gierschik P (2008) Constitutive serum response factor activation by the viral chemokine receptor homologue pUS28 is differentially regulated by Galpha(q/11) and Galpha(16). Cell Signal 20:1528–1537
35. Slinger E, Maussang D, Schreiber A, Siderius M, Rahbar A, Fraile-Ramos A, Lira SA, Soderberg-Naucler C, Smit MJ (2010) HCMV-encoded chemokine receptor US28 mediates proliferative signaling through the IL-6-STAT3 axis. Sci Signal 3:ra58
36. Soroceanu L, Matlaf L, Bezrookove V, Harkins L, Martinez R, Greene M, Soteropoulos P,

Cobbs CS (2011) Human cytomegalovirus US28 found in glioblastoma promotes an invasive and angiogenic phenotype. Cancer Res 71:6643–6653

37. Boomker JM, The TH, de Leij LF, Harmsen MC (2006) The human cytomegalovirus-encoded receptor US28 increases the activity of the major immediate-early promoter/enhancer. Virus Res 118:196–200
38. McLean KA, Holst PJ, Martini L, Schwartz TW, Rosenkilde MM (2004) Similar activation of signal transduction pathways by the herpesvirus-encoded chemokine receptors US28 and ORF74. Virology 325:241–251
39. Maussang D, Verzijl D, van Walsum M, Leurs R, Holl J, Pleskoff O, Michel D, van Dongen GA, Smit MJ (2006) Human cytomegalovirus-encoded chemokine receptor US28 promotes tumorigenesis. Proc Natl Acad Sci U S A 103:13068–13073
40. Maussang D, Langemeijer E, Fitzsimons CP, Stigter-van Walsum M, Dijkman R, Borg MK, Slinger E, Schreiber A, Michel D, Tensen CP, van Dongen GA, Leurs R, Smit MJ (2009) The human cytomegalovirus-encoded chemokine receptor US28 promotes angiogenesis and tumor formation via cyclooxygenase-2. Cancer Res 69:2861–2869
41. Baryawno N, Rahbar A, Wolmer-Solberg N, Taher C, Odeberg J, Darabi A, Khan Z, Sveinbjornsson B, FuskevAg OM, Segerstrom L, Nordenskjold M, Siesjo P, Kogner P, Johnsen JI, Soderberg-Naucler C (2011) Detection of human cytomegalovirus in medulloblastomas reveals a potential therapeutic target. J Clin Invest 121:4043–4055
42. Casarosa P, Menge WM, Minisini R, Otto C, van Heteren J, Jongejan A, Timmerman H, Moepps B, Kirchhoff F, Mertens T, Smit MJ, Leurs R (2003) Identification of the first nonpeptidergic inverse agonist for a constitutively active viral-encoded G protein-coupled receptor. J Biol Chem 278:5172–5178
43. Stropes MP, Miller WE (2008) Functional analysis of human cytomegalovirus pUS28 mutants in infected cells. J Gen Virol 89:97–105
44. Waldhoer M, Casarosa P, Rosenkilde MM, Smit MJ, Leurs R, Whistler JL, Schwartz TW (2003) The carboxyl terminus of human cytomegalovirus-encoded 7 transmembrane receptor US28 camouflages agonism by mediating constitutive endocytosis. J Biol Chem 278:19473–19482
45. Miller WE, McDonald PH, Cai SF, Field ME, Davis RJ, Lefkowitz RJ (2001) Identification of a motif in the carboxyl terminus of beta -arrestin2 responsible for activation of JNK3. J Biol Chem 276:27770–27777
46. Maussang D, Vischer HF, Schreiber A, Michel D, Smit MJ (2009) Pharmacological and biochemical characterization of human cytomegalovirus-encoded G protein-coupled receptors. Methods Enzymol 460:151–171
47. Brune W, Messerle M, Koszinowski UH (2000) Forward with BACs: new tools for herpesvirus genomics. Trends Genet 16:254–259
48. Datsenko KA, Wanner BL (2000) One-step inactivation of chromosomal genes in Escherichia coli K-12 using PCR products. Proc Natl Acad Sci U S A 97:6640–6645
49. Warming S, Costantino N, Court DL, Jenkins NA, Copeland NG (2005) Simple and highly efficient BAC recombineering using galK selection. Nucleic Acids Res 33:e36
50. Tischer BK, von Einem J, Kaufer B, Osterrieder N (2006) Two-step red-mediated recombination for versatile high-efficiency markerless DNA manipulation in Escherichia coli. Biotechniques 40:191–197
51. Sinzger C, Hahn G, Digel M, Katona R, Sampaio KL, Messerle M, Hengel H, Koszinowski U, Brune W, Adler B (2008) Cloning and sequencing of a highly productive, endotheliotropic virus strain derived from human cytomegalovirus TB40/E. J Gen Virol 89:359–368
52. Dunn W, Chou C, Li H, Hai R, Patterson D, Stolc V, Zhu H, Liu F (2003) Functional profiling of a human cytomegalovirus genome. Proc Natl Acad Sci U S A 100:14223–14228
53. Yu D, Silva MC, Shenk T (2003) Functional map of human cytomegalovirus AD169 defined by global mutational analysis. Proc Natl Acad Sci U S A 100:12396–12401
54. Margulies BJ, Gibson W (2007) The chemokine receptor homologue encoded by US27 of human cytomegalovirus is heavily glycosylated and is present in infected human foreskin fibroblasts and enveloped virus particles. Virus Res 123:57–71
55. Mokros T, Rehm A, Droese J, Oppermann M, Lipp M, Hopken UE (2002) Surface expression and endocytosis of the human cytomegalovirus-encoded chemokine receptor US28 is regulated by agonist-independent phosphorylation. J Biol Chem 277: 45122–45128
56. Fraile-Ramos A, Pelchen-Matthews A, Kledal TN, Browne H, Schwartz TW, Marsh M (2002) Localization of HCMV UL33 and US27 in endocytic compartments and viral membranes. Traffic 3:218–232
57. Droese J, Mokros T, Hermosilla R, Schulein R, Lipp M, Hopken UE, Rehm A (2004) HCMV-encoded chemokine receptor US28 employs

multiple routes for internalization. Biochem Biophys Res Commun 322:42–49

58. Penfold ME, Schmidt TL, Dairaghi DJ, Barry PA, Schall TJ (2003) Characterization of the rhesus cytomegalovirus US28 locus. J Virol 77:10404–10413
59. Pleskoff O, Casarosa P, Verneuil L, Ainoun F, Beisser P, Smit M, Leurs R, Schneider P, Michelson S, Ameisen JC (2005) The human cytomegalovirus-encoded chemokine receptor US28 induces caspase-dependent apoptosis. FEBS J 272:4163–4177
60. Pleskoff O, Treboute C, Alizon M (1998) The cytomegalovirus-encoded chemokine receptor US28 can enhance cell-cell fusion mediated by different viral proteins. J Virol 72:6389–6397
61. Pleskoff O, Treboute C, Brelot A, Heveker N, Seman M, Alizon M (1997) Identification of a chemokine receptor encoded by human cytomegalovirus as a cofactor for HIV-1 entry. Science 276:1874–1878
62. Vomaske J, Melnychuk RM, Smith PP, Powell J, Hall L, DeFilippis V, Fruh K, Smit M, Schlaepfer DD, Nelson JA, Streblow DN (2009) Differential ligand binding to a human cytomegalovirus chemokine receptor determines cell type-specific motility. PLoS Pathog 5:e1000304
63. Casarosa P, Waldhoer M, LiWang PJ, Vischer HF, Kledal T, Timmerman H, Schwartz TW, Smit MJ, Leurs R (2005) CC and CX3C chemokines differentially interact with the N terminus of the human cytomegalovirus-encoded US28 receptor. J Biol Chem 280:3275–3285
64. Tschische P, Tadagaki K, Kamal M, Jockers R, Waldhoer M (2011) Heteromerization of human cytomegalovirus encoded chemokine receptors. Biochem Pharmacol 82:610–619
65. Farrell HE, Abraham AM, Cardin RD, Sparre-Ulrich AH, Rosenkilde MM, Spiess K, Jensen TH, Kledal TN, Davis-Poynter N (2011) Partial functional complementation between human and mouse cytomegalovirus chemokine receptor homologues. J Virol 85:6091–6095
66. Redwood AJ, Messerle M, Harvey NL, Hardy CM, Koszinowski UH, Lawson MA, Shellam GR (2005) Use of a murine cytomegalovirus K181-derived bacterial artificial chromosome as a vaccine vector for immunocontraception. J Virol 79:2998–3008
67. Messerle M, Crnkovic I, Hammerschmidt W, Ziegler H, Koszinowski UH (1997) Cloning and mutagenesis of a herpesvirus genome as an infectious bacterial artificial chromosome. Proc Natl Acad Sci U S A 94:14759–14763
68. Lee SH, Girard S, Macina D, Busa M, Zafer A, Belouchi A, Gros P, Vidal SM (2001) Susceptibility to mouse cytomegalovirus is associated with deletion of an activating natural killer cell receptor of the C-type lectin superfamily. Nat Genet 28:42–45
69. Webb JR, Lee SH, Vidal SM (2002) Genetic control of innate immune responses against cytomegalovirus: MCMV meets its match. Genes Immun 3:250–262
70. Upton JW, Kaiser WJ, Mocarski ES (2010) Virus inhibition of RIP3-dependent necrosis. Cell Host Microbe 7:302–313
71. Ghazal P, Messerle M, Osborn K, Angulo A (2003) An essential role of the enhancer for murine cytomegalovirus in vivo growth and pathogenesis. J Virol 77:3217–3228
72. Pollock JL, Virgin HW IV (1995) Latency, without persistence, of murine cytomegalovirus in the spleen and kidney. J Virol 69:1762–1768
73. Okada M, Minamishima Y (1987) The efficacy of biological response modifiers against murine cytomegalovirus infection in normal and immunodeficient mice. Microbiol Immunol 31:45–57
74. Kaptein SJ, Beisser PS, Gruijthuijsen YK, Savelkouls KG, van Cleef KW, Beuken E, Grauls GE, Bruggeman CA, Vink C (2003) The rat cytomegalovirus R78 G protein-coupled receptor gene is required for production of infectious virus in the spleen. J Gen Virol 84:2517–2530
75. Miller WE, Zagorski WA, Brenneman JD, Avery D, Miller JL, O'Connor CM (2012) US28 is a potent activator of phospholipase C during HCMV infection of clinically relevant target cells. PLoS One 7(11):e50524

# Chapter 11

# Methods for the Detection of Cytomegalovirus in Glioblastoma Cells and Tissues

Charles S. Cobbs, Lisa Matlaf, and Lualhati E. Harkins

## Abstract

An increased awareness of the potential oncomodulatory properties of human cytomegalovirus (HCMV) has evolved over the last decade. We first reported the presence of HCMV in human glioblastomas, and subsequently these findings have been corroborated by other groups. However, some controversy has been associated with the immunohistochemical and in situ hybridization techniques used, since standard immunohistochemical and in situ hybridization techniques have been insufficient to detect low level HCMV antigens and nucleic acids in some tumor tissues. Here, we present detailed methods that can be used for the sensitive detection of low level HCMV antigens and nucleic acids in human glioblastoma specimens. Using these techniques, HCMV is frequently detected in frozen and formalin fixed paraffin-embedded tissue specimens. Furthermore, we demonstrate how human primary glioblastoma cells can be cultured in vitro, and how these cells can be used for detection of HCMV by immunofluorescence, in situ hybridization, western blot, and RT-PCR.

**Key words** Human cytomegalovirus, Glioblastoma, Immunostaining, PCR

## 1 Introduction

Human cytomegalovirus (HCMV) is a betaherpesvirus that persistently infects the majority of adult humans. HCMV is a well-known cause of disease during fetal infections and in the setting of immunosuppression. In the last decade, an emerging association between HCMV infection and glioblastoma has occurred. HCMV infection in the majority of glioblastomas has been reported by multiple groups, although others have not been able to detect HCMV in these tumors. Evidence suggests that low level HCMV infection is correlated with glioblastoma disease progression, and that antiviral strategies aimed at HCMV may play a novel role in therapy to gliomas [1]. The goal of this chapter is to provide a detailed methodology for the immunohistochemical and in situ hybridization techniques that we have used in order to consistently identify

Andrew D. Yurochko and William E. Miller (eds.), *Human Cytomegaloviruses: Methods and Protocols*, Methods in Molecular Biology, vol. 1119, DOI 10.1007/978-1-62703-788-4_11, 

HCMV infections in human tumors. We hope that these techniques will lead to some increased understanding of the role HCMV plays in human malignancies.

## 2 Materials

Standard materials and equipment for molecular and cell biological experiments (e.g., Refrigerators, Freezers, PCR machines, incubators)

### 2.1 Cell Culture

1. 1× Neurobasal Medium, serum free, 500 mL (Invitrogen #21103-049) supplemented with 5 mL penicillin–streptomycin (100 U/mL final concentration), 5 mL 100× GlutaMAX (Invitrogen #35050-061), and 5 mL 100× N-2 supplement (Invitrogen #17502-048).
2. Epidermal Growth Factor (Sigma-Aldrich #E9644-.2MG) and Fibroblast Growth Hormone, basic human (Sigma-Aldrich #F0291-25UG) are resuspended in ultrapure water, aliquoted into small volumes and stored at −20 °C. Growth factors are added to medium just before culturing at a 20 ng/mL final concentration.
3. Laminin from Engelbreth-Holm-Swarm murine sarcoma (Sigma-Aldrich #L2020-1MG) is added to complete NB medium at a 1 μg/mL final concentration to establish adherent cultures.
4. Papain from papaya latex (Sigma-Aldrich #P5306-25MG).
5. Tissue culture plates and flasks. 24-well multiwell tissue culture plates with inserted sterile 12 mm cover glass circles (Fisher #12-545-80).
6. Hemocytometer.

### 2.2 Immunostaining

1. Cold methanol.
2. Protein free (TBS) blocking buffer (Pierce #37570).
3. Tris buffered saline (50 mM Tris, 150 mM NaCl) with 0.05%Tween-20.
4. Primary antibodies:
   (a) Mouse anti-cytomegalovirus antibody MAB810 against IE1 and IE2 (Millipore #MAB810), 1:200 dilution.
   (b) Mouse anti-glycoprotein B antibody 2 F12 (Virusys #CA005-100), 1:1,000 dilution.
   (c) Goat anti-US28 vC-17 antibody (Santa Cruz Biotechnology #sc-28042), 1:50 dilution.
   (d) Mouse anti-pp65 antibody 3A12 (Virusys #CA003-100), 1:500 dilution.

   (e) Mouse anti-pp71 antibody 10G11 (courtesy of Dr. Tom Shenk, Princeton University), 1:2 dilution.

5. Secondary antibodies, all used at a 1:1,000 dilution:
   (a) Alexa Fluor 488 donkey anti-mouse IgG (Invitrogen #A-21202).
   (b) Alexa Fluor 488 donkey anti-goat igG (Invitrogen #A-11055).
6. Mounting medium containing either DAPI (SlowFade gold antifade reagent with DAPI, Invitrogen #S36938) or propidium iodide (VECTASHIELD mounting medium with propidium iodide, Vector Labs #H-1300).
7. Nail polish.
8. Superfrost Plus glass microscope slides, white 75×25 mm (Fisher #12-550-15).

### 2.3 Extraction of Protein and RNA from GBM Tissue

1. Qiazol lysis reagent (Qiagen #79306).
2. Chloroform, without additives.
3. Isopropanol.
4. 100 % ethanol.
5. 70 % ethanol.
6. 75 % ethanol.
7. DEPC water.
8. 0.8 mM NaOH (FW:40).
9. 0.1 M sodium citrate in 10 % ethanol.
10. 1 % SDS solution.
11. 0.3 M Guanidine hydrochloride in 0.05 % ethanol (FW 95.53).
12. TissueRuptor homogenizer (Qiagen #9001271).
13. TissueRuptor disposable probes (Qiagen #990890).
14. Qiashredder columns (Qiagen #79654).
15. RNeasy Lipid Tissue Mini Kit (Qiagen #74804).
16. Cold centrifuge.

### 2.4 Western Blotting

1. DC protein assay reagents package (#500-0166).
2. 2.1 % SDS solution.
3. Plate reader.
4. Criterion XT precast 4–12 % Bis–Tris SDS-PAGE gels and electrophoresis system (Bio-Rad #345-0123, #165-6001, and #170-4071).
5. 5.20× XT MOPS Running Buffer, 500 mL (Bio-Rad #161-0788).
6. 4× XT sample buffer, 10 mL (Bio-Rad #161-0791).

7. 20× XT reducing agent, 1 mL (Bio-Rad #161-0792).
8. Protein standard.
9. BupH Tris–Glycine transfer buffer packs (Thermo Fisher #28380).
10. Immun-Blot PVDF membrane roll (Bio-Rad #162-0177).
11. Thick blot paper, 9.5 × 15.2 cm, pack of 50 (Bio-Rad #170-4085).
12. Tupperware or plastic blotting chamber.
13. Ampac sealpack pouches, 8 × 12″, pack of 40 (Fisher #01-812-25H).
14. Ponceau S solution, 1 L (Sigma Aldrich #P7170-1 L).
15. Lyophilized Bovine Serum Albumin, 50 g (Sigma Aldrich #A9418-50G).
16. Tris buffered saline (50 mM Tris, 150 mM NaCl) with 0.05 %Tween-20.
17. 10× blocker BSA in TBS, 125 mL (Thermo Fisher #37520).
18. Primary antibodies:
    (a) Mouse anti-cytomegalovirus antibody MAB810 against IE1 and IE2 (Millipore #MAB810), 1:1,000 dilution.
    (b) Mouse anti-glycoprotein B antibody 2 F12 (Virusys #CA005-100), 1:5,000 dilution.
    (c) Mouse anti-pp65 antibody 3A12 (Virusys #CA003-100), 1:2,000 dilution.
    (d) Mouse anti-pp71 antibody 2H10-9 (courtesy of Dr. Tom Shenk, Princeton University), 1:10 dilution.
    (e) Rabbit anti-actin antibody (Sigma Aldrich #A2066), 1:1,000 dilution.
    (f) Mouse anti-tubulin antibody (Abcam #ab7291), 1:5,000 dilution.
19. Secondary antibodies:
    (a) Immunopure goat anti-mouse IgG (H + L), peroxidase conjugated, 2 mL (Thermo Fisher #31430), 1:10,000 dilution.
    (b) Immunopure goat anti-rabbit IgG (H + L), peroxidase conjugated, 2 mL (Thermo Fisher #31460), 1:10,000 dilution.
20. Super signal west pico chemiluminescent substrate, 500 mL kit (Thermo Fisher #34080).
21. Super signal west femto chemiluminescent substrate, 100 mL kit (Thermo Fisher #34095).
22. Restore western blot stripping buffer, 500 mL (Thermo Fisher #21059).

### 2.5 RT-PCR

1. iScript cDNA synthesis kit (Bio-Rad #170-8890).
2. Filter pipette tips.
3. Taq PCR core kit (Qiagen #201223).
4. Nuclease-free water, 10 × 50 mL (Qiagen #129114).
5. PCR primers:
   (a) pp71F 5′-AGAAACACGCTGGTCGGCGG-3′.
   pp71R 5′-CGCGGCGGCGAAGAAAATCG-3′.
   (b) IE1F 5′-AGCACCATCCTCCTCTTCCTCTG-3′.
   IE1R 5′-AAGCGGCCTCTGATAACCAAGCC-3′.
   (c) IE2F 5′-CGCCACTTCGGGTGGGTGTG-3′.
   IE2R 5′-GGTGAGCCGCATGTTCCGCA-3′.
   (d) US28N-F 5′-ATGACACCGACGACGACGG-3′.
   US28N-R 5′-GCTAGGGAGTTGTGATCTAG-3′.
   (e) US28C-F 5′-TCGCGCCACAAAGGTCGCAT-3′.
   US28C-R 5′-GACGCGACACACCTCGTCGG-3′.
   (f) UL56F 5′-GAGTTGTTTCCCGAAAGTTTCATTAT-3′.
   UL56R 5′-CCTCTCTCACAATGTGGACATG-3′.
   (g) UL84F 5′-GCGCCCGGCCTTCTCTCTCT-3′.
   UL84R 5′-CCGTTGACTCCGCGGCATCG-3′.
   (h) gBF-external 5′- TCCAACACCCACAGTACCCGT-3′.
   gBR-external 5′- CGGAAACGATGGTGTAGTTCG-3′.
   (i) gBF-internal 5′-CGCCGCCCGCCCCGCGCCCGCCG CGGCAGC.
   ACCTGGCT-3′.
   gBR-internal 5′- GTAAACCACATCACCCGTGGA-3′.
   (k) Rab14F 5′-GCAGATTTGGGATACAGCAGG-3′.
   Rab14R 5′-CAGTGTTTGGATTGGTGAGATTC-3′.
6. MinElute PCR purification kit (Qiagen #28004).
7. Ultrapure agarose, 500 g (Invitrogen #16500500).
8. Agarose gel electrophoresis chamber and power supply.
9. 10× TAE pH 8.0, 5 L cube (Bio-Rad #161-0773).

### 2.6 Fixation of Human Glioblastoma Frozen Tissue Sections

1. FrozFix Fixative (www.newcomersupply.com # 1096).
2. FrozFix® Conditioning Set (www.newcomersupply.com # 1097).
3. Tris Buffered Saline + Tween, 10× (# 140305). Discard used buffer solution after using ten slides for every 50 mL of buffer.
4. Tissue-Tek staining racks or equivalent.
5. Tissue-Tek staining trays or equivalent.

6. Slide box.
7. Aluminum foil.
8. Slide marking pen.
9. Superfrost Plus slides or equivalent (Fisher # 12-550-15 or Aminosilane #5050, #5070).
10. Pepsin (BioGenex Laboratories #HK-054-5K).
11. Trypsin (BioGenex Laboratories #EK-0015K or EK-001 10K).
12. Thermocycler or moist heating block.
13. Cryostat set to −20 °C.
14. Forceps.
15. Fine tip painters brush.
16. Liquid nitrogen.

#### 2.7 Immunohistochemistry on Frozen Sections

1. Incubation tray, slide show 30, with black lid (New Comer Supply # 6844-30 BL).
2. 30 % Hydrogen peroxide, reagent grade (Sigma # H-1009)—dilute to 3 % in distilled water prior to use.
3. Avidin/Biotin Block (BioGenex # HK 154-20× or DAKO # X0590).
4. FC Receptor block (Innovex Biosciences # NB-309, available through www.newcomersupply.com).
5. PAP Pen (www.newcomersupply.com #6505A).
6. CMV IE1/IE2 mAb (Millipore #MAB810), 1:40 dilution.
7. CMV early/late cocktail (Innovex Biosciences #MAB337C), 1:40–1:60 dilution.
8. CMV late antigen mAb (Millipore #MAB8127), 1:35 dilution.
9. CMC pp65 antigen (www.leica-microsystems.com #NCL-CMVpp65), 1:40 dilution.
10. Goat anti mouse secondary antibody (BioGenex Laboratories #HK-325 UM), 1:18 dilution.
11. Peroxidase labeled streptavidin (BioGenex Laboratories #HK-320-UK), 1:18 dilution.
12. Common antibody diluent (BioGenex Laboratories # HK156-5K).
13. Streptavidin/Peroxidase diluent (BioGenex Laboratories # HK 157-5K).
14. DAB (Innovex Biosciences # NB-314SB, available through www.newcomersupply.com).
15. DAB enhancer (Innovex Biosciences # NB-308, available through www.newcomersupply.com).

16. Aqueous hematoxylin (Innovex Biosciences #NB 305 or NB 305 A, available through www.newcomersupply.com).
17. Advantage mount or equivalent (www.newcomersupply.com #NB300 or NB 300A).
18. Glass coverslips.

### 2.8 Preparation and Pretreatment of Paraffin-Embedded Sections

1. Formalin fixed, paraffin-embedded glioblastoma specimens.
2. Positive control (CMV infected tissue such CMV infected lung or any tissue from transplant patients), or commercially available CMV positive control (www.newcomersupply.com #3240A or 3249B).
3. Microtome designated for paraffin sections.
4. Water bath set to 37–40 °C.
5. Distilled water.
6. Paper towels.
7. Tissue-Tek plastic staining racks or equivalent.
8. Laboratory oven set to 45–50 °C.
9. Microwave oven.
10. Plastic Tissue-Tek staining jars or equivalent.
11. Superfrost Plus slides or equivalent (Fisher # 12-550-15).
12. 10 % Neutral buffered formalin (New Comer Supply #1090 N).
13. Xylene, ACS or reagent grade (Fisher # X5p-1 GAL (Un1307)).
14. Ethanol, 200 proof, reagent grade (AAper Alcohol and Chemical Co or Pharmco Products Inc.).
15. Tris buffer (pH 7.6) Trizma (Sigma #T- 4253 pre-ph crystals, 6.96 g/L distilled water, and 8.5 g NaCl/L distilled water or ready-to-use Tris buffered saline with Tween www.newcomersupply.com #140305 A).
16. Pepsin (HK-054–5 K) (BioGenex Laboratories, San Ramon, California, USA 1-800-421-4149).
17. Trypsin (EK-0015KOREK-001 10K) (BioGenex Laboratories, San Ramon, California, USA 1-800-421-4149).
18. 1× Citra Plus buffer (pH 7.0) BioGenex Laboratories, cat. # HK080-5Kor HK-090-9K.
19. Thermocycler or moist heating block.

### 2.9 Immunohistochemistry Using Paraffin-Embedded Sections

1. Incubation tray, Slide show 30, with black lid, (cat. # 6844-30 BL, New Comer Supply, Middleton, WI USA).
2. 30 % Hydrogen peroxide (reagent grade, Sigma cat. # H-1009) (dilute to 3 % in distilled water, prior to use).
3. Avidin/Biotin Block (BioGenex cat. # HK 154-20X OR DAKO Catalog # X0590).

4. FC Receptor block (cat. # NB-309, Innovex Biosciences), available through www.newcomersupply.com.
5. PAP Pen (New Comer Supply, cat. # 6505A).
6. Chemicon CMV IE1/IE2 mAb (Millipore Corp, MAB810) 1:40.
7. Innovex Biosciences CMV early/late cocktail (Innovex Biosciences, cat. # MAB337C) 1:40–1:60.
8. Chemicon CMV late Ag mAb (Millipore, MAB8127) 1:35.
9. Cytomegalovirus, PP65 antigen, cat. # NCL-CMVpp65, www.leica-microsystems.com (1:40).
10. Goat Anti Mouse secondary antibody (1:18), cat. # HK-325 UM, BioGenex Laboratories, San Ramon, CA USA 1-800-421-4149.
11. Peroxidase Labeled Streptavidin (1:18), cat. # HK- 320 –UK, BioGenex Laboratories, San Ramon, CA USA 1-800-421-4149.
12. Common Antibody Diluent, BioGenex Laboratories cat. # HK156-5K.
13. Streptavidin/Peroxidase diluents, BioGenex Laboratories, cat. # HK 157-5K.
14. DAB, Innovex Biosciences, cat. # NB-314SB, New Comer Supply, Middleton, WI, USA.
15. DAB Enhancer (Innovex Biosciences, cat. # NB-308, New Comer Supply, Middleton, WI, USA).
16. Aqueous Hematoxylin, Innovex Biosciences, cat. # NB 305 OR NB 305 A, available through www.newcomersupply.com.
17. Advantage mount or equivalent (www.newcomersupply.com, cat. # OR NB 300A).
18. Glass coverslips.
19. Bright field microscope.
20. Slide folders.

### 2.10 DNA/RNA In Situ Hybridization

1. Formalin fixed, paraffin-embedded glioblastoma specimens.
2. Positive control (CMV infected tissue such CMV infected lung or any tissue from transplant patients) or commercially available CMV positive control (www.newcomersupply.com #3240A or 3249B).
3. Microtome designated for paraffin sections.
4. Water bath set to 37–40 °C.
5. Distilled water.
6. Paper towels.
7. 4×4 gauze sponges.
8. Tissue-Tek plastic staining racks or equivalent.

9. Laboratory oven set to 45–50 °C.
10. Microwave oven.
11. Plastic Tissue-Tek staining jars or equivalent.
12. Superfrost Plus slides or equivalent (Fisher #12-550-15).
13. 10 % Neutral buffered formalin (www.newcomersupply.com #1090 N).
14. Xylene, ACS or Reagent Grade (Fisher # X5p-1 GAL (Un1307)).
15. Ethanol, 200 proof, reagent grade (AAper Alcohol and Chemical Co. or Pharmco Products Inc).
16. Pepsin (BioGenex Laboratories #HK-054-5K).
17. 1× Citra Plus buffer, pH 7.0 (BioGenex Laboratories # HK080-5Kor HK-090-9K).
18. Thermocycler or moist heating block (e.g., MISHA thermocycler, Shandon-Lipshaw).
19. Tris buffer (pH 7.6) Trizma (Sigma #T- 4253 pre-ph crystals, 6.96 g/L distilled water, and 8.5 g NaCl/L distilled water or ready-to-use Tris buffered saline with Tween www.newcomersupply.com #140305 A).
20. ISH universal hybridization kit (www.leica-microsystems.com #ISH-DK or Biogenex ISH core kit for fluorescein labeled probe #DF-132-60K).
21. CMV RNA fluoresceinated probe (www.leica-microsystems.com #NCL-CMV).
22. POLY d(T) probe, fluorescein conjugated (www.leica-microsystems.com).
23. Negative control probe (#ISH 5950 A).
24. 1× SSC Buffer.
25. Alkaline phosphatase enhancing buffer (Innovex Biosciences #NB302 OR NB302S, available from www.newcomersupply.com).
26. Fc receptor block (Innovex Biosciences #NB309 A, NB309-15).
27. Permanent BCIP/NBT, single solution (Innovex Biosciences #NB321NBT, available from www.newcomersupply.com).
28. Nuclear red counterstain (Innnovex Biosciences #NB326, available from www.newcomersupply.com).
29. Advantage permanent mounting medium (Innovex Biosciences #NB 300 OR NB 300A, available from www.newcomersupply.com).
30. Glass coverslips.
31. Brightfield microscope.

## 3 Methods

### 3.1 Detection of HCMV in Primary Tumor Cells

Primary glioblastoma derived cell cultures are generated from fresh surgically resected tissue according to IRB approved protocols. The resulting cultures can be propagated as unattached neurospheres or grown with laminin to allow attachment for immunostaining protocols. Once a primary culture is established, it is possible to perform many standard in vitro experiments such as RNA interference, cell sorting, drug treatment, or ELISA analysis for secreted factors.

#### 3.1.1 Establishing Primary Glioblastoma Derived Cell Cultures

1. Working in a tissue culture hood, place the piece of GBM tissue into a 10 cm sterile tissue culture dish.
2. Using two sterile scalpels, cut tissue into small pieces (*see* **Note 1**).
3. Add 10 mg of papain to 4 mL neurobasal medium (NB) without growth factors and swirl to dissolve.
4. Add NB + papain to the tissue and pipette up and down several times to mechanically disrupt tissue.
5. Place plate in 37 °C incubator and swirl every 10 min for 1 h to facilitate enzymatic digestion of the tissue.
6. Move the cells in NB + papain to a 15 mL conical tube and spin down 5 min at 200 × *g*.
7. Discard supernatant and resuspend in 2 mL NB medium + growth factors.
8. Mechanically disrupt cells (*see* **Note 2**).
   (a) Pass through a 5 mL pipette several times.
   (b) Pass through a glass Pasteur pipette several times.
   (c) Melt the opening of the glass Pasteur pipette with a flame to make a smaller bore, then pass cells through the pipette several times.
9. At this point you should have a single-cell suspension.
10. Seed cells to allow propagation as neurospheres or plate directly into 24-well dishes with coverslips and laminin for immunofluorescence:
    (a) Count the number of cells with a hemacytometer.
    (b) Approximate seeding densities:

        T-75 or 10 cm dish = $2 \times 10^6$ cells.

        6-well plate = $3 \times 10^5$ cells.

        24-well plate = $5 \times 10^4$cells.

    (c) Fill to appropriate volume with NB + growth factors, + laminin for adherent conditions (*see* **Note 3**).

11. Grow cells at 37 °C.
12. Feed neurosphere cells every 2–3 days with NB + growth factors until spheres 1 mm in diameter form. At this point spheres can be mechanically disrupted and replated or frozen down (*see* **Note 4**).
13. For adherent cells, replace medium 1–2 days after initial plating to remove any dead cells or necrotic debris from the medium. The viability of GBM cells in culture is different from tumor to tumor, so the number of cells that attach relative to the seeding density will vary.

#### 3.1.2 Immunostaining of Primary GBM Cells

1. Check to make sure your primary GBM cells are attached to coverslips in a 24-well dish.
2. Aspirate medium from cells and rinse 1× in sterile PBS (about 500 μL/well).
3. Add 500 μL of cold methanol to each well and fix for 10 min.
4. Rinse the cells 2× in PBS.
5. Aspirate PBS and overlay in 500 μL/well protein free blocking buffer.
6. Place on rotator and block for 30 min at room temperature.
7. Rinse the wells 3× with 500 μL TBS-T.
8. Aspirate TBS-T and add 25 μL/well of primary antibody diluted in protein free blocking buffer.
9. Place on rotator for 30 min at room temperature, or incubate overnight at 4 °C (scale up primary antibody to 100 μL for overnight incubation).
10. Rinse wells 3× with 500 μL TBS-T.
11. Aspirate TBS-T and add 25 μL/well appropriate secondary antibody diluted in protein free blocking buffer.
12. Cover plate with foil and rotate for 30 min at room temperature.
13. Always include a secondary antibody only negative control (no primary antibody) to account for any non-specific binding (*see* **Note 5**, Fig. 1).
14. Rinse wells 3× with 500 μL TBS-T.
15. Continue with co-immunostaining for other viral or cellular proteins or rinse 2× in PBS before mounting.
16. Make sure glass slides are appropriately labeled and put a drop of mounting medium (about 10 μL) on slide.
17. Leave second PBS wash on cells and remove coverslips from well by lifting the edge of the coverslip with a pipette tip and removing with tweezers.

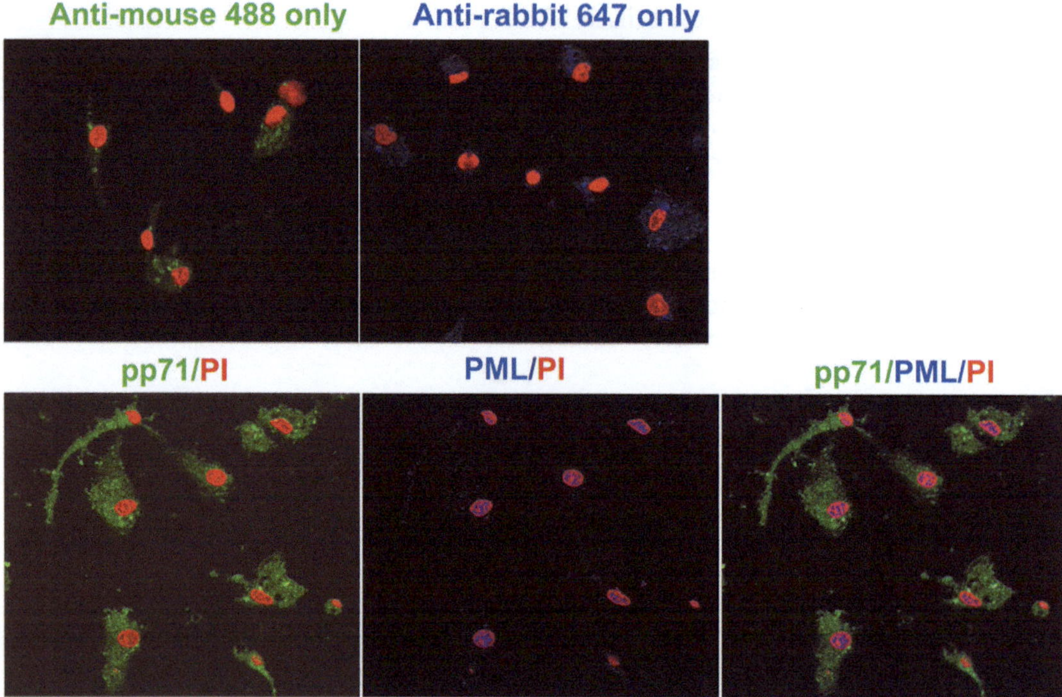


**Fig. 1** Primary tissue from GBM#4134 was homogenized and cultured as described above. Cells were then fixed and immunostained for pp71 (mouse antibody 10G11) and Promyelocytic Leukemia Protein PML (rabbit antibody ab53773, Abcam). Secondary antibodies were Alexa Fluor 488 anti-mouse antibody and Alexa Fluor 647 anti-rabbit antibody, respectively. Immunostained cells were then mounted with VECTASHIELD mounting medium with propidium iodide, sealed, and visualized by confocal fluorescence microscopy using a Nikon Eclipse C1 Confocal microscope (Nikon TE2000-U) fitted with a "Cool Snap" Photometrix camera (Roper Scientific). Images were acquired using EZ-C1 v2.20 software and further processed using Adobe Photoshop CS4

18. Place coverslip onto the drop of mounting medium with the cells facing downward into the medium.
    (a) Optional: Seal the edges of the coverslip with nail polish and allow to dry for several hours in the dark.
19. Visualize cells with a fluorescent microscope, and compare anti-CMV immunostained cells with the secondary-only negative control.

### 3.2 Isolation and Analysis of Protein and RNA from GBM Tissue

This protocol allows simultaneous extraction of protein, RNA, and DNA from a single piece of either fresh or frozen primary GBM tissue. The initial steps of tumor processing are performed with RNAse-free reagents, equipment, and protocols. For western blot analysis we prefer to use the Bio-Rad Criterion XT precast Bis–Tris gel system to maintain a neutral pH during electrophoresis, but any SDS-PAGE system is acceptable.

*3.2.1 Homogenization of Tissue*

1. Put tissue sample in a sterile petri dish and weigh (this protocol works best for 50–100 mg of tissue).
2. In a fume hood, add 1 mL Qiazol to the tissue and dice into small pieces using a sterile scalpel.
3. Transfer the tissue and Qiazol to a 50 mL conical tube with a wide bore pipette tip.
4. Homogenize tissue for 2–3 min with a TissueRuptor probe—avoid making bubbles.
5. Incubate homogenized sample at room temperature for 5 min to dissociate the protein.
6. Divide the sample into two Qiashredder columns (500 μL per column) and homogenize 2×.
7. Combine homogenized sample into one tube and add 200 μL chloroform. Cap the tube and vortex for 15 s.
8. Incubate at room temperature for 3 min.
9. Centrifuge the tube for 15 min, 9,500 rpm, at 4 °C.
10. Pipette off upper clear, aqueous phase to a clean microfuge tube (~400 μL) being careful not to disrupt the interphase or organic phase.
11. Proceed with RNA extraction, the interphase and organic phase containing protein and DNA can be stored overnight at 4 °C.

*3.2.2 Extraction of RNA*

1. Add an equivalent volume of 70 % ethanol to the aqueous RNA phase and proceed with RNA purification using the RNeasy lipid tissue mini kit according to the manufacturer's protocol.
2. Elute RNA from RNeasy column and quantify by spectrophotometry.
3. Verify integrity of the RNA by resolving about 1 μg of RNA on a 1 % agarose gel to visualize intact ribosomal RNA bands.
4. If the RNA is of highly quality, proceed with cDNA synthesis (see below).
5. Store remaining RNA at −80 °C.

*3.2.3 Extraction of DNA*

1. Add 300 μL 100 % ethanol to interphase/organic phase and mix by inverting.
2. Incubate for 2–3 min at room temperature to allow DNA to precipitate.
3. Centrifuge for 5 min, 4,000 rpm, at 4 °C.
4. Remove supernatant (protein) to a new microfuge tube and put on ice.
5. Add 1 mL 0.1 M sodium citrate in 10 % ethanol to the DNA pellet and incubate with occasional mixing at room temperature.

6. Centrifuge for 5 min, 4,000 rpm, at 4 °C.
7. Discard supernatant and repeat **steps 21** and **22** two more times.
8. Add 1 mL 75 % ethanol to DNA pellet and incubate at room temperature with occasional mixing for 10–20 min.
9. Centrifuge for 5 min, 4,000 rpm at 4 °C.
10. Discard supernatant, air-dry for ~15 min.
11. Redissolve the pellet in TE buffer (pH 8.0) to achieve desired concentration (typically addition of 300–600 μL of TE buffer to DNA isolated from 50 to 70 mg tissue will result in a DNA concentration of 0.2–0.3 μg/L).
12. Incubate at 37 °C for 1–2 h to help dissolve gel-like DNA.
13. Pass sample through pipette.
14. Optional: centrifuge for 10 min, 9500 rpm at 4 °C and transfer supernatant (dissolved DNA) to a new microfuge tube.
15. Long-term storage at 4 °C or −20 °C.

#### 3.2.4 Extraction of Protein

1. Add 500 μL isopropanol to protein supernatant and mix by inverting.
2. Incubate for 10 min at room temperature.
3. Centrifuge for 10 min, 9,500 rpm at 4 °C.
4. Discard supernatant.
5. Resuspend pellet in 1 mL 0.3 M guanidine hydrochloride in 95 % ethanol.
6. Mix sample and incubate at room temperature for 20 min.
7. Centrifuge the sample for 5 min, 7,500 rpm at 4 °C.
8. Discard the supernatant and repeat **steps 5–7** two more times.
9. Add 1 mL 100 % ethanol to pellet, vortex, and incubate at room temperature for 20 min.
10. Centrifuge for 5 min, 7,500 rpm at 4 °C.
11. Discard supernatant and let pellet dry.
12. Add 1 % SDS to the pellet (~400 μL) and break up pellet with a sterile toothpick.
13. Incubate tube in a 50 °C heat block for 10 min to help solubilize the protein.
14. Centrifuge for 10 min, 8,500 rpm at 4 °C.
15. Transfer the supernatant to a new tube and quantify protein using the DC protein assay reagents pack according to the manufacturer's instructions (*see* **Note 6**).

*3.2.5 Western Blot Analysis of Protein Lysates*

1. Prepare protein samples by combining equivalent amounts of protein with 4× XT sample buffer and 20× reducing agent to give a 1× final concentration (e.g., For 50 μL total volume, combine 35 μL protein lysate, 12.5 μL 4× buffer, and 2.5 μL 20× reducing agent).
2. Boil samples for 5 min in microfuge tubes with lid-lock clips.
3. Vortex briefly and spin down, and then put the tubes on ice.
4. Prepare 1× MOPS running buffer by diluting the 20× stock solution in ultrapure water. Mix well before adding to the gel chamber (*see* **Note 7**).
5. Load protein samples and electrophorese at 150 V until dye front runs off the gel, or longer if a greater separation is required.
6. Prepare 1.5 L transfer buffer by dissolving 3 BupH Tris-Glycine transfer buffer packs in 300 mL methanol and 1.3 L ultrapure water (20 % methanol final concentration). Store in cold room until ready to transfer.
7. Arrange wet blot transfer according to standard protocols and transfer at 90 V for 90 min in cold room, or transfer overnight at 22 V.
8. Remove PVDF membrane from blotter and discard gel and filter paper. Place membrane in plastic blotting chamber and cover with Ponceau S.
9. Incubate for 5 min on rotator then rinse three times with ultrapure water. Protein bands should be distinct and equivalent between lanes. The membrane can also be trimmed or cut at this point as desired.
10. Prepare a blocking solution of 5 % BSA in TBS-T (e.g., 1 g BSA dissolved in 20 mL TBS-T).
11. Rinse membrane one time in TBS-T for 5 min then incubate in blocking solution for 1 h at room temperature on a rotator.
12. Rinse three times with TBS-T for 15′, 5′, and 5′.
13. Dilute primary antibody in 1× BSA in TBST (diluted from 10× blocker BSA in TBS) and incubate with membrane for 1 h at room temperature or overnight at 4 °C on a rotator. For smaller volumes (2–3 mL), put membrane and antibody in sealpack pouches and heat seal (*see* **Note 8**).
14. Rinse membrane three times in TBS-T for 15′, 5′ and 5′.
15. Dilute secondary antibody in 1× BSA in TBST and incubate at room temperature for 1 h on rotator.
16. Rinse membrane three times in TBS-T for 15′, 5′, and 5′.
17. Blot excess moisture from membrane and place face up on a piece of plastic or saran wrap.

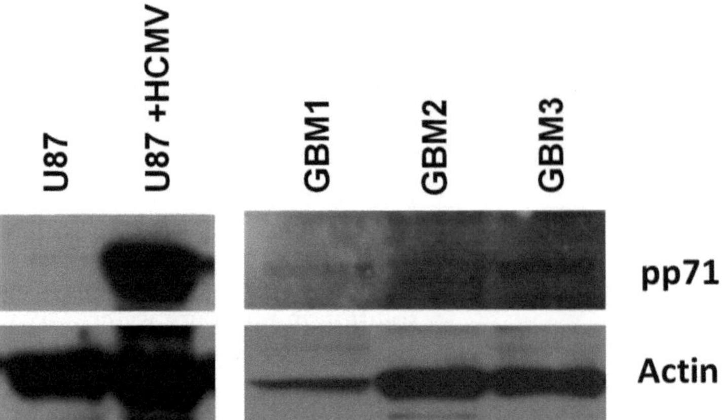


**Fig. 2** Protein lysates were prepared from a U87 glioma cell line that was either mock infected or HCMV infected by harvesting directly in RIPA buffer (Thermo Fisher) with protease and phosphatase inhibitors. Protein lysates from primary glioma tissues were prepared as described in Subheading 3. Equivalent volumes for each lysate were loaded on 4–12 % Bis–Tris precast gels, electrophoresed, and blotted according the described protocols. The membrane was first probed with anti-pp71 antibody 2H10-9 (1:10 dilution) and then re-probed with anti-actin antibody (1:1,000 dilution)

18. Mix chemiluminescent substrate 1:1 according to the manufacturer's instructions, vortex, and cover membrane in a thin layer of substrate.
19. Make sure membrane is completely covered in substrate and cover with another piece of plastic or wrap completely in saran wrap, avoiding creases. Wipe away any excess substrate so membrane pouch is completely dry.
20. Tape membrane into exposure cassette and proceed to the dark room.
21. Expose membrane to film, optimizing the length of exposure to adjust for the strength of the signal. Develop film (*see* **Note 9**).
22. In addition to blotting for viral proteins, always re-blot or do a parallel blot for a cellular loading control like actin or tubulin (Fig. 2).
23. To strip and re-blot the membrane, incubate for 10 min in Restore western blot stripping buffer, rinse three times, re-block, and re-probe.
24. Blots can be stored at 4 °C for several months.

#### *3.2.6 RT-PCR Analysis of GBM RNA*

1. Reverse transcribe 1 μg of RNA using the iScript cDNA synthesis kit according to the manufacturer's instructions (*see* **Note 10**).

2. Prepare a PCR master mix with the Taq PCR core kit reagents, 49 μL per PCR reaction (*see* **Note 11**).
   (a) 5 μL CoralLoad PCR buffer (10×).
   (b) 10 μL Q-solution (5×).
   (c) 2 μL Forward primer (10 μM).
   (d) 2 μL Reverse primer (10 μM).
   (e) 1 μL dNTP mix (10 mM each).
   (f) 0.5 μL Taq polymerase (5 units/μL).
   (g) 28.5 μL nuclease-free water.
3. Vortex master mix for 10 s then spin down.
4. Aliquot 49 μL per PCR tube then add 1 μL cDNA (50 μL total PCR reaction volume).
5. Always include a negative control containing 49 μL of the PCR master mix with 1 μL of nuclease-free water only.
6. Put PCR tubes in the thermocycler and run the following PCR reaction:

| | |
|---|---|
| 1 cycle: | 94° for 5′ |
| 50 cycles: | 94° for 30″ |
| | 58/60° for 30″ (*see* **Note 12**) |
| | 72° for 30″ |
| 1 cycle: | 72° for 10′ |
| | 4° hold |

7. Run 20 μL of each reaction on a 1 % agarose gel (Fig. 3) and purify remaining PCR reaction with the MinElute PCR purification kit.
8. Sequence PCR reactions to verify that the PCR product was not the result of a laboratory strain contaminant (Viral genes from patient samples generally have unique nucleotide changes which distinguish them from sequenced laboratory strains.)

### 3.3 Immunohistochemical Detection of HCMV Antigens in Human Glioblastoma Using Frozen Sections

#### 3.3.1 Freezing Histology Tissues with Liquid Nitrogen

1. Cut a sample of the biological tissue to approximate dimensions of 1.5 by 1.5 by 0.5 cm. Larger tissue samples can warp during freezing; cutting tissue samples to these approximate dimensions can help minimize or avoid warping the tissue.
2. Pour approximately 2 mL of an OCT frozen tissue media (e.g., Sakura Finetek or an equivalent media) onto a suitable support, such as a plastic basemold. The tissue media facilitates cutting of the biological tissue sample.
3. Place the embedded tissue sample into a small zipper-locking plastic bag or equivalent and close the seal, expelling excess air prior to closure.

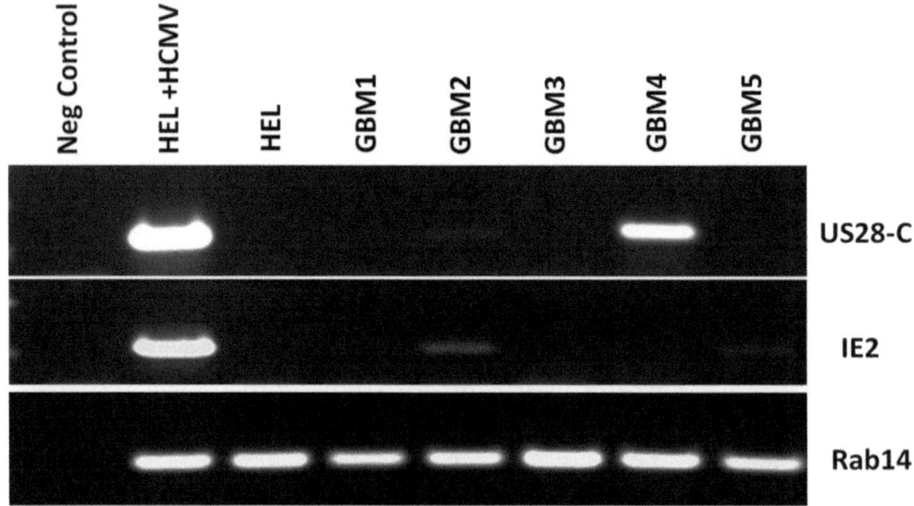


**Fig. 3** RNA was extracted from human embryonic lung fibroblast cells (HEL) that were either mock infected or HCMV infected using the RNeasy mini kit according to the manufacturer's protocol (Qiagen). RNA from primary GBM tissue was isolated as described in Subheading 3. 1 μg of each RNA sample was reverse transcribed, PCR amplified using primers to US28, IE2 and the cellular gene Rab14, and resolved on a 1 % agarose gel. All PCR products were purified and confirmed by sequencing. The negative control is a PCR reaction with water instead of cDNA

4. Immerse the plastic bag with the trimmed and embedded tissue sample into liquid nitrogen for 15–30 s, rapidly freezing the tissue sample to about −20 °C. Rapid freezing in liquid nitrogen helps prevent ice crystal formation and consequential damage to the tissue sample.
5. If the frozen tissue sample is to be stored prior to subsequent processing, remove the frozen tissue sample from the plastic bag, and place into a cassette. Store in a freezer maintained between −70 and −140 °C. For long-term storage (greater than 2 months), tightly wrap the cassette in aluminum foil before storing in the freezer.

#### *3.3.2 Freezing Histology Tissues Without Liquid Nitrogen (see **Note 13**)*

1. Cut the tissue sample to approximate dimensions of 1 by 1 by 0.2–0.5 cm.
2. Pour approximately 2 mL of an OCT frozen tissue media (e.g., Sakura Finetek, or an equivalent media) onto a suitable support, such as a plastic base mold. The tissue media facilitates cutting of the biological tissue sample.
3. Lay the biological tissue sample onto the plastic base mold in the desired orientation, and layer an additional amount (approximately 2 mL) of the tissue media on top of the tissue, so all exposed surfaces of the tissue sample are covered with the frozen sectioning media.

4. Place the tissue sample on the freezing platform in a cryostat maintained at −20 °C for 5 min.
5. To flatten the tissue, position the "cold press" for 2 min (The cold press is a small flat metallic instrument used in the cryostat that keeps the tissue cold.)
6. Spray the tissue sample with a rapid freezing aerosol (e.g., Cytocool II, Richard Allen Scientific or equivalent) for approximately 1 min.
7. Allow the tissue and media to completely freeze, (approximately 2–5 min).
8. If frozen tissue sample is stored prior to subsequent processing, immediately wrap the frozen sample with aluminum foil, and store in a freezer maintained between −20 and −140 °C.

*3.3.3 Sectioning of Frozen Human Glioblastoma Tissue*

1. Immediately transfer frozen tissue, embedded in an OCT media, into a −20 °C cryostat for sectioning.
2. Section the frozen, OCT embedded tissues at desired thickness (4–10 μm) or recommended thickness depending on tissue type and intended application: 7–10 μm for Human or Animal Sections for IHC, Tissues thicker than 10 μm will require longer times for fixation.
3. Pick up sections with positive charge glass slides (silane coated slides or equivalent) and keep at −20 °C until ready for use.
4. Proceed to the "conditioning" step or store cut sections in properly sealed slide boxes. Store at −20 °C for a period of less than 2 months or −70 °C or lower for longer storage.

*3.3.4 Conditioning of Frozen Sections Prior to Fixation (see **Note 14**)*

1. If sections were stored at −70 °C, transfer the sections to −20 °C for at least 2 h prior to conditioning steps.
2. Acclimate tissue sections by either laying the slides containing the frozen sections at room temperature (RT) for 30 s to 5 min, or warming the bottom of slide with natural heat from index finger for 15 s/slide.
3. Sequentially immerse frozen section slides into the following conditioning solutions with continuous, very slow dips (1 s intervals) 20 times with no swish action.

   *Solution A*. 95 % ethanol: 20 dips (1 s intervals × 20 dips)

   *Solution B*. 1 part 95 % ethanol + 1 part acetone: 20 dips (1 s intervals × 20 dips). Continue to dip if a gelatinous substance is still visible until it clears. Discard used FrozFix® Conditioning Solutions after 20 slides for every 50 mL. Do not let slides dry out at any point from **step 6** through the chosen staining application.

*3.3.5 Fixing the Frozen Sections*

1. Immerse conditioned slides into fresh FrozFix® using a coplin jar or other appropriate container at a recommended volume of at least 5 mL/slide with total fluid level above the tissue section.
2. Continue immersion at RT. The fixation times suggested are only guidelines. The user may optimize the time to reflect the tissue thickness, type of tissue, type of antigen, and intended applications. 10–20 min for fresh frozen sections in FrozFix® for immunohistochemistry, immunofluorescence, or in situ hybridization is usually sufficient. Sections thicker than 10 μm will require longer fixation times. For sections stored frozen more than 36 h at −20, −70, or −80 °C, these will require 25–45 min. Discard used FrozFix® after fixing 20 slides for every 50 mL of FrozFix®
3. If a gelatinous substance appears, transfer the sections to fresh Conditioning Solution A, then gently dip until clear, and continue with fixation.

*3.3.6 Rinsing the Frozen Sections*

1. After fixation, transfer fixed tissue sections to unused 1× Tris Buffer Saline + Tween (TBST) pH 7.6, at RT (ten quick dips/rinse). Rinse two times.
2. Continue with immunohistochemistry procedure.

*3.3.7 Procedure for Immunohistochemistry Using Histology Sections Fixed in FrozFix®*

1. Sections stored in TBST at 2–8 °C for 9–48 h must be acclimated to RT.
2. Re-incubate in unused FrozFix® for 20 min at RT. Rinse twice in unused 1× TBST pH 7.6 at RT (15 s/rinse).
3. Due to FrozFix® hybrid formulation, enzymatic pretreatment using a 10× dilution of pepsin, trypsin, or pronase may be necessary to obtain superior IHC results. If using Chemicon IE antibody, digest sections with 1:10 pepsin at 37 °C for 8 min. If using pp65 antibody, digest sections with 1:10 pronase 14 at 37 °C for 8 min. For late antigen antibody, digest with 1:10 pepsin or trypsin for 4 min at 37 °C. If enzyme pretreatment is not desired, proceed to **step 5**.
4. Rinse twice in 1× TBST pH 7.6, at RT (15 s/rinse).
5. Block endogenous peroxidase in tissue sections for 5–10 min with 3 % aqueous $H_2O_2$ (for most IHC antigens) or 0.3 % (for CD and other sensitive antigens). DI water or Tris buffer may be used to dilute $H_2O_2$. Store diluted $H_2O_2$ in a plastic bottle at 2–8 °C and discard after 3 weeks.
6. Rinse twice in 1× TBST pH 7.6, at RT (15 s/rinse).
7. Perform avidin biotin block for biotin rich tissues.
   (a) Block with avidin at RT for 15 min. Rinse twice with 1× TBST pH 7.6, at RT for 2 min each.
   (b) Perform biotin block at RT for 15 min. Rinse twice with unused 1× TBST pH 7.6, at RT for 2 min each.

8. Rinse twice in fresh 1× TBST pH 7.6, at RT (15 s/rinse).
9. Blot around edge of slide without touching the tissue, and then apply Fc Receptor Blocker at RT for 10–30 min depending on tissue type. Fc Receptor Blocker is strongly recommended to avoid nonspecific staining for all tumors and normal non-lymphoid tissues.
10. Blot around edge of slide without touching the tissue. Apply enough CMV antibody (Chemicon IE) at 1:80–1:160, depending on length of time the tissue taken from the subject to liquid nitrogen, or if previously frozen the average storage temperature and length of storage (higher concentration antibody is needed for more poorly fixed tissue). Perform the same step for positive control section and negative control sections.
11. Incubate primary antibody based on recommended established working conditions (antibody dilution, temperature and time). For some antibodies that require overnight or longer incubation times, use 2–8 °C incubation.
12. After incubation, rinse twice in 1× TBST pH 7.6, at RT for 2 min each. Let acclimate at RT for 45 min prior to **step 13**.
13. Blot around the edge of slide without touching the tissue. Apply goat anti mouse biotinylated secondary antibody at 1:18 dilution for 20–30 min at room temperature.
14. Rinse two times in unused 1× TBST pH 7.6, at RT for 2 min each. Apply enough volume of HRP enhancing buffer to cover entire sections and incubate for 2 min at room temperature.
15. Blot around the slide edges without touching the tissue. Apply HRP Streptavidin Label for Peroxidase System at 1:18 for 20–30 min at room temperature. For Alkaline Phosphatase System, apply Streptavidin/Alkaline Phosphatase at 1:18 for 30 min at room temperature.
16. Rinse two times in 1× TBST pH 7.6, at RT for 2 min each. Apply enough HRP Enhancing buffer to cover entire sections and incubate for 2 min at room temperature.
17. Blot around slide edges without touching the tissue. Apply chromogen, e.g., DAB if HRP for 1–5 min or Fast Red if Alkaline Phosphatase for 5–10 min at RT.
18. Examine the tissue under the microscope while developing. Stop color development by rinsing with tap water (2× 15 s/rinse).
19. Counterstain with aqueous hematoxylin (5–10 dips), depending on contrast desired.
20. Rinse off excess hematoxylin in tap water.
21. For Alkaline Phosphatase and Peroxidase Systems (AEC Chromogen), do not dehydrate in alcohols. Proceed to mount with an aqueous mounting media (e.g., Advantage Permanent

Aqueous Mounting Medium, Innovex Biosciences). For immediate viewing with Advantage Permanent Aqueous Mounting Medium, quick-set the coverslip by placing the mounted slides in an oven for 5 min at 37 °C. Do not leave slides at 37 °C for more than 1 h. Refer to Advantage Permanent Mounting Media protocol.

22. Apply coverslip with an aqueous mounting media and let the slides sit at RT for 10–15 min.
23. For Peroxidase System (DAB Chromogen), dehydrate in ascending series of ethanol by dipping 20 times in each step in series: 50, 75, 95, 100 % (2×), xylene (2×) each. Mount and coverslip with Permaslip mount or equivalent solvent based mounting media.
24. Examine slides under the microscope for reactivity.

### 3.4 Immunohistochemical Detection of HCMV Antigens in Human Glioblastoma Using Paraffin-Embedded Sections

#### 3.4.1 Preparation of Paraffin-Embedded Sections for Immunohistochemistry

1. Cut paraffin blocks at 6 μm for brain sections (sections thicker or thinner do not stain optimally for immunohistochemistry).
2. Float the sections in water bath with distilled or deionized water only (Please note that histology additives such as gelatin, stay-on, or any other similar protein additives cause additional nonspecific or background staining).
3. Pick up the sections onto Plus slides or equivalent Superfrost microscope slides.
4. Drain the slides vertically onto absorbent paper towels at room temperature for at least 1 h.
5. Transfer the slides into staining tray, then allow to dry in 50 °C oven for at least 4 h, preferably overnight.
6. Deparaffinize in xylene (leave at room temp for 45 min).
7. Transfer sections in two changes of fresh xylene (same type as above), 5 min/change.
8. Hydrate in graded alcohols starting with 100, 95, 70, and 50 % ETOH. For each series of ETOH, dip the slides 20 times in two separate baths each (Ethanol used in these must be reagent grade 200 proof nondenatured.)
9. Rinse in running tap water quickly to remove alcohol.
10. Transfer into TBST pH 7.6.
11. Because of variability of fixation (from autolysis from specimen to specimen and block to block, age of tissue blocks, depth of cuts from block to block within the same case, etc.) conditioning of tissues before immunostaining steps is required. This will allow unfixed areas to be fixed. Fixative in general works by penetrating the immediate 1 mm area of the tissue and prolonged exposure to the fixative causes over fixation of the periphery and does not necessarily penetrate through the center of the section. *See* **Note 15** for specific guidelines.

12. Rinse quickly with running water (1 min) to remove fixative, and then transfer to TBST. Soak tissue sections for 2 h or overnight at 4–8 °C. If sections were soaked overnight at 4 °C, allow sections to reach room temperature for 1 h before proceeding with enzyme digestion.
13. Mark tissue sections with Pap pen to contain liquid without drying sections. Add enough drops of enzyme onto the sections then incubate as recommended below. Do not substitute any enzyme—must not be over 3 weeks old after reconstitution. Enzymes from other vendors must be optimized before use.
14. For CMV-IE, CMV–Late (Chemicon), and CMV cocktail (Innovex Biosciences), Pepsin is recommended at 37 °C for 4–6 min. For pp65, pepsin or trypsin is recommended at 37 °C for 4–6 min. This step is critical and if modified, the entire protocol will not work. All enzyme digestion must be conducted in a humidified environment and sections are not allowed to dry, otherwise, no signal will be detected. Use MISHA thermocycler or equivalent—if not available, optimize the temperature setting to simulate 37 °C as actual heating temperature on the slide, not the oven. The bottom of the slides with section facing upwards must be completely dry before lying onto the hot surface. Moisture trapped on the bottom causes irregular heating and boiling over of the enzyme. Due to differences in fixation as well as autolysis and over-fixation, this step must be optimized to fit the status of fixation of tissues (*see* **Note 16**).
15. Rinse quickly in distilled water and then transfer to TBST.
16. Retrieve the sections as follows: Sections are retrieved in 1× Citra Plus buffer by preheating the Citra buffer (without the slides) in microwave at high setting for 2–4 min depending on the number of slides/bath. The resulting buffer temperature must be at least 85–90 °C. The slides are then immersed on the preheated Citra and transferred to 45–50 °C water bath for 2.5 h.
17. Allow the slides to cool at room temperature for 5 min. then rinse with running tap water for 2 min then transfer to TBST (pH 7.6) for minimum 2 h at room temperature or overnight at 4–8 °C (do not shorten this step). If stored overnight at 4–8 °C, let the slides reach room temperature before proceeding to next step (at least 1 h at room temp).

#### 3.4.2 Immunohistochemistry on Paraffin-Embedded Tissue Sections

1. Block endogenous peroxidase by incubating slides in freshly prepared 3 % $H_2O_2$ for 12 min at room temperature. Use plastic staining jars only, and do not use glass.
2. Rinse quickly in distilled water, and then transfer to TBST. All incubations are done using incubation tray, or equivalent to prevent drying of sections.

3. Blot dry around edges with gauze or tissue paper. Mark or circle around tissue sections with Pap Pen. Apply avidin block to sections for 15 min at room temperature.
4. Rinse 2× TBST, 3 min/rinse. Immediately follow by adding biotin block for 15 min at room temperature. Rinse 2× TBST, 3 min/rinse.
5. With a Pap Pen mark a circle around the sections, then apply Fc receptor block using enough volume to adequately cover the sections.
6. Carefully blot the Fc receptor block solution, then without letting sections dry, add primary antibody. Incubate in humidified container overnight at 4–8 °C (*see* **Notes 17** and **18**).
7. Rinse with TBST (3 × 10 min each) at RT. Let soak for 2 h at RT or overnight at 2–8 °C.
8. The following steps can be automated in any immuno automated system set up for room temperature conditions.
9. Apply goat anti-mouse secondary antibody at 1:18; in the BioGenex link diluents (*see* **Note 19**).
10. Incubate for 30–45 min RT.
11. Rinse with TBST 3× (10 min each) RT.
12. Apply Peroxidase-Labeled Streptavidin diluted in Streptavidin-Peroxidase Diluent (1:18).
13. Incubate for 30–45 min at RT.
14. Rinse with TBST 3× (10 min. each) at RT.
15. Develop with DAB under microscope for 1–5 min (*see* **Note 20**).
16. Transfer to tap water when signals are optimum. Signals weaker than 2 plus staining can be enhanced by using DAB enhancer. If background is problem, soak developed slides in fresh TBST for 24 h or up to 72 h (2–8 °C).
17. Counterstain with Harris hematoxylin (Richard Allan) by dipping five times or more depending on intensity of counterstain desired.
18. Rinse in running tap water, and then remove excess hematoxylin by dipping in bluing solution 5–10 times. Rinse in running tap water.
19. Dehydrate in descending series of alcohol (50, 75, 95, 100 % (2×)) by dipping 10× each, then two changes of xylene (reagent grade).
20. Coverslip and mount in Permount.

### 3.5 DNA/RNA In Situ Hybridization Methods for Paraffin-Embedded Tissue Sections and Primary Cell Cultures

#### 3.5.1 In Situ Hybridization Protocol for DNA/RNA Detection in Paraffin Sections: Fluorescein or Biotinylated Probes

1. Cut Sections as per Immunohistochemistry protocol and use same glass slides (**steps 1–5**). Deparaffinize tissue sections in xylene and graded alcohols using RNAse-free (with RNAse inhibitor) ddH$_2$O (**step 6–10**, Immuno protocol).
2. Post fix and pretreatment—refer to immunohistochemistry protocol **steps 11–15**.
3. Rinse in DNAse/RNAse-free H$_2$O ethanol.
4. Dehydrate to 100 % ETOH in graded alcohols with RNAse inhibitor.
5. Dry sections at room temperature or 50 °C oven.
6. Add 20 μL HCMV nucleic acid probe or sufficient amount to cover the size of sections (biotinylated or fluorescein labeled probes for RNA or DNA detection).
7. On separate slides add positive control and no probe (diluent only) or negative control probe.
8. Denature at 90 °C for 15–20 min (normal and low grade tumor) to 25 min (high grade) on MISHA thermocycler or equivalent chamber, with moist environment to prevent evaporation of probe.
9. Incubate overnight in humidified chamber at 37 °C. Soak in TBST pH 7.6 until coverslips come off.
10. Continue with **step 11** for DNA probes. For RNA probes, continue with **step 12**.
11. Stringency wash 40 °C 1× SSC for 20 min. Perform this step with DNA probes only.
12. Rinse 2× with TBS then incubate in Alkaline phosphatase enhancing buffer for 1 min.
13. Apply Fc block for 15 min RT.
14. Blot around edges.
15. Apply link 1 mouse anti-biotin or mouse anti-fluorescein 45 min RT.
16. Wash 2× with TBS, then incubate in alkaline phosphatase enhancing wash buffer for 1 min.
17. Blot around edges, then apply link 2 (goat anti-mouse biotinylated FAB2 fragment) 45 min RT. **Steps 16–20** can be substituted with rabbit F(ab′) conjugated to alkaline phosphatase diluted in 1:200 TBS containing 3 % w/v BSA and O.1 % v/v Triton x-100. Incubate for 30–35 min depending on type of tumor. Skip **steps 19** and **20**, proceed to **step 21**.
18. Wash 2× with TBST then incubate with alkaline phosphatase enhancing buffer for 1 min.
19. Blot around edges and apply alkaline phosphatase-streptavidin label 45 min RT.

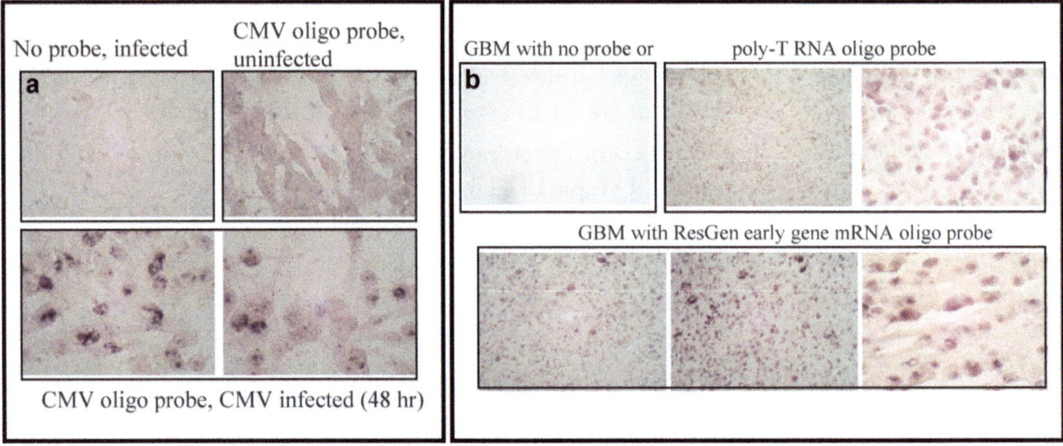


**Fig. 4** In situ hybridization(ISH) for CMV. (**a**) U251 glioblastoma (GBM) cells were grown in tissue culture and ISH was performed for CMV RNA with negative controls (no probe—CMV infected, and CMV probe-uninfected). Cells infected for 48 h have blue staining after CMV oligo probe hybridization (*bottom*). (**b**) Frozen section of human GBM hybridized with no probe (negative control), poly-T RNA oligo probe (positive control), or CMV probe for CMV early gene mRNA. GBM cells are positive for CMV by ISH on *bottom* panel (low and high power)

20. Wash 2× with TBST then incubate with alkaline phosphatase enhancing buffer for 1 min.
21. Blot dry, then add chromogen substrate (BCIP/NBT) RT 5 min or until signal develops.
22. Wash in distilled $H_2O$.
23. Counterstain with Innovex Nuclear Red and rinse with tap water.
24. Wipe around edges with gauze or equivalent.
25. Mount in permanent aqueous mounting medium.
26. Dry in 45–50 °C oven for 45 min.
27. Examine (Example shown in Fig. 4).

*3.5.2 In situ Hybridization Protocol for DNA/RNA Detection in Primary Glioblastoma Derived Cell Cultures Using Fluorescein or Biotinylated Probes*

1. Rinse slides with RNAse-free $ddH_2O$.
2. Fix in –20 °C Methanol 20 min and let air-dry.
3. Add HCMV nucleic acid probe.
4. Denature in Misha thermocycler 90 °C, 15 min.
5. Hybridize at 37 °C for 1–1.5 h in MISHA humidified chamber with coverslips.
6. Soak in TBST to remove coverslips and then rinse in TBST ×2.
7. Probe wash 40 °C 1× SSC 10 min, skip this step if using RNA probes.
8. Hybridization wash RT for 10 min.
9. Fc receptor block 10 min at RT.

10. Blot around edges.
11. Apply link 1 (mouse anti-biotin or mouse anti-fluorescein) 15–20 min at RT.
12. Wash ×2 TBST, then incubate in alkaline phosphatase enhancing wash buffer for 1 min.
13. Blot around edges, apply link 2 (goat anti-mouse biotinylated FAB2 fragment) 15–20 min at RT.
14. Wash ×2 TBST, incubate in alkaline phosphatase enhancing wash buffer for 1 min.
15. Blot around edges, apply alkaline phosphatase-streptavidin label 15–20 min at RT. **Steps 11–15** can be substituted with rabbit F(ab′) conjugated to alkaline phosphatase diluted in 1:200 TBS containing 3 % w/v BSA and O.1 % v/v Triton x-100. Incubate for 30–35 min depending on type of tumor. Proceed to **step 16**.
16. Wash ×2 TBST, then incubate in alkaline phosphatase enhancing wash buffer for 1 min.
17. Blot around edges, add chromogen substrate (BCIP/NBT) RT 5 min or until signal develops.
18. Wash in distilled $H_2O$.
19. Counterstain with nuclear red and rinse with tap water.
20. Wipe around edges with gauze.
21. Mount in Advantage permanent mounting medium.
22. Dry in 45–50 °C oven for 45 min.
23. Examine (Example shown in Fig. 4).

## 4 Notes

1. Try to remove red blood vessel tissue from tumor tissue before homogenization.
2. If the tissue is still in large pieces and is not breaking up easily, then place back in the 37 °C incubator for hour-long intervals. Larger tumors or tumors with sinewy necrotic regions may need longer digestion times.
3. All cultures can be grown in the presence of laminin, but it is really only absolutely necessary when cells are going to be subjected to immunostaining.
4. Neurospheres do not have to be enzymatically detached from plates for passaging. Just collect the culture medium in a 15 mL conical tube and spin the spheres down gently. Aspirate the supernatant, resuspend in a small volume of NB + growth factors, and pass through a small bore glass Pasteur pipette several times to dissociate the spheres before re-seeding.

5. This negative control is necessary because the fluorescently labeled secondary antibodies often bind non-specifically to primary GBM cells, which appear as small very bright spots in the cytoplasm. This control is necessary to ascertain if the anti-HCMV antigen immunostaining observed is specific. HCMV infected cells can also be used as a positive control, but the pattern and localization of staining seen in acutely infected tissue culture cells is often different from what is observed in endogenously infected primary GBM cells.
6. Protein lysates prepared from primary tumor tissues contain some insoluble material and necrotic debris, which often results in artificially high protein concentration values in the Lowry assay. Lysates can be centrifuged and the pellet discarded to improve the quantification accuracy. Alternatively, when quantitative analysis is not necessary, equivalent volumes of protein from tumor specimens of comparable size can be loaded and qualitatively analyzed.
7. The Bio-Rad criterion gel system can be used with various combinations of gel composition, gel percentage, and running buffer to achieve a range of separations. We use the 4–12 % gradient Bis-Tris gels with XT MOPS buffer which is optimal for midsize proteins. (Refer to http://www.bio-rad.com/webroot/web/pdf/lsr/literature/Bulletin_4110001E.pdf.)
8. A 1 h incubation time at room temperature is usually sufficient for monoclonal antibodies. Overnight incubation improves the binding efficiency of some polyclonal antibodies, goat antibodies in particular.
9. Depending on the binding specificity of the primary antibody and the amount of antibody loaded, the intensity of the signal may vary. For a strong signal, the west pico substrate is usually sufficient to detect a signal in less than a minute exposure. For weaker signals, the Thermo Fisher dura or femto chemiluminescent substrates can be used to improve detection.
10. There are many reagents that can be used for cDNA synthesis, though they differ in their ability to detect low abundance transcripts. For best results, test several different enzymes to determine the optimal reagents for your experiments.
11. The PCR conditions used to detect low levels of viral transcripts in whole tumor tissues requires a high number of PCR cycles. Therefore, it is extremely important to prevent contamination of PCR reagents with viral DNA. We have established a remote PCR bench with a dedicated set of pipettes, tubes, and filter tips to prevent contamination. It is also essential to include a negative control to ensure that any positive PCR signals detected are the result of amplification of genuine HCMV transcripts in the cDNA sample and not due to contamination. All PCR products are also sequenced to

verify their identity and sequence variation from laboratory strains.

12. The UL56 and Rab14 primers are used with a 58° annealing temperature. All others use a 60° annealing temperature.
13. This method of freezing biological tissues may be employed in place of liquid nitrogen for rapid freezing. Since the freezing rate will be slower, the possibility of ice crystal formation is more likely to occur with this method, contributing to tissue damage.
14. Conditioning of sections and the sequence of steps is critical to the performance of FrozFix®. Modifications instituted by user may result in different performance of the fixative.
15. Guidelines to conditioning of tissue sections: For paraffin blocks originally fixed in 10 % neutral buffered formalin (10 % NBF) or other fixative, aged 6 months to 2 years, paraffin sections from these blocks should be post-fixed for 1.5–2 h in 10 % NB. For blocks 2–4 year old, post-fix for 3–4 h in 10 % NBF, and for blocks over 4 year old, postfix for 6–8 h in 10 % NBF. (For further reference to immunohistochemistry, *see* refs. 2–4.)
16. Positive control for detection must be selected from tissues that most closely reflect the fixation and other tissue handling as the GBM. It is known that HIV infected tissues, as well as infected transplant tissues are very good sources of cytomegalovirus positive controls. However, due to the significant difference in terms of viral copies per cell population, direct correlation as to antibody dilution, incubation time and detection parameters is not feasible. Therefore *optimization of all reagents with a known infected tissue serves as a guideline only and therefore provides information as to the suitability or potency of reagents.* We have found that the antibody concentration varies within lots and therefore optimization is a necessity before final immunostaining steps. We use the following guideline: If primary antibody dilution of 1:50 is optimum for positive control tissues using HIV infected and transplant tissues, the starting dilution for GBM will be approximately 1:20 (with consideration of the initial protein concentration of primary antibody if known). Difference in thickness of sections utilized needs to be considered for length of incubation for the components of detection system. For example, in the case of *GBM and other brain tissue* components cut at 6 μm require incubation of the secondary antibody and the Peroxidase Streptavidin Label for 40–45 min each at room temperature whereas *lung, kidney or other tissues cut at 4–5*μm require 30 min incubation on the detection system components. As a working guideline, the positive control immunostaining for HIV infected lung must be an intensity of at least 4+ to observe an intensity of 2+ for the GBM. Based on

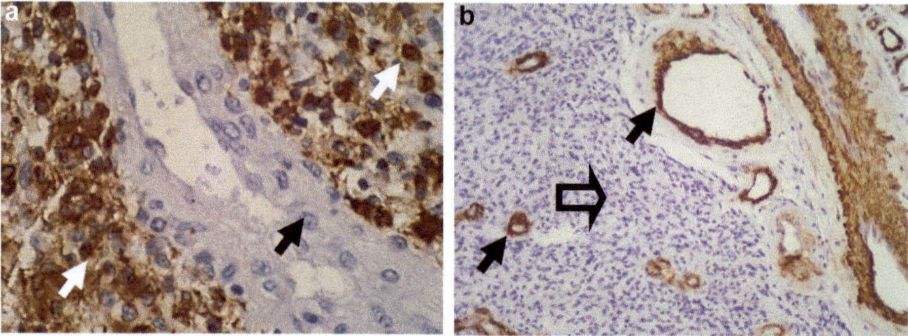


**Fig. 5** Paraffin sections of human glioblastoma immunostained with monoclonal antibody specific for CMV IE1-72 (**a**), and IgG control monoclonal antibody specific for smooth muscle actin (SMA) specific for blood vessels (**b**). Tumor cells stain in cytoplasm and nucleus (*white arrows*) in glioblastoma, but area of vascular endothelial proliferation is negative (*black arrow*). IgG control SMA antibody is specific for blood vessels (*black arrows*) but negative in areas of tumor (*open arrow*)

our sensitive immunostaining techniques, immunostaining of lung epithelium is observed in addition to lung pneumocytes using this technique (Figs. 5 and 6).

17. The following antibody concentrations are recommended, but proper optimization must be conducted before the final run due to variability from lot to lot. We recommend using Chemicon CMV IE1/IE2 mAb as screening antibody.
    (a) Chemicon CMV IE1/IE2 mAb (MAB810) 1:40.
    (b) Innovex Biosciences CMV early/late cocktail (MAB337C) 1:40–1:60.
    (c) Chemicon CMV late Ag mAb (MAB8127) 1:35.
    (d) Novocastra pp65 mAbs (two clones) 1:35–1:45.
    (e) Biogenex CMV IE mAb (MU254-UC) 1:15–1:25.
18. All antibodies are diluted fresh and used immediately or no more than 24 h after dilution with common antibody TBST with BSA. We prepare corresponding volume for all the slides at around 150–250 μL/slide/antibody (depending on the size of the sections) to allow consistency. We have observed a significant decrease in reactivity 24 h after diluting the antibody.
19. For utmost specificity and sensitivity, we optimized the system for brain tumors with this secondary antibody system. Our results indicate that Multilink systems (i.e., secondary antibodies specific to mouse and rabbit antibodies) could possibly cause non-specificity because the secondary component are designed to detect polyclonal antibodies also. Other kits such as polymer based and dextran did not provide the comparable sensitivity or specificity. Detection kits other than Biogenex Supersensitive Monoclonal kit must be optimized before use.

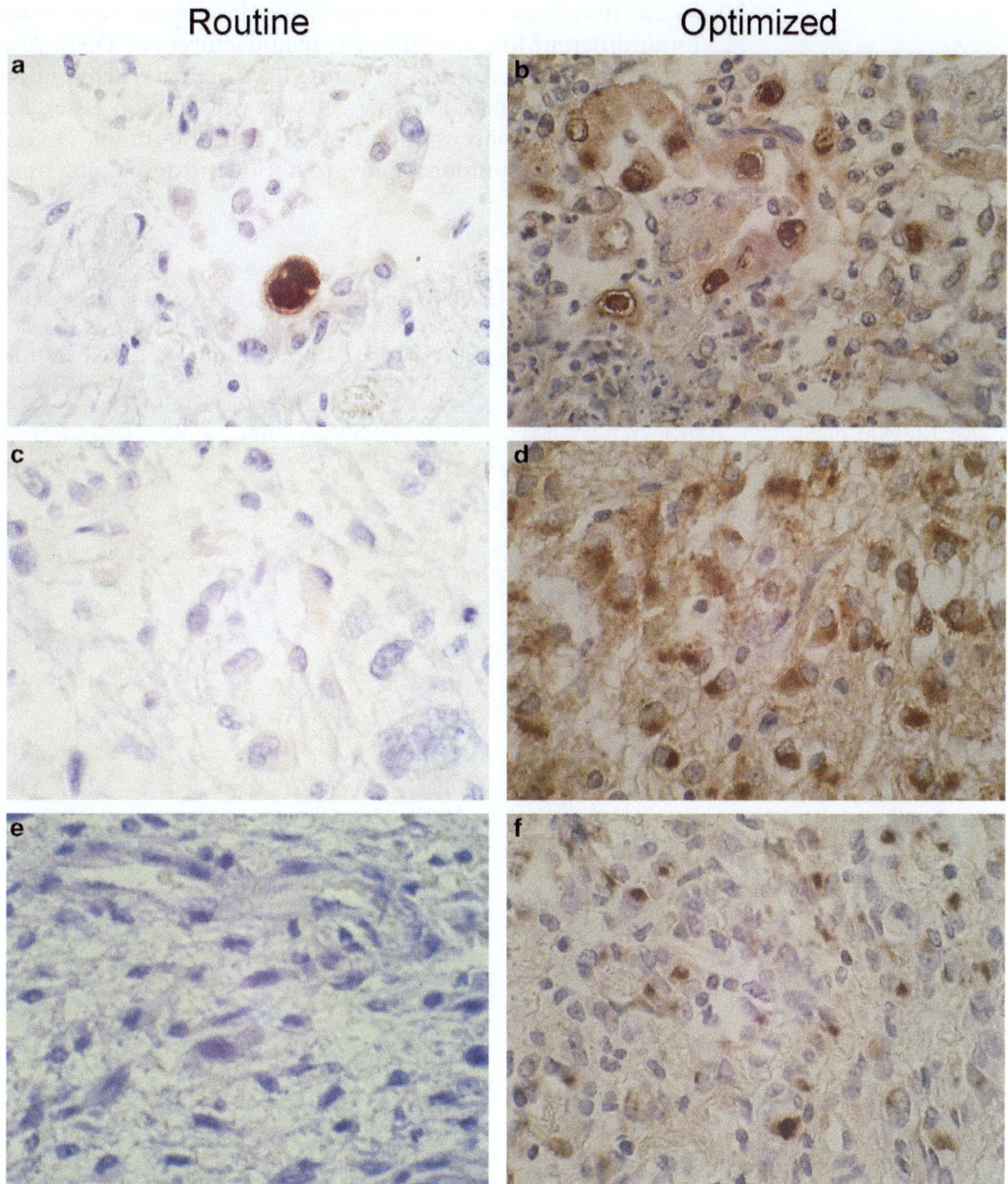


**Fig. 6** Routine versus optimized immunostaining of paraffin sections of CMV-infected lung (**a**, **b**) and two different glioblastomas (**c** and **d**, and **e** and **f**) with anti-IE1-72 mAb

20. Please note that there is variation in developing time when using different lots and different manufacturers of DAB. Also note that positive controls by nature have high and active infection and more viral copies per cell and will develop faster and require weaker concentration of DAB than the tumor. Do not allow automated immunostainer to decide the developing time.

## References


1. Dziurzynski K, Chang SM, Heimberger AB, Kalejta RF, McGregor Dallas SR, Smit M, Soroceanu L, Cobbs CS (2012) Consensus on the role of human cytomegalovirus in glioblastoma. Neuro Oncol 14(3):246–255
2. Elias J (2003) Immunohistopathology, a practical approach to diagnosis, 2nd edn. ASCP, Chicago, IL
3. Sheehan DC (1980) Theory and practice of histotechnology. CV Mosby Company, St. Louis
4. Freida CL (1997) Histology & self instructional text, 2nd edn. ASCP p 4. Detailed technical bulletin for FrozFix. www.newcomersupply.com. 2505 ParviewRoad, Middleton, WiI53562

# Chapter 12

# Methods to Study the Nucleocytoplasmic Transport of Macromolecules with Respect to Their Impact on the Regulation of Human Cytomegalovirus Gene Expression

Marco Thomas, Barbara Zielke, Nina Reuter, and Thomas Stamminger


## Abstract

One defining feature of eukaryotic cells is their compartmentalization into nucleus and cytoplasm which provides sophisticated opportunities for the regulation of gene expression. Accurate subcellular localization is crucial for the effective function of most viral macromolecules, and nuclear translocation is central to the function of herpesviral proteins that are involved in processes such as transcription or DNA replication. Human cytomegalovirus (HCMV) encodes several transactivator proteins which stimulate viral gene expression either on the transcriptional or posttranscriptional level. In this chapter, we focus on nucleocytoplasmic transport mechanisms of either proteins or RNA that are utilized during HCMV infection. We describe commonly used assays to determine the subcellular localization of a protein, its nucleocytoplasmic shuttling activity, its capacity to export unspliced RNA from the nucleus, and its association with RNA in vivo.


**Key words** Nucleus, Cytoplasm, Immunofluorescence analysis, Shuttling, Heterokaryon, CAT mRNA export assay, Coimmunoprecipitation, RNA-immunoprecipitation

## 1 Introduction

*Herpesviridae* encode a conserved family of regulatory proteins, which are known to function as posttranscriptional activators to facilitate nuclear export of intronless mRNAs via recruitment of the cellular mRNA export machinery [1–3]. The HCMV-encoded representative of this protein family is pUL69, a multifunctional regulatory protein that acts as a pleiotropic transactivator by associating with the cellular transcription elongation factor hSPT6 [4]. In addition, pUL69 possesses properties of a viral mRNA export factor, since it binds to RNA [5], shuttles from the nucleus to the cytoplasm [6], and recruits the cellular mRNA export machinery via interaction with UAP56/URH49 [7]. The closely related cellular DExD/H-box RNA helicases UAP56 and URH49 play an

Andrew D. Yurochko and William E. Miller (eds.), *Human Cytomegaloviruses: Methods and Protocols*, Methods in Molecular Biology, vol. 1119, DOI 10.1007/978-1-62703-788-4_12, © Springer Science+Business Media New York 2014

essential role in cellular and viral mRNA export by recruiting the adaptor protein REF to spliced and unspliced mRNAs in order to facilitate their nuclear export. To gain further insight into their mode of action, we aimed to characterize these cellular RNA helicases in more detail by examining their subcellular localization via indirect immunofluorescence analysis and by investigating UAP56 function, with respect to nucleocytoplasmic shuttling via an interspecies heterokaryon assay. Next, coimmunoprecipitation assays were performed in order to confirm the association of HCMV pUL69 with its cellular interaction partners UAP56 or hSPT6 in eukaryotic cells. To elucidate the betaherpesviral mRNA export in a more global sense, we investigated whether the closely related pUL69-homolog M69 of murine cytomegalovirus could similarly function as a posttranscriptional transactivator protein and determined if it is capable of exporting unspliced CAT-reporter mRNAs. Finally, RNA-immunoprecipitation assays were performed to confirm a direct interaction of pUL69 and the before mentioned CAT-reporter RNA in transfected cells. The experiments described in this chapter can be easily adapted to examine related proteins for their subcellular localization, nucleocytoplasmic shuttling, association with known or suspected interaction partners, and RNA-binding capacity. Collectively, these methods will help to further define the activities of cytomegalovirus-encoded transcriptional or posttranscriptional regulatory factors.

## 2 Materials

Prepare all solutions using ultrapure water and analytical grade reagents. Unless indicated otherwise, prepare and store all reagents at room temperature.

### 2.1 Buffers and Solutions

*2× BES* (BES-buffered saline): 50 mM BES, 280 mM NaCl, 1.5 mM $Na_2HPO_4$, pH 6.95.

*Coimmunoprecipitation (CoIP) buffer*: 10 ml 1 M Tris–HCl, pH 8.0, 6 ml 5 M NaCl, 2 ml 0.5 M EDTA, 1 ml Nonidet P-40 (NP-40), adjusted to 200 ml with sterile $H_2O$. Protease inhibitors were added as follows: 2 ml 100 mM PMSF and 400 μl each of 1 mg/ml aprotinin, leupeptin, and pepstatin (Sigma-Aldrich).

*CHX*, cycloheximide.

*DAPI*(4′,6-diamidino-2-phenylindole)-containing VECTASHIELD mounting medium.

*ECL solution A*: 50 mg luminol (Sigma-Aldrich) dissolved in 200 ml 0.1 M Tris–HCl, pH 8.6.

*ECL solution B*: 11 mg *p*-hydroxycoumarin acid (Sigma-Aldrich) was dissolved in 10 ml DMSO.

Glycogen (Roche).

Hoechst 33258 (Sigma-Aldrich).

*LMB*, leptomycin B (Sigma-Aldrich).

*4 % paraformaldehyde solution*: 4 % paraformaldehyde dissolved in PBSo.

*PBSo* (phosphate-buffered saline without $CaCl_2$ and $MgCl_2$): 138 mM NaCl, 2.7 mM KCl, 6.5 mM $Na_2HPO_4$, 1.5 mM $KH_2PO_4$.

*PEG*, polyethylene glycol—PEG8000.

Protein A Sepharose (GE Healthcare).

Protein A-Dynabeads (Invitrogen).

*Ribonucleoprotein-immunoprecipitation (RIP) buffer*: 10 ml 1 M Tris–HCl, pH 8.0, 12 ml 5 M NaCl, 2 ml 0.5 M EDTA, 1 ml NP-40, adjusted to 200 ml with sterile $H_2O$. Protease inhibitors were added as follows: 2 ml 100 mM PMSF and 400 μl each of 1 mg/ml aprotinin, leupeptin, and pepstatin. In addition, 20 μl of the RNase-inhibitor RNasin was included.

4× *SDS protein loading buffer*: 125 mM Tris–HCl, pH 6.8, 2 mM EDTA, 20 % glycerol, 4 % SDS, 10 % β-mercaptoethanol, 0.01 % bromophenol blue.

*10× SDS-PAGE buffer*: 286 g of glycine, 60.6 g of Tris base, and 20 g of SDS were dissolved in $H_2O$ with the volume adjusted to 2 l.

*0.2 % Triton X-100*: 0.2 % Triton X-100 dissolved in PBSo.

*Western blotting buffer*: 75 g of glycine, 15.1 g of Tris base, and 1 l of ethanol were dissolved in $H_2O$ with the volume adjusted to 5 l.

### 2.2 Cell Culture

*DMEM* (Dulbecco's modified Eagle medium): a ready-to-use mixture purchased from Gibco/BRL dissolved in sterile water and adjusted to pH 7.0.

*FCS*, fetal calf serum (Sigma-Aldrich).

*MEM* (Eagle's minimal essential medium): a ready-to-use mixture purchased from Gibco/BRL dissolved in sterile water and adjusted to pH 7.0.

*Trypsin/EDTA*: 0.25 % trypsin, 140 mM NaCl, 5 mM KCl, 0.56 mM $Na_2HPO_4$, 25 mM Tris–HCl, pH 7.5, 5 mM D(+) glucose, 0.01 % EDTA, pH 7.0.

### 2.3 Reporter and Effector Constructs

*pCFN-βGal*: eukaryotic expression plasmid encoding β-galactosidase fused to the nuclear localization signal (NLS) of SV40 T antigen [8].

*pCFNrev-βGal*: eukaryotic expression plasmid encoding β-galactosidase in fusion with the NLS of SV40 large T antigen and the nuclear export signal of the HIV-1 Rev protein [8].

*pcRev*: eukaryotic expression plasmid encoding HIV-1 Rev protein [9].

*pcRex*: eukaryotic expression plasmid encoding HTLV-1 Rex protein [10].

*pDM128/CMV/RRE*: CAT-reporter plasmid which contains an intronic CAT-reporter gene along with the HIV-1 Rev-responsive element [11].

*pDM128/CMV/RxRE*: CAT-reporter plasmid which contains an intronic CAT-reporter gene along with the HTLV-1 Rex-responsive element [11].

*pHM829*: NLS-mapping vector, inserts can be fused simultaneously to the C-terminus of β-galactosidase and the N-terminus of GFP [12].

*pHM830*: NLS-mapping vector, inserts can be fused simultaneously to the C-terminus of GFP and the N-terminus of β-galactosidase [12].

*FLAG-pUL69* (pHM2098): eukaryotic expression plasmid encoding an N-terminally FLAG-tagged version of the pleiotropic transactivator pUL69 of HCMV [7].

*Myc-pUL69* (pHM2235): eukaryotic expression plasmid encoding an N-terminally Myc-tagged version of the pleiotropic transactivator pUL69 of HCMV [13].

*UL69* (pHM160): eukaryotic expression plasmid encoding full-length HCMV pUL69 in pCB6 vector [14].

*FLAG-UL69ΔR1* (pHM2322): eukaryotic expression plasmid encoding N-terminally FLAG-tagged pUL69 with internal deletion of aa17–30 [5].

*FLAG-UL69ΔR1ΔRS* (pHM2325): eukaryotic expression plasmid encoding N-terminally FLAG-tagged pUL69 with internal deletion of aa17–30 and aa123–139 [5].

*FLAG-UL69ΔR2ΔRS* (pHM2326): eukaryotic expression plasmid encoding N-terminally FLAG-tagged pUL69 with internal deletion of aa36–50 and aa123–139 [5].

*FLAG-UL69mutUAP56* (pHM2316): eukaryotic expression plasmid encoding the amino acid replacement mutant pUL69 R22,23,25,26A unable to interact with UAP56 [7].

### 2.4 Antibodies

*M2*: antibody directed against the FLAG epitope (DYKDDDDK) (Sigma-Aldrich).

*Anti-β-galactosidase*: antibody to *E. coli* β-galactosidase (Roche).

*Anti-FLAG*: purified serum directed against FLAG epitope (Sigma-Aldrich).

*Anti-Myc*: Hybridoma 1–9E10.2, directed against the Myc epitope (EQKLISEEDL) (ATCC).

*MAb 69-66*: mouse monoclonal antibody for detection of the viral tegument protein pUL69 [14].

*Secondary antibodies*: HRP (horseradish peroxidase)-coupled or FITC/Cy3-conjugated goat anti-rabbit and anti-mouse IgG (H+L), respectively (Dianova).

### 2.5 Kits and Ready-to-Use Products

CAT-ELISA (Roche).

Dynabeads-ProteinA-Sepharose (Invitrogen).

DynaMag-2magnet (Invitrogen).

Lipofectamine 2000 (Invitrogen).

RNase-free DNase I (Roche).

RNase-free Water (Invitrogen).

RNasin (Roche).

Transcriptor One-Step RT-PCR system (Roche).

TRIZOL (Invitrogen).

X-tremeGENE HP (Roche).

## 3 Methods

Carry out all procedures at room temperature unless otherwise specified.

### 3.1 Indirect Immunofluorescence Analysis

One defining feature of eukaryotic cells is their compartmentalization into the nucleus and the cytoplasm, which from a molecular standpoint provides a number of different opportunities for the regulation of protein expression. To gain insight into the function of a particular protein, it is important to first determine its subcellular localization, and this can be easily achieved by performing indirect immunofluorescence experiments.

1. One day prior to transfection, seed $3.0 \times 10^5$ HeLa (alternatively use $3.0 \times 10^5$ HFF for infection studies) cells on cover slips in 2 ml medium (MEM, 5 % (v/v) FCS, 350 μg/ml L-glutamine, and 10 μg/ml gentamicin) per six well.
2. For ectopic expression analysis, transfect 2 μg of a plasmid encoding your protein of interest by either using standard calcium phosphate precipitation methods or liposome-based reagents like Lipofectamine 2000 or X-tremeGENE HP according to the manufacturer's protocols.
3. Approximately 10–15 h post-transfection, wash the cells with PBSo and add fresh medium (*see* **Notes 1** and **3**).
4. Two days post-transfection, wash cells on cover slips twice with PBSo.
5. Next, fix the cells by adding 1 ml of 4 % paraformaldehyde and incubate for 10 min at room temperature (*see* **Note 2**).
6. Wash cells twice with PBSo.

7. Permeabilize the cells with 0.2 % Triton X-100 in PBSo for 20 min at 4 °C.
8. Wash cells three times with PBSo over a total time period of 5 min.
9. Incubate the cells with the appropriate primary antibody directed against your protein of interest diluted in PBSo/1 % FCS for 30–45 min at 37 °C (*see* **Note 4**).
10. Wash cells twice with PBSo to remove unbound primary antibody.
11. Incubate the cells with the respective fluorescence-conjugated secondary antibodies (e.g., FITC or Cy3) (*see* **Note 4**).
12. Mount the cells by using DAPI (4′,6-diamidino-2-phenylindole)-containing VECTASHIELD mounting medium for staining of genomic DNA to visualize cell nuclei.
13. Analyze the subcellular localization of the desired protein using a fluorescence microscope (e.g., confocal laser scanning microscope) that is equipped with the appropriate filter sets necessary for the detection of fluorescent dyes given by the secondary antibodies or fluorophores like GFP coupled to your protein of interest (*see* Subheading 3.2).
14. For further processing of images, Adobe Photoshop can be used, although one should carefully follow approved journal parameters for any changes to collected images.

### 3.2 Mapping of Nuclear Localization Signals by Simultaneous Fusion to Green Fluorescent Protein and β-Galactosidase

Having determined that the protein of interest is predominantly localized to the cell nucleus as determined by indirect immunofluorescence analysis (*see* Subheading 3.1), the identification of a nuclear localization signal (NLS) within its primary amino acid sequence can provide important additional evidence supporting the ability of a protein to exhibit nuclear patterning. To this end, we highly recommend the reporter constructs initially described by Sorg and Stamminger [12]. These vectors allow for the expression of proteins that are simultaneously fused to green fluorescent protein (GFP) and to β-galactosidase (β-Gal) (*see* Fig. 1). The GFP part of the fused protein encoded by these plasmids functions as a fluorescent tag, whereas the β-Gal part of the fused protein increases the molecular weight of the fusion protein, thus preventing that proteins <40 kDa passively diffuse into the nucleus. As demonstrated previously, this assay can be used for the unambiguous identification of the NLS activity of a protein domain due to the strictly cytoplasmic staining of a GFP–β-Gal fusion without NLS [15].

1. Perform experiment essentially as described in Subheading 3.1 by transfecting your protein of interest fused in-frame into either of the GFP–β-Gal reporter plasmids (*see* Fig. 1).

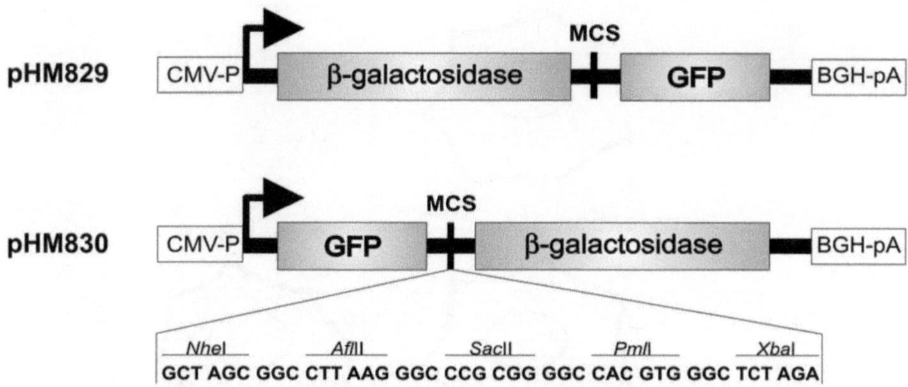


**Fig. 1** Schematic representation of the NLS-mapping vectors pHM829 and pHM830. *Inserts* can be cloned in-frame into the multiple cloning site (MCS) of pHM829 to generate proteins that are simultaneously fused to the C-terminus of β-Gal and the N-terminus of GFP. When pHM830 is used for cloning, fusion proteins contain an N-terminal GFP and a C-terminal β-Gal domain [12]. *CMV-P* CMV-promoter, *BGH-pA* BGH-polyA signal

Localization of your protein is determined by examining for appropriate GFP fluorescence or β-Gal expression using an anti-β-galactosidase antibody.

### 3.3 Interspecies Heterokaryon Assay

An interspecies heterokaryon assay is used to analyze whether a nuclear protein possesses the capacity for nucleocytoplasmic shuttling within a heterokaryon (cell with multiple nuclei formed during fusion of distinct cells). The experiment is performed as initially described by Pinol-Roma and colleagues [16] and relies on the ability of a shuttling protein to translocate from the nucleus of the human cell into the nucleus of the murine cell within one heterokaryon (*see* Fig. 2a). Each experiment should include the internal control plasmids pCFNrev-βGal and pCFN-βGal, respectively (*see* Fig. 2b, [8]). The positive control pCFNrev-βGal encodes β-Gal fused to the NLS of the SV40 T antigen and the nuclear export signal (NES) of the HIV-1 Rev protein, which enables the fusion protein to shuttle between the nucleus and the cytoplasm of transfected cells. The pCFN-βGal plasmid, in contrast, lacks the NES of HIV-1 Rev and is therefore retained in the nucleus, hence serving as the internal shuttling-negative control. Alternatively, HCMV pUL69 can likewise be used as a shuttling-positive control. However, in contrast to the classical leucine-rich NES of Rev, which targets the export receptor CRM1 (exportin1) and can specifically be blocked by the antibiotic LMB, HCMV pUL69 export is CRM1-independent as it cannot be blocked by LMB [6].

1. For each experimental condition to be tested, plate $2.8 \times 10^5$ HeLa cells in duplicate in 2 ml medium (MEM, 5 % (v/v) FCS, 350 μg/ml L-glutamine, and 10 μg/ml gentamicin) per well of a six-well dish.

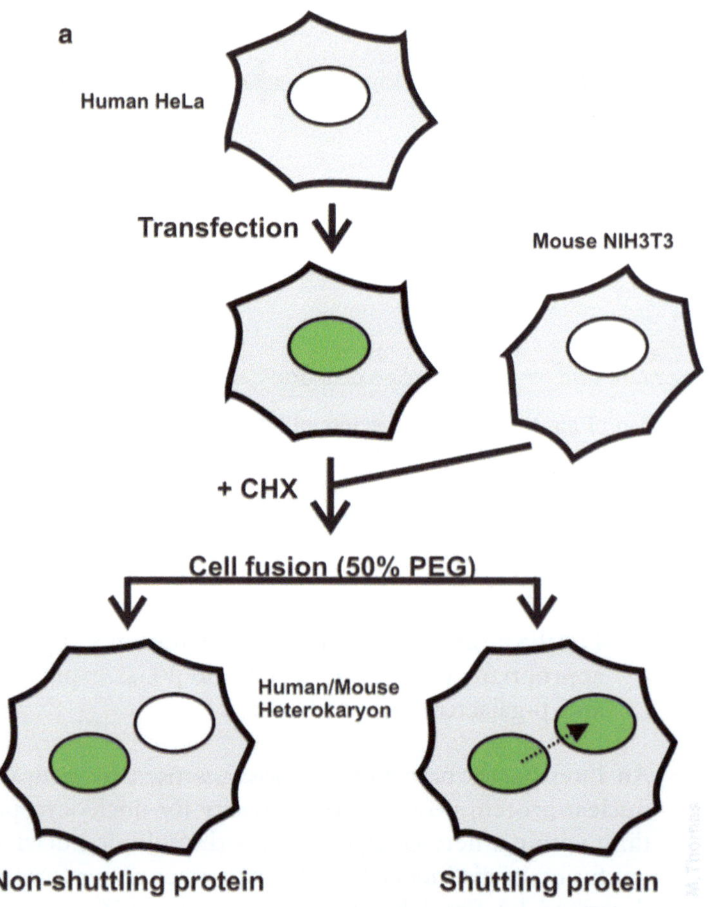


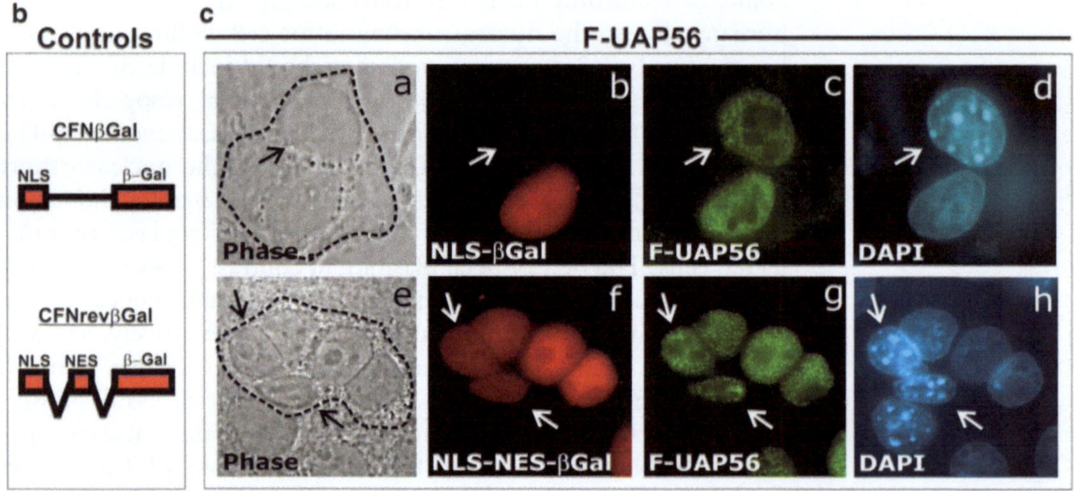


**Fig. 2** Interspecies heterokaryon assay. (**a**) Flow chart of a heterokaryon experiment *as described in* Subheading 3.3. (**b**) Internal controls used in heterokaryon experiments: the shuttle-positive pCFNrev-βGal and the shuttle-negative pCFN-βGal plasmid. (**c**) Nucleocytoplasmic shuttling of UAP56. HeLa cells were cotransfected with expression plasmids for either of the internal control plasmids (as indicated in **b**) (*a* to *d*+ CFN-βGal; *e* to *f*+ CFNrev-βGal) and FLAG-UAP56. The transfected cells were cocultivated with murine NIH3T3 cells

2. Transfect 2 μg of DNA (including shuttling positive and negative controls) by using the liposome-based reagent Lipofectamine 2000 according to the manufacturer's protocol.
3. The next day, wash the transfected HeLa cells with PBSo and trypsinize them by using trypsin/EDTA. After extensive resuspension of the HeLa cells, mix them thoroughly with $6.6 \times 10^5$ NIH3T3 cells. Thereafter, seed the co-culture of transfected HeLa and NIH3T3 cells on glass cover slips in six-well dishes containing 2 ml of medium (DMEM, 8 % (v/v) FCS, 350 μg/ml L-glutamine, and 10 μg/ml gentamicin), and incubate the cell mixture overnight at 37 °C.
4. Block protein synthesis 30 min prior to the fusion step (**step 5**) and throughout the experiment by adding 50 μg/ml CHX diluted in DMEM without supplements to the medium (*see* **Note 5**).
5. For fusion, wash cells once with PBSo and form heterokaryons by incubating the cover slips for 2 min with 200 μl pre-warmed 50 % PEG diluted in DMEM without supplements (*see* **Note 6**). For this, take the glass cover slip out of the medium, invert it, and put it on the PEG drop.
6. After exactly 2 min, immediately remove the cover slip, invert and return to the six-well dish.
7. Wash the cover slips extensively four to six times with pre-warmed PBSo (*see* **Note 7**).
8. Incubate cells in fresh medium containing 50 μg/ml CHX, and allow proteins to shuttle for 1–4 h at 37 °C in a cell incubator (*see* **Note 8**).
9. Thereafter, fix the cells with 4 % paraformaldehyde, and subject the heterokaryons to standard indirect immunofluorescence analysis as described above (*see* Subheading 3.1, **steps 6–14**).
10. For each experiment, analyze at least 20 heterokaryons. Nuclear shuttling should only be scored positive when a minimum of 80 % of the heterokaryons show a localization of the investigated protein within the murine nucleus. Human and murine nuclei can easily be discriminated by DAPI or Hoechst 33258 staining since murine nuclei display a punctate chromatin pattern compared to HeLa nuclei which are more diffusely stained.

**Fig. 2** (continued) and fused with PEG as described in Subheading 3.3 (compare **a**). Heterokaryon formation was controlled by light microscopy (panels *a* and *e*: Phase, phase contrast image of the heterokaryon; *arrows* indicate the murine nuclei). In order to detect and differentiate between the respective ectopically expressed proteins, double immunofluorescence analysis was performed using monoclonal anti-β-Gal antibody (*b* and *f*) and polyclonal anti-FLAG serum (*c* and *g*). DAPI counterstaining of the nuclei (*d* and *h*) revealed that FLAG-UAP56 was able to target the murine nuclei (panels *c* and *g*) as it was demonstrated for the shuttling-positive control NLS-NES-βGal (panel *f*), while the shuttle-negative control NLS-βGal was detected exclusively in the human nuclei (panel *b*) [21]

### 3.4 CAT mRNA Export Reporter Assay

Eukaryotic gene expression is tightly regulated and can be stimulated by *trans*-acting cellular or viral proteins, the so-called transactivator proteins. The transactivation activity of a protein can be investigated by the fact that the respective protein enhances the expression of reporter genes like luciferase or chloramphenicol acetyltransferase (CAT). Because proteins encoded by these reporter constructs catalyze the enzymatic conversion of their appropriate substrate, reporter gene expression can be easily detected and quantified by light-sensitive instruments like a microscope or a luminometer.

Here, we describe a method that can be used to analyze the transactivation capacity of a protein specifically on the posttranscriptional level, i.e., mRNA export. This assay, initially developed by Hope and colleagues [17], is based on the observation that mRNAs containing an intron are generally retained within the nucleus until splicing is completed. This technique utilizes a reporter construct (pDM128/CMV/RRE or pDM128/CMV/RxRE) containing a CAT gene within the *tat–rev* intron of the HIV-1 *env* gene driven by an HCMV promoter (*see* Fig. 3a). In addition to the CAT gene, both reporter encode for either of the *cis*-active RNA sequences: the HIV-1 Rev or HTLV-1 Rex-responsive element (RRE or RxRE), respectively, also contained within the intron of the construct. Since the *tat–rev* intron is inefficiently spliced due to suboptimal HIV-1 splice sites, unspliced CAT mRNAs are retained within the nucleus, and thus only marginal CAT protein levels are detectable. However, cotransfection of pDM128/CMV/RRE or pDM128/CMV/RxRE with plasmids expressing proteins like the well-characterized mRNA export factors HIV-1 Rev (binds to the Rev-responsive element RRE contained within the intron sequence of pDM128/CMV/RRE) or HTLV Rex (binds to the Rex-responsive element RxRE contained within the intron sequence of pDM128/CMV/RxRE) results in the nuclear export of unspliced pre-mRNAs, leading to an increase in CAT protein expression (for pDM128/CMV/RRE and HIV-1 Rev, *see* Fig. 3b, data not shown for pDM128/CMV/RxRE and HTLV Rex). In this context, it is of note that the HCMV-encoded mRNA export factor pUL69 facilitates the cytoplasmic accumulation of both reporter constructs, thus serving as an alternative positive control (for pDM128/CMV/RRE and pUL69, *see* Fig. 3b). However, in contrast to the mRNA export factors Rev or Rex, the pUL69-mediated mRNA export is CRM1-independent and therefore cannot be blocked by the antibiotic LMB [7].

**Fig. 3** (continued) CAT protein are given as percentages relative to that of cotransfected HIV-1 Rev. Standard deviations from at least three experiments are shown as *bars*. (**c**) Western blot analyses of lysates used for CAT mRNA export assays (**b**) were conducted to ensure that the effector proteins were expressed in comparable amounts by using an anti-Myc antibody (αMyc, *upper panel*). Staining of the cellular housekeeping gene β-actin served as loading control (α-actin, *lower panel*). In conclusion, while HCMV pUL69 and HIV-1 Rev facilitate the cytoplasmic accumulation of CAT mRNA, murine pM69 fails to do so [7, 22]

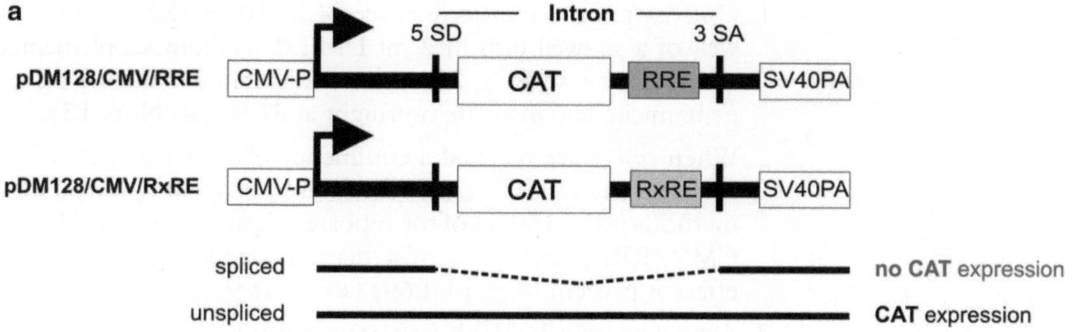


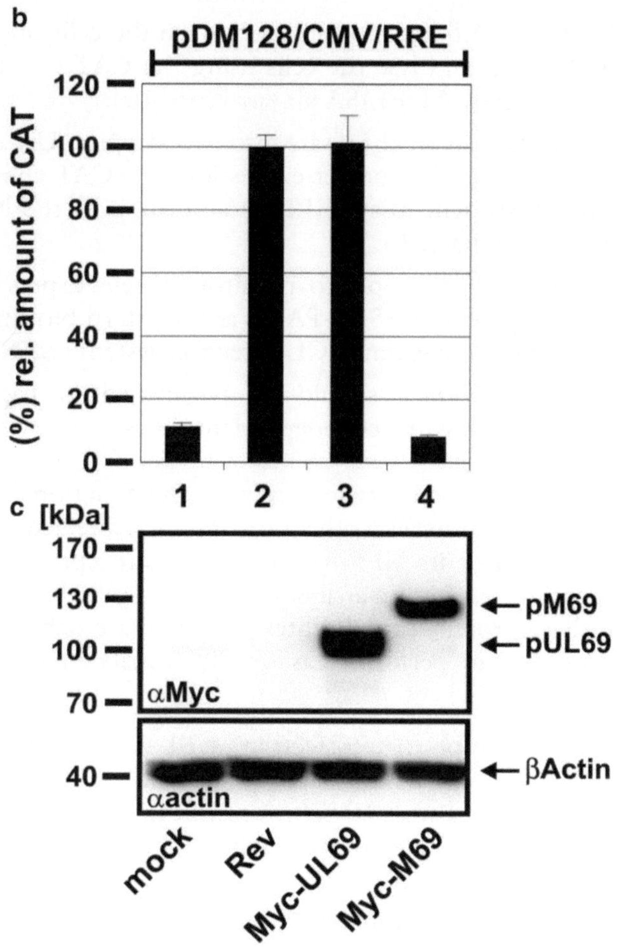


**Fig. 3** CAT mRNA export assay. (**a**) Schematic representation of the mRNA export reporter pDM128/CMV/RRE and pDM128/CMV/RxRE, respectively, *as described in* Subheading 3.4. *CAT* chloramphenicol acetyltransferase gene, *CMV-P* CMV-promoter, *SD* and *SA* splice donor and acceptor sites, *SV40PA* SV40 late poly(A) signal, *RRE* HIV-1 Rev-responsive element or *RxRE*, HTLV-1 Rex-responsive element. (**b**) CAT mRNA export assay of HCMV pUL69 and MCMV pM69 that was performed as described in Subheading 3.4. The diagram shows the relative amounts of CAT protein measured 36 h post-transfection of HEK293T cells with a mixture of the reporter pDM128/CMV/RRE and plasmids expressing either Myc-pUL69 (*lane 3*) or Myc-pM69 (*lane 4*). HIV-1 Rev served as a positive control (*lane 2*) and pcDNA3 as a negative control (mock, *lane 1*). The relative amounts of

1. One day prior to transfection, seed $4.8 \times 10^5$ HEK293T into one well of a six-well dish in 2 ml DMEM medium supplemented with 10 % (v/v) FCS, 350 μg/ml L-glutamine, and 10 μg/ml gentamicin, and incubate overnight at 37 °C (*see* **Note 13**).
2. When cells have reached a confluency of approximately 80 %, transfect the cells by standard calcium phosphate precipitation methods using 150 ng of the reporter plasmid (e.g., pDM128/CMV/RRE) and 2 μg of a plasmid encoding the putative effector protein (e.g., pUL69) (*see* **Note 9**).
3. Approximately 10–15 h post-transfection, wash the cells with PBSo and add fresh medium (*see* **Notes 1** and **3**).
4. At about 36 h post-transfection, wash the cells once with ice-cold PBSo, and lyse the cells using the CAT lysis buffer supplied in the CAT-ELISA kit (*see* **Note 10**).
5. Thereafter, divide the lysate into two samples. Use one sample to quantify CAT-reporter expression by a CAT enzyme-linked immunosorbent assay (ELISA) according to the instructions of the manufacturer.
6. Use the other sample to monitor protein expression of the effector protein by SDS-PAGE and Western blotting by using the chemiluminescent ECL detection system (*see* **Note 11**).
7. Each transfection should be performed in triplicate and repeated at least three independent times.

### *3.5 Coimmunoprecipitation Assay (CoIP)*

The aforementioned methods rely on the functional interaction of the transactivators with other proteins (e.g., interaction of its NLS with importin or its NES with the export receptor) (*see* Fig. 4). Coimmunoprecipitation analyses represent a valuable tool to investigate these protein–protein interactions on a biochemical level in cells, and they are performed as initially described by Bannister and Kouzarides [18].

1. One day prior to transfection, seed $4.8 \times 10^5$ HEK293T in duplicate wells of a six-well plate in 2 ml DMEM medium supplemented with 10 % (v/v) FCS, 350 μg/ml L-glutamine, and 10 μg/ml gentamicin, and incubate overnight at 37 °C (*see* **Notes 12** and **13**).
2. When cells have reached a confluency of approximately 80 %, transfect the cells by standard calcium phosphate precipitation methods using 2 μg of each plasmid for ectopic expression of your protein(s) of interest.
3. Approximately 10–15 h post-transfection, wash the cells with PBSo and add fresh medium (*see* **Note 1**).
4. Two days post-transfection, wash the cells twice with ice-cold PBSo and harvest them by collecting the cells of two wells in 1 ml PBSo. During this harvesting step, cells are pooled from

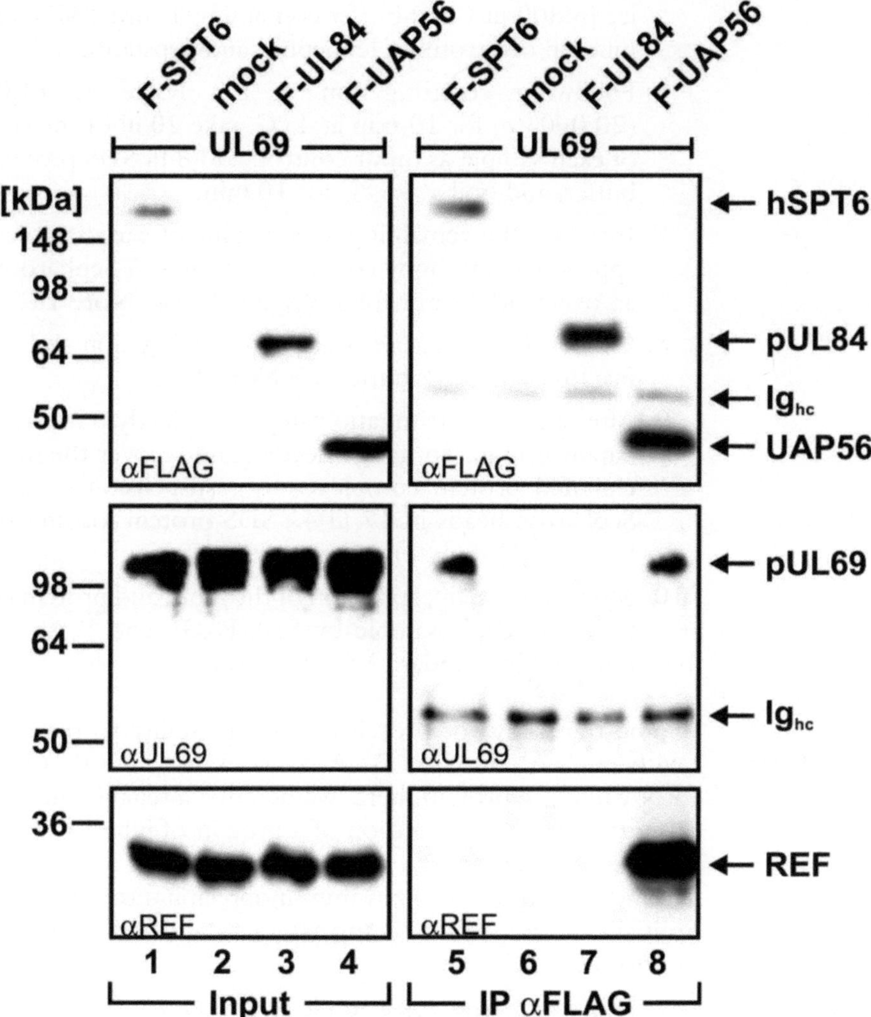


**Fig. 4** Coimmunoprecipitation analyses for the in vivo detection of the association of HCMV pUL69 and its cellular interaction partners hSPT6 and UAP56. Coimmunoprecipitation experiments were performed as described in Subheading 3.5. Briefly, HEK293T cells were cotransfected with UL69 (*lanes 1–8*) and FLAG-hSPT6 (*lanes 1* and *5*), FLAG-pUL84 (*lanes 3* and *7*), FLAG-UAP56 (*lanes 4* and *8*), or an empty vector (*lanes 2* and *6*). 36 h post-transfection, cells were lysed, and immunoprecipitations were performed, using anti-FLAG antibody (IP αFLAG *lanes 5–8*). The amount of protein in the input (*lanes 1–4*) as well as the coprecipitated samples (*lanes 5–8*) was visualized by Western blotting using anti-pUL69 (*upper panel*) or anti-REF (*lower panel*) antibodies. Immunoglobulin heavy chain (Ig$_{hc}$) served as an internal control for the presence of the precipitating antibody. This experiment demonstrates that HCMV pUL69 not only interacts with the cellular transcription elongation factor hSPT6 [4], but can also be found in complexes containing the cellular mRNA export factors UAP56 and REF [7]

duplicate (multiple) independent transfections. Via this procedure transfection rates and hence the amount of protein within the input are increased, and transfection variability is decreased.

5. Centrifuge for 5 min at 4,000 rpm (1,700 × *g*) and 4 °C, remove the supernatant, and lyse the cell-pellet for 20 min on

ice in 800 μl CoIP buffer containing 1 mM PMSF and 2 μg/ml each of aprotinin, leupeptin, and pepstatin.

6. Following centrifugation of the lysate at 14,000 rpm (20,000 × *g*) for 10 min at 4 °C, take 20 μl of the supernatant of each sample as input control, add 8 μl SDS protein loading buffer, and boil at 95 °C for 10 min.
7. Incubate the remaining supernatant of each sample with the appropriate antibody coupled to protein A Sepharose beads in an overhead rotator for 1.5 h at 4 °C (*see* **Note 14**).
8. Collect the Sepharose beads by centrifugation and wash them five times in CoIP buffer (*see* **Note 15**).
9. After a final centrifugation step, completely remove the supernatant using a 26-gauge needle, and recover the immunoprecipitated protein complexes by resuspension of the pelleted Sepharose beads in 17 μl 4× SDS protein loading buffer followed by boiling for 10 min at 95 °C.
10. Monitor protein expression of the input and protein complexes from the CoIP sample by SDS-PAGE and Western blotting followed by standard ECL reaction.

### 3.6 RNA-Immunoprecipitation Assay (RIP)

Some transactivator proteins have the ability to directly interact with nucleic acids like DNA or RNA. As HCMV pUL69 is an RNA-binding protein, here, we describe a reliable method to analyze the physical association of a protein of interest with RNAs in vivo and to identify the associated RNAs (Fig. 5). This technique, termed ribonucleoprotein-immunoprecipitation assay (RIP), was initially developed by Niranjanakumari and colleagues [19] and is performed as follows:

1. Plate $5.0 \times 10^6$ HEK293T cells per 10 cm dish in 10 ml medium (DMEM, 10 % (v/v) FCS, 350 μg/ml L-glutamine, and 10 μg/ml gentamicin), and incubate overnight at 37 °C (*see* **Notes 12** and **13**).
2. For ectopic expression analysis, transfect 10 μg of a plasmid encoding an RNA-binding protein of interest (e.g., FLAG-UL69 or mutants) and 1.5 μg of a reporter plasmid, coding for a putative target RNA (e.g., pDM128/CMV/RRE) using standard calcium phosphate precipitation protocols.
3. Approximately 10–15 h after transfection, wash the cells with PBSo and add fresh medium (*see* **Note 1**).
4. Two days post-transfection, wash cells twice with 10 ml ice-cold PBSo.
5. Harvest the cells by collecting them in 1 ml PBSo.
6. Centrifuge for 5 min at 4,000 rpm (1,700 × *g*) 4 °C, remove the supernatant, and lyse the cell-pellet for 20 min on ice in

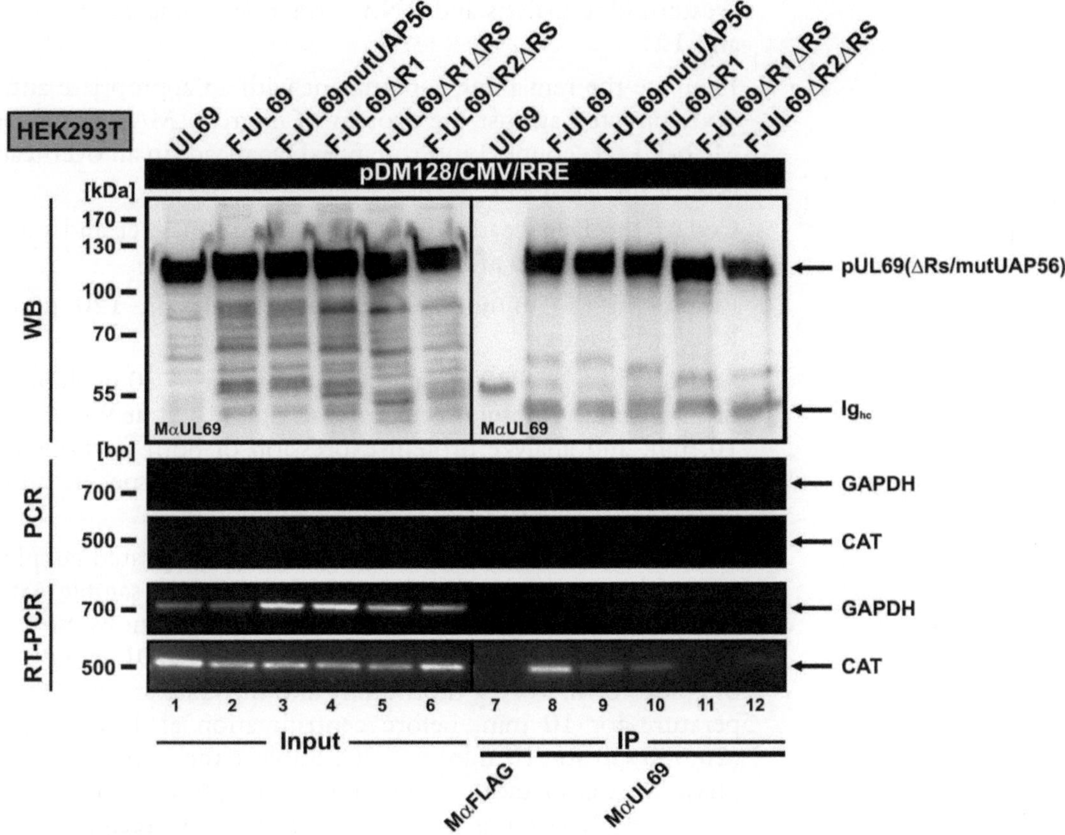


**Fig. 5** Ribonucleoprotein-immunoprecipitation analysis to investigate the association of pUL69 and mutants with CAT-reporter mRNA in vivo. Ribonucleoprotein-immunoprecipitation analyses were performed as described in Subheading 3.6. To this end, pUL69 and mutants were immunoprecipitated from HEK293T cells that were transfected with the CAT-reporter plasmid pDM128/CMV/RRE (*see* Fig. 3a) and pUL69 (*lanes 1* and *7*), FLAG-pUL69 (*lanes 2* and *8*), or any of the FLAG-tagged pUL69 mutants as indicated (*lanes 3–6* and *9–12*), followed by Western blotting using a monoclonal anti-pUL69 antibody (MαUL69, *upper panel*, *lanes 1–6*). Immunoprecipitation of pUL69 and mutants was performed using either an anti-FLAG antibody (MαFLAG, *lane 7* using MAb-FLAG) or the monoclonal anti-pUL69 antibody (MαUL69, *lanes 8–12* using Hybridoma MAb-UL69 69-66 [4]). RNAs were extracted prior to and after immunoprecipitation and subjected to RT-PCR using sequence-specific primers for the intronic CAT-reporter gene and GAPDH (RT-PCR, *lanes 1–12*). In parallel, PCRs were performed to exclude any DNA contamination in the RNA samples (PCR, *lanes 1–12*). RT-PCR and PCR products were monitored by standard agarose gel electrophoresis. This experiment confirms that HCMV pUL69 selectively binds to CAT mRNA but not to GAPDH, thereby implying selective mRNA binding of HCMV pUL69 in vivo ([5] and Zielke et al. in preparation). Moreover, these results indicate that both the RNA-binding and the UAP56-interaction capacity of pUL69 are essential for the association with the reporter mRNA in vivo

800 μl RIP buffer containing 1 mM PMSF and 2 μg/ml each of aprotinin, leupeptin, pepstatin and 20 units/ml of RNase-inhibitor RNasin.

7. After centrifugation at 14,000 rpm (20,000×*g*) for 10 min at 4 °C, take aliquots (120 μl) of each sample as input controls for

Western blot analysis and RNA extraction (compare **steps 11** and **12**).

8. Incubate the remaining supernatant with an appropriate antibody directed against the protein of interest (MAb-FLAG or MAb-UL69) coupled to Protein A-Dynabeads in an overhead rotator for 1.5 h at 4 °C.
9. Collect the Dynabeads using a DynaMag-2magnet and wash five times with 500 μl of RIP buffer (*see* **Note 15**).
10. Resuspend the immunoprecipitated samples in 120 μl of RNase-free water.
11. Add 10 μl of 2× SDS protein loading buffer to 20 μl aliquots of input and immunoprecipitated samples, boil at 95 °C for 10 min, and analyze protein expression of both fractions by SDS-PAGE and Western blotting using target-specific antibodies directed against the proteins of interest.
12. Isolate RNA from both input and immunoprecipitated samples by incubation of the remaining 100 μl of each sample with TRIZOL, according to the manufacturer's instructions. Briefly, mix 100 μl of each sample with 300 μl of TRIZOL and 80 μl of chloroform. Vortex thoroughly and incubate at room temperature for 10 min, before centrifugation at 14,000 rpm (20,000 × *g*) for 10 min at 4 °C. Subject the upper aqueous phase to isopropanol precipitation in the presence of 1 μl of 20 μg/μl glycogen as a carrier. Collect RNA precipitates by centrifugation at 14,000 rpm (20,000 × *g*) for 10 min, wash with 70 % ethanol, air-dry, and resuspend in 50 μl of RNase-free water. In addition, to exclude any DNA contamination within RNA precipitates, treat each sample with 20 units of recombinant RNase-free DNase I for 1 h at 37 °C.
13. Reverse transcribe the RNAs from input and immunoprecipitated samples using the Transcriptor One-Step RT-PCR system according to the manufacturer's instruction. Briefly, use 5 μl of template RNA and target-specific primers to amplify known or suspected RNA targets (e.g., the CAT-reporter RNA) and a non-associated reference RNA (e.g., GAPDH or other abundant cellular housekeeping genes). In addition, to assess possible DNA contamination of RNA samples, perform PCR reactions in parallel by skipping reverse transcription (*see* **Note 16**).
14. Analyze the RT-PCR products by agarose gel electrophoresis for semiquantitative determination of the mRNA levels prior and after RNA-immunoprecipitation (*see* **Note 17**).

## 4 Notes

1. When standard calcium phosphate precipitation methods are used for transfection of cells, replace medium within 12–15 h after transfection. Avoid longer incubation times as calcium phosphate precipitates can be toxic to cells.
2. Alternatively, cells can be fixed and permeabilized by incubation with ice-cold methanol for 5 min. After removal of the methanol, let the cells air-dry until the methanol is completely evaporated before continuing Subheading 3.1, **step 8**. Note that methanol fixation cannot be used when localization of GFP-tagged fusion proteins is to be determined, as alcohol fixation rapidly destroys GFP fluorescence.
3. To determine the impact of inhibitors (e.g., protein kinase inhibitors) on protein localization/activity, remove the entire medium, and feed cells with fresh medium containing the particular drug (e.g., 10 μM roscovitine for inhibition of cyclin-dependent kinase activity or 2 μM Gö6976 to block pUL97 (and PKC) activity, respectively [20]) starting the day after transfection and maintaining this medium throughout the experiment until harvesting.
4. It is important to plan your experiment thoroughly in advance. For co-staining experiments, it is necessary to use primary antibodies from different species and secondary antibodies with clearly distinguishable emission maxima in order to allow for a discrete staining of your respective protein.
5. If inhibition of CRM1-dependent protein export has to be achieved, treat the cells additionally with 2.5 ng/ml LMB, 3 h before cell fusion and throughout the whole experiment.
6. Prepare 50 % PEG solution as follows: weigh 10 g PEG into a glass vessel and autoclave the dry powder. Then add 10 ml of DMEM without supplements, and let the PEG dissolve in a 65 °C water bath for several hours. Once dissolved, the PEG solution can be stored at room temperature for about 4 weeks.
7. Do not exceed 2 min PEG incubation as this leads to the formation of artificially large heterokaryons which can impede the detection of your protein.
8. As soon as heterokaryons are formed, proteins start to shuttle. However, they also can be targeted for degradation and might therefore be hard to detect as de novo protein synthesis is blocked by CHX. Thus, shuttling time is highly dependent on the stability of the protein of interest and has to be determined empirically for each individual protein.

9. When performing transactivation assays using luciferase or CAT reporters, it is absolutely necessary to guarantee that each sample is transfected with the same amount of DNA. Therefore, differences in DNA quantities have to be adjusted with empty vector DNA in order to exclude secondary effects due to the amount of transfected DNA.
10. If CRM1-dependent export has to be blocked, treat the cells additionally with 2.5 ng/ml leptomycin B at 6 h before harvesting the cells.
11. In order to make an unambiguous conclusion on the transactivation capacity of your protein of interest, it is absolutely essential to monitor the expression levels of effector proteins by Western blotting. Equal effector protein levels can be achieved by adjusting the amount of transfected DNA based on protein levels as determined by immunoblotting.
12. The indicated number of cells is a recommended amount based on our previous experience. If protein expression is too low (as is often the case for endogenous proteins), we recommend to increase the number of cells that are used in each experiment.
13. Because proper protein expression is required for CAT mRNA export assays and CoIP and RIP experiments, we do highly recommend HEK293T cells, but you can also perform these experiments with HeLa cells. In this case seed $4.0 \times 10^5$ HeLa cells per well of a six-well dish in 2 ml MEM medium supplemented with 5 % (v/v) FCS, 350 μg/ml L-glutamine, and 10 μg/ml gentamicin, and continue the method as described. To study protein–protein or protein–RNA interactions in infected cells, seed $3.0 \times 10^6$ HFF cells per 10 cm dish in 10 ml MEM medium supplemented with 5 % (v/v) FCS, 350 μg/ml L-glutamine, and 10 μg/ml gentamicin. One day later, infect the cells with the respective virus (-mutant) and the desired multiplicity of infection, and perform the experiment at those timepoints postinfection when your protein of interest is synthesized to appropriate amounts and continue the method at the washing step of Subheadings 3.5 and 3.6.
14. Alternatively, use Protein A-Dynabeads according to the instructions of the manufacturer.
15. To avoid coprecipitation of nonspecific proteins that can occur for a number of reasons, one can increase the number of washing steps. If this does not solve the problem, one can increase the salt concentration within the lysis buffer stepwise until specificity is achieved, as controlled by analysis of noninteracting proteins.
16. To ensure that the amplification reaction is within the linear range, either dilute the RNA sample or decrease the number of the PCR cycles within the RT-PCR.

17. Alternatively, quantification of the RNA levels can be achieved by quantitative RT-PCR (e.g., *SYBR-Green* or *TaqMan*) using target-specific oligonucleotides for amplification. To normalize the expression level of individual mRNAs, quantification of cellular housekeeping genes (e.g., GAPDH or RPL4) needs to be determined in parallel in order to compare the respective mRNA expression levels.

## Acknowledgments

The authors thank past and present members of the laboratory whose work has contributed to studies that are described in the review. Research in the Stamminger laboratory was supported by the DFG (SFB796), the IZKF Erlangen, and the graduate school BIGSS.

### References

1. Sandri-Goldin RM (2001) Nuclear export of herpes virus RNA. Curr Top Microbiol Immunol 259:2–23
2. Toth Z, Stamminger T (2008) The human cytomegalovirus regulatory protein UL69 and its effect on mRNA export. Front Biosci 13:2939–2949
3. Sandri-Goldin RM (2008) The many roles of the regulatory protein ICP27 during herpes simplex virus infection. Front Biosci 13:5241–5256
4. Winkler M, aus Dem ST, Stamminger T (2000) Functional interaction between pleiotropic transactivator pUL69 of human cytomegalovirus and the human homolog of yeast chromatin regulatory protein SPT6. J Virol 74:8053–8064
5. Toth Z, Lischka P, Stamminger T (2006) RNA-binding of the human cytomegalovirus transactivator protein UL69, mediated by arginine-rich motifs, is not required for nuclear export of unspliced RNA. Nucleic Acids Res 34:1237–1249
6. Lischka P, Rosorius O, Trommer E, Stamminger T (2001) A novel transferable nuclear export signal mediates CRM1-independent nucleocytoplasmic shuttling of the human cytomegalovirus transactivator protein pUL69. EMBO J 20:7271–7283
7. Lischka P, Toth Z, Thomas M, Mueller R, Stamminger T (2006) The UL69 transactivator protein of human cytomegalovirus interacts with DEXD/H-Box RNA helicase UAP56 to promote cytoplasmic accumulation of unspliced RNA. Mol Cell Biol 26:1631–1643
8. Roth J, Dobbelstein M (1997) Export of hepatitis B virus RNA on a Rev-like pathway: inhibition by the regenerating liver inhibitory factor IkappaB alpha. J Virol 71:8933–8939
9. Malim MH, Hauber J, Le SY, Maizel JV, Cullen BR (1989) The HIV-1 rev trans-activator acts through a structured target sequence to activate nuclear export of unspliced viral mRNA. Nature 338:254–257
10. Rimsky L, Hauber J, Dukovich M, Malim MH, Langlois A, Cullen BR, Greene WC (1988) Functional replacement of the HIV-1 rev protein by the HTLV-1 rex protein. Nature 335:738–740
11. Farjot G, Buisson M, Duc DM, Gazzolo L, Sergeant A, Mikaelian I (2000) Epstein-barr virus EB2 protein exports unspliced RNA via a crm-1-independent pathway. J Virol 74:6068–6076
12. Sorg G, Stamminger T (1999) Mapping of nuclear localization signals by simultaneous fusion to green fluorescent protein and to beta-galactosidase. Biotechniques 26:858–862
13. Lischka P, Thomas M, Toth Z, Mueller R, Stamminger T (2007) Multimerization of human cytomegalovirus regulatory protein UL69 via a domain that is conserved within its herpesvirus homologues. J Gen Virol 88:405–410
14. Winkler M, Rice SA, Stamminger T (1994) UL69 of human cytomegalovirus, an open reading frame with homology to ICP27 of herpes simplex virus, encodes a transactivator of gene expression. J Virol 68:3943–3954
15. Lischka P, Sorg G, Kann M, Winkler M, Stamminger T (2003) A nonconventional nuclear localization signal within the UL84 protein of human cytomegalovirus mediates

nuclear import via the importin alpha/beta pathway. J Virol 77:3734–3748

16. Pinol-Roma S, Dreyfuss G (1992) Shuttling of pre-mRNA binding proteins between nucleus and cytoplasm. Nature 355:730–732
17. Hope TJ, Huang XJ, McDonald D, Parslow TG (1990) Steroid-receptor fusion of the human immunodeficiency virus type 1 Rev transactivator: mapping cryptic functions of the arginine-rich motif. Proc Natl Acad Sci U S A 87:7787–7791
18. Bannister AJ, Kouzarides T (1996) The CBP co-activator is a histone acetyltransferase. Nature 384:641–643
19. Niranjanakumari S, Lasda E, Brazas R, Garcia-Blanco MA (2002) Reversible cross-linking combined with immunoprecipitation to study RNA-protein interactions in vivo. Methods 26:182–190
20. Thomas M, Rechter S, Milbradt J, Auerochs S, Muller R, Stamminger T, Marschall M (2009) Cytomegaloviral protein kinase pUL97 interacts with the nuclear mRNA export factor pUL69 to modulate its intranuclear localization and activity. J Gen Virol 90:567–578
21. Thomas M, Lischka P, Muller R, Stamminger T (2011) The cellular DExD/H-box RNA-helicases UAP56 and URH49 exhibit a CRM1-independent nucleocytoplasmic shuttling activity. PLoS One 6:e22671
22. Zielke B, Thomas M, Giede-Jeppe A, Muller R, Stamminger T (2011) Characterization of the betaherpesviral pUL69 protein family reveals binding of the cellular mRNA export factor UAP56 as a prerequisite for stimulation of nuclear mRNA export and for efficient viral replication. J Virol 85:1804–1819

# Chapter 13

# Fluorescence-Based Laser Capture Microscopy Technology Facilitates Identification of Critical In Vivo Cytomegalovirus Transcriptional Programs

**Craig N. Kreklywich, Patricia P. Smith, Carmen Baca Jones, Anda Cornea, Susan L. Orloff, and Daniel N. Streblow**

## Abstract

Cytomegalovirus gene expression in highly permissive, cultured fibroblasts occurs in three kinetic classes known as immediate early, early, and late. Infection of these cells results in a predictable transcriptional program leading to high levels of virus production. Infection of other, so-called, nonpermissive cell types results in a transcriptional program that either fails to produce virus particles or production is substantially reduced compared to fibroblasts. We have found that CMV gene expression profiles in tissues from infected hosts differ greatly from those observed in infected tissue culture cells. The number of viral genes expressed in tissues is much more limited, and the number of highly active genes does not correlate with viral DNA load. Additionally, viral gene expression in vivo is tissue selective with no two tissues expressing the exact same viral gene profile. Thus, in vivo CMV gene expression appears to be governed by mechanisms that are still uncharacterized. Cytomegalovirus remains in a persistent phase for the lifetime of the host. During this phase only a limited number of host cells are infected, and it is very difficult to detect CMV gene expression in whole tissues without sub-fractionating infected vs. uninfected cells. Herein, we describe the development of a fluorescence-based laser capture microscopy technique coupled with small sample size microarray analysis to determine the viral gene expression in 50–100 infected cells isolated from frozen RCMV-infected tissue sections.

**Key words** Laser capture microscopy, Cytomegalovirus, Green fluorescence protein, Microarray analysis

## 1 Introduction

Cytomegaloviruses (CMVs) are ubiquitous β-herpesviruses that establish lifelong persistence following primary infection. In asymptomatic individuals, CMV persistence is typically maintained in a latent state, which is associated with a general lack of virus production and limited viral gene expression. However, CMV reactivation is considered to be the major source of virus in immunocompromised individuals, leading to significant disease in transplant patients or to congenital disease, such as deafness [1]. CMV

Andrew D. Yurochko and William E. Miller (eds.), *Human Cytomegaloviruses: Methods and Protocols*, Methods in Molecular Biology, vol. 1119, DOI 10.1007/978-1-62703-788-4_13, © Springer Science+Business Media New York 2014

is the largest human herpesvirus containing over 200 potential open reading frames, and many of these genes have been implicated in CMV disease [2, 3]. During productive infections, CMV gene transcription is controlled temporally giving three kinetic categories of viral proteins: immediate early (IE), early (E), and late (L). Expression of the IE genes does not require de novo protein synthesis, and these viral proteins are potent transactivators of both viral and cellular gene transcription. The E class of viral proteins functions in a number of different processes including cell cycle control, replication, and immune evasion. Expression of the E genes requires the synthesis of the IE proteins and is also dependent upon certain cellular factors. The L genes encode mostly structural proteins involved in virion assembly and egress, and expression of the L viral genes requires viral DNA production. Therefore, antiviral drugs that block viral DNA synthesis, such as ganciclovir, inhibit viral L gene expression but not the IE or E classes of viral genes. While much is known about viral gene transcription during lytic infections in vitro, very little is known about the specific gene transcription profiles that are associated with in vivo reactivation from latency or during persistence. It is now thought that CMV persistence may proceed via a nonclassical gene expression profile that involves E and/or L gene expression without IE.

Typically, CMV gene expression studies have been limited to the analysis of only a few viral genes of interest. However, since the adoption of microarray technology, several studies have reported the global transcription profiles associated with infection of cultured cells. Chambers et al. were the first to publish the viral transcriptional analysis from human-(H)CMV-infected human foreskin fibroblasts utilizing microarrays and were able to kinetically classify the HCMV AD169 transcriptome [4]. Goodrum et al. used microarrays to study HCMV acute infection and latency transcription programs in CD34+ cells infected in vitro. They discovered that the pattern of viral gene expression in this progenitor cell type was different than that found in fibroblasts [5–7]. In addition, the same group has reported that one of the genes expressed in the CD34+ stem cells, HCMV UL138, controls the ability of infected progenitor cells to become latently infected [8]. Using microarray technology, we have compared rat CMV (RCMV) gene expression profiles of infected cultured endothelial cells, fibroblasts, and smooth muscle cells to the viral transcriptomes detected in tissues derived from infected rats [9]. In cultured cells, RCMV expresses about 95 % of the known viral open reading frames (ORFs) with few differences between these three cell types. In contrast, we observed that RCMV gene expression in virus-infected rats is highly restricted with significantly lower expression of replication-essential viral genes ($1 \times 10^2$-copies/gram tissue) and high expression ($1 \times 10^6$-copies/gram tissue) of viral genes with unknown function or those with immune modulator phenotypes [9]. In addition, we have also observed that the RCMV transcription profiles are tissue specific. Recently, next-generation sequencing techniques have been employed to rapidly sequence whole CMV genomes, as well as to identify the complete HCMV transcriptome including the characterization of the viral noncoding RNAs [10, 11].

CMV infections persist for the life of the infected individual and are hypothesized to be involved in a number of chronic inflammatory diseases including atherosclerosis, graft rejection, and cancer [12–16]. However, the mechanisms and the viral genes involved in viral persistence and chronic disease are unknown. A major hurdle in determining which viral genes are expressed in tissues during persistence is the paucity of infected cells, making viral gene expression detection in whole tissue samples extremely difficult, to nearly impossible. Isolation of CMV-infected cells from whole tissues is the only way to increase the viral to cellular gene signal to noise ratio and to reliably determine viral gene expression. We have thus developed a method to isolate RCMV-infected cells from tissues utilizing a recombinant RCMV expressing green fluorescence protein (GFP) under the constitutive cellular EF1α promoter [17]. This virus produces high levels of GFP even during nonproductive infection conditions. Using fluorescence-based laser capture microscopy (LCM), we are able to isolate infected GFP+ cells from infected rat tissues, and this process has greatly enhanced our ability to detect viral gene expression. We have tested a number of different tissue fixation conditions in order to optimize GFP retention and RNA extraction, and these are described herein. We have also optimized the microarray conditions for these small-sized total RNA samples. To identify viral gene expression, we have employed both printed glass-slide and semiconductor microarrays containing oligonucleotide primers that represent the full complement of RCMV genes. While the techniques outlined here work well for isolating GFP+ cells from tissues, future directions include the characterization of HCMV gene expression in samples from human tissues using immunostaining techniques to identify and capture HCMV-infected cells.

## 2 Materials

### 2.1 Rat Cytomegalovirus Expressing GFP

1. Recombinant RCMV Maastricht strain expressing GFP (RCMV-GFP) was constructed using homologous recombination by replacing the RCMV ORFs r145, r146, and a portion of r147 with a GFP expression cassette under the constitutive cellular EF1α promoter [17] (*see* **Note 1**).

### 2.2 RCMV Purification and Titration Components

1. Sorbitol cushion underlay: 20 % D-sorbitol, 0.05-M Tris–HCl pH 7.4, and 1-mM $MgCl_2$. Chemicals are from Sigma. Filter the reagent to sterilize.
2. Titration overlay: Minimal Essential Medium (MEM; Life Technologies) supplemented with 10 % fetal bovine serum (FBS), nonessential amino acids, 100-units/ml penicillin, 100-mg/ml streptomycin (pen/strep), 20-mM L-glutamine, and 0.5 % (w/v) carboxymethyl cellulose (CMC). Store at 4 °C and preheat to 37 °C prior to use.
3. Carboxymethyl cellulose (CMC): weigh 0.75 g of low-viscosity CMC and 0.75 g of high-viscosity CMC from Sigma. To a 500-ml bottle, add CMC and 60 ml of PBS and 40 ml of $dH_2O$. Do not mix. Autoclave the liquid to solubilize CMC and sterilize. Store at room temperature. To each bottle, add 200 ml of DMEM medium containing pen/strep and 5 % fetal bovine serum.

### 2.3 Laser Capture Microscopy

1. Microscope slides used for laser capture microscopy were 1.4-μm Pet-membrane slides (MMI-membrane slides #50102).

### 2.4 RCMV-Specific Microarrays and Reagents

1. Glass RCMV microarray slides were printed at the Spotted Microarray Core (SMC) at the Vaccine and Gene Therapy Institute, Oregon Health and Sciences University. Recent studies have also utilized semiconductor microarray slides (4x2k custom microarray chips) purchased from Custom Array Inc. (Bothell, WA).
2. Spotted RCMV microarray slides contain two unique 70mer antisense oligos for each of the 159 predicted viral ORFs [4] and an additional 2,925 rat cellular genes. The oligos were chosen with a 3′ bias and blasted against the NCBI database for alignment to the RCMV Maastricht strain and for possible cross-hybridization to cellular sequences. Custom Array slides contain two unique 30–50mers antisense probes to each of the predicted RCMV ORFs. The Custom Array slides contain an additional 1,400 unique cellular genes.
3. Custom Array's Hybridization solution: 6× SSPE, 0.05 % Tween-20, 20-mM EDTA, 0.04 % SDS, and 0.1-μg/ml Salmon sperm DNA.
4. Prehybridization buffer: 6× SSPE, 0.05 % Tween-20, 20-mM EDTA, 5× Denhardt's solution, and 0.05 % SDS.
5. Eukaryotic Total RNA Pico chip and RNA Ladder: Agilent RNA 6000 Pico Kit #5067-1573.

## 3 Methods

### 3.1 Production of Rat Cytomegalovirus RCMV Expressing GFP

1. Infect rat lung fibroblasts (RFL-6) with RCMV-GFP at multiplicity of infection (MOI) equal to 0.5. Incubate cells at 37 °C until full cytopathic effect is observed.
2. Scrape infected cells, and then collect culture medium containing scraped cells into 250-ml bottles.
3. Centrifuge at 7,000 rpm (7,500 × *g*; Beckman JA-14 rotor) for 20 min to remove cell debris. Transfer cell-free supernatants to an additional clean tissue culture flask and set aside for further processing as described below in **step 5**.
4. Lyse the infected cell pellets by freeze/thaw a total of three times in the 250-ml bottles.
5. Pellet lysed cell debris at 7,000 rpm (7,500 × *g*; Beckman JA-14 rotor) for 20 min, and add the clarified supernatant to the culture supernatants collected above in **step 3**.
6. Transfer cell-free supernatants to SW28 centrifugation tubes containing a 5 ml sorbitol cushion underlay.
7. Centrifuge at 22,000 rpm (87,000 × *g*; Beckman ultracentrifuge; rotor SW28) for 1 h at 8 °C.
8. Resuspend the virus pellets in Minimal Essential Medium [18] culture media.
9. Store virus at −80 °C.

### 3.2 Titration of RCMV-GFP

1. Prepare triplicate tenfold serial dilutions of viral stocks in culture medium using at least two independent vials of stock virus.
2. Infect 24-well plates of RFL-6 rat fibroblasts with 0.2 ml of tenfold viral dilutions.
3. Incubate the infected cells at 37 °C for 3 h.
4. Overlay wells with titration overlay material 1-ml MEM supplemented with 10 % fetal bovine serum (FBS), nonessential amino acids, 100-units/ml penicillin, 100-mg/ml streptomycin (pen/strep), 20-mM L-glutamine, and 5 % carboxymethyl cellulose (CMC).
5. Incubate for 7 days at 37 °C.
6. Fix cells for 10 min at room temperature with 3.7 % formalin prepared in 1× PBS.
7. Wash fixed cells with $H_2O$.
8. Stain cells for 30 min with 0.05 % aqueous methylene blue (Sigma).
9. Decant stain and wash with $H_2O$.

10. Count virus plaques in triplicate wells.
11. Calculate the average virus titer as the number of plaque-forming units per ml (pfu/ml) taking into account the dilution factor.

### 3.3 Infection of Rats with RCMV-GFP

1. Irradiate Lewis rats (γ-irradiated with 600 Gy).
2. Infect irradiated rats with $5 \times 10^5$ pfu of RCMV-GFP within 24 h of irradiation treatment. Mock-infected animals served as negative controls (*see* **Note 2**).
3. Euthanize rats at 3, 5, 7, 10, and 21 days postinfection (dpi) via $CO_2$ asphyxiation. Harvest tissues (heart, liver, kidney, spleen, bone marrow, and salivary glands) and blood.
4. Place a small portion of each tissue into a Cryo-safe tube and snap freeze in liquid nitrogen. Store at −80 °C until used for nucleic acid analyses.
5. Embed a second portion of each tissue in optimal cutting temperature (OCT) medium and snap freeze and stored at −80 °C. This portion will be cut and utilized in microscopic analyses as described below (Subheading 3.6).

### 3.4 RCMV-GFP Infection of Cardiac Allograft Recipient Rats

We also infected a set of Lewis rats that received F344 rat donor heart allografts to study the effect that the immune environment present during allograft rejection plays on RCMV gene expression. Rat transplant operations were performed as previously described [19–23]. Described first is the donor operation (**steps 1**–7) and then transplant operation (**steps 8–12**).

1. Anesthetize the rat with rat cocktail (1 ml/kg, i.p.).
2. Open the abdomen midline.
3. Inject 100 units (0.3 ml) of aqueous heparin into the inferior vena cava.
4. Open the anterior chest wall via the sternum.
5. Place gauze containing iced saline on the heart.
6. Ligate the superior and inferior vena cava as well as the confluence of pulmonary veins and divide them.
7. Place heart in UW solution during recipient operation and euthanize the donor by exsanguination followed by bilateral thoracotomy.
8. Anesthetize the recipient with inhaled isoflurane mixed with oxygen using a anesthesia machine.
9. Open the abdomen at the midline.
10. Isolate the infrarenal aorta and inferior vena cava.
11. Sew donor heart to the recipient aorta and donor pulmonary artery to the recipient inferior vena cava.

12. Irrigate the abdomen with warm saline with 1-mg/ml gentamicin sulfate. Provide analgesics as necessary.
13. Treat rats with low-dose cyclosporine A (5 mg/kg/day) for 10 days following transplantation to prevent acute cellular rejection.
14. Infect rats with $5 \times 10^5$ pfu of RCMV-GFP within 24 h of transplantation. Mock-infected animals served as negative controls.
15. Euthanize rats at 7 and 21 days posttransplantation via $CO_2$ asphyxiation. Harvest tissues (graft and native hearts, liver, kidney, spleen, bone marrow, and salivary glands) and blood.
16. Place a small portion of each tissue into a Cryo-safe tube and snap freeze in liquid nitrogen. Store at −80 °C until used for nucleic acid analyses.
17. Embed a second portion of each tissue in optimal cutting temperature (OCT) medium and snap freeze and stored at −80 °C. This portion will be cut and utilized in microscopic analyses as described below (Subheading 3.6).

### 3.5 Isolation and Analysis of Rat PBMC and Bone Marrow Cells

At each of the time points described above in Subheading 3.2, whole blood and bone marrow cells were isolated and analyzed by flow cytometry for the presence of GFP positivity.

1. Flush rat femurs and tibias from the infected rats with a 23-gauge needle and syringe containing RPMI-1640 medium.
2. Strain bone marrow cells through a 70-μm filter.
3. Wash bone marrow cells twice with 50 ml of RPMI-1640 medium containing 10 % v/v fetal bovine serum, pen/strep, and 2-mM L-glutamine.
4. Isolate rat peripheral blood cells (PBMC) from 5 ml of non-clotted whole blood.
5. Overlay whole blood over 5 ml of Ficoll hypaque in a 15-ml conical tube.
6. Centrifuge at 2,000 rpm ($800 \times g$) in a Beckman tabletop centrifuge for 30 min.
7. Remove buffy coat and place cells into a fresh 50-ml conical.
8. Wash cells twice with complete RPMI-1640 medium.
9. Analyze BM cells and PBMC for the presence of GFP on a FACS Calibur (Becton Dickinson) utilizing Cell Quest software. Analyze data using FlowJo software (Tree Star, Inc.). As shown in Fig. 1, GFP expression is very high in the PBMC from RCMV-GFP-infected rats. The data shown in Table 1 represents an analysis of BM and PBMC cells isolated from three infected rats at 3, 5, 7, and 10 days postinfection. GFP+ cells are expressed as the percentage of total cells, which are

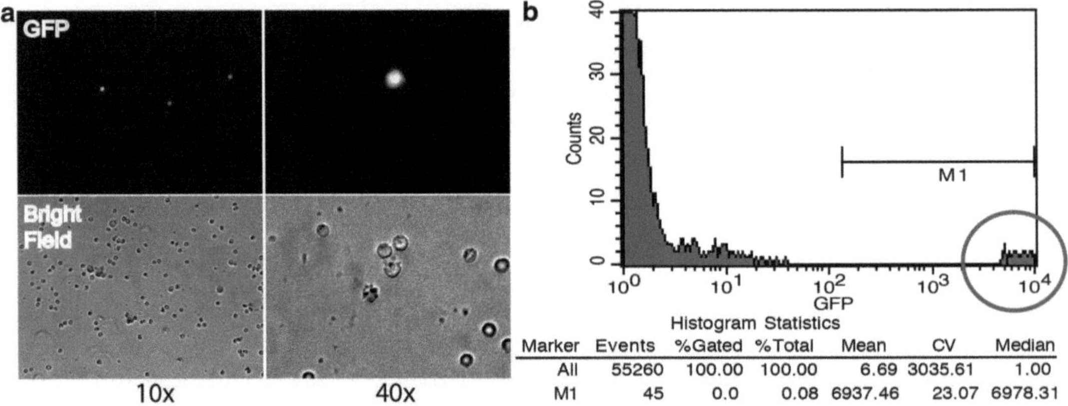


**Fig. 1** GFP detection in peripheral blood cells isolated from Lewis rats infected with RCMV-GFP at 10 dpi. (**a**) GFP expression in peripheral blood cells at 10× and 40× using fluorescence microscopy. (**b**) Total peripheral blood cells were analyzed using a BD FACScan flow cytometer. The number of RCMV-GFP+ peripheral blood cells was determined to be about 1 in 1,000 cells

**Table 1**
**Detection of RCMV-GFP in infected rat tissues[a]**

| Organ | Day 3 | Day 5 | Day 7 | Day 10 | Day 21 | Uninfected |
|---|---|---|---|---|---|---|
| Salivary glands | – | + | + | ++ | ++ | – |
| Heart | + | ± | ± | + | – | – |
| Lung | ++ | ++ | + | ++ | + | – |
| Spleen | ++ | ++ | ++ | ++ | – | – |
| Liver | ± | ++ | + | + | – | – |
| Kidney | ± | ++ | + | ± | – | – |
| PBMC[b] | 0.021 | 0.077 | 0.027 | 0 | 0 | 0 |
| BMC | 0.005 | 0.006 | 0 | 0 | 0 | 0 |

[a]Groupings: no GFP+ foci in more than five tissue sections (–); 1–5 GFP+ foci per five tissue sections (±); multiple GFP+ foci per tissue section (+); and multiple GFP+ foci per 20× microscope field (++). $n = 5$ rats

[b]GFP positivity of PBMC and BMC was determined by flow cytometry. Listed are positive cells as a percent of total cells

typically present at 3, 5, and 7 dpi but peak at 5 dpi. Accordingly, there is approximately 1 infected cell per 1,000 total PBMC and approximately 1 infected cell per 10,000 total bone marrow cells (Table 1). Thus, RCMV-GFP+ BM cells and PBMC can be sorted using a FACS Calibur flow cytometer making it possible to isolate the infected cell populations.

### 3.6 Cryosectioning of Frozen Rat Tissues

1. Cut 8-μm-thick tissue sections from the OCT-embedded RCMV-GFP-infected rat tissues using a Leica CM3050S Cryostat.

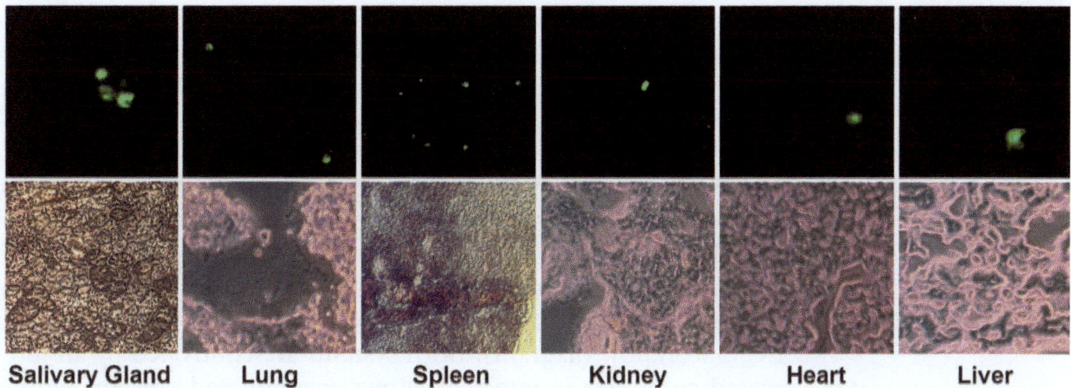


**Fig. 2** Fluorescence microscopic detection of RCMV-GFP-infected tissues at 7 dpi. Frozen tissues (salivary gland, lung, spleen, kidney, heart, and liver) were cryosectioned (8 μm) and mounted on microscope slides. The thin sections were viewed using a Deltavision Deconvolution microscope, and pictures were taken at 60× magnification. Fluorescence images are shown above their corresponding phase image

2. Mount cut tissue sections for immunofluorescence microscopy onto glass microscope slides.
3. Mount cut tissue sections for Laser Capture Microscopy (LCM) onto 1.4-μm Pet-membrane slides (MMI-membrane slides #50102).
4. Store tissue sections frozen at −80 °C until visualization or visualize immediately.

### 3.7 Fluorescence Microscopy

Tissue sections can be visualized immediately upon thawing or first fixed and then visualized by fluorescence microscopy (*see* **Note 3**). Herein, we will outline the processing and visualization of fixed tissue sections, which was used to determine the level of infection (RCMV-GFP+ cells) in rat tissues at 3, 5, 7, 10, and 21 dpi (Table 1). The following fixing and washing steps can be performed in Coplin jars or in a humidified chamber.

1. Fix slides containing tissue sections from RCMV-GFP-infected rats with 2 % paraformaldehyde diluted in 1× PBS for 10 min at room temperature in the dark.
2. Wash twice with 1× PBS.
3. Visualize GFP fluorescence using a Delta Vision RT microscope by Applied Precision. Representative photomicrographs shown in Fig. 2 were obtained at 60× magnification.
4. Count the RCMV-GFP+ cells from three infected rats per time point and tabulate data using excel, and display as shown in Table 1. Virus-infected cells are readily detectable in the lung and spleen at 3 dpi. At 5, 7, and 10 dpi, the virus was observed in multiple tissues including the heart, lung, spleen, liver, and kidney. The peak of GFP detection for most tissues occurred at 5 dpi. However, the peak in salivary gland tissues was highest at day 21.

### 3.8 Fluorescence-Based Laser Capture Microscopy

Our goal with LCM was to differentially collect RCMV-GFP-infected vs. uninfected cells from tissue sections and extract intact viral and cellular RNA for expression analysis. LCM allows excision of samples as small as 100 μm$^2$ by using a software-directed microscope stage and ultraviolet cutting laser focused through the objective. Infected cells can be easily visualized by fluorescence of constitutively expressed GFP without permeabilizing or staining of cells. For our study, we employed two different laser capture microscopes. Initially, we used a PALM laser pressure catapulting system. We then switched to a newer ArcturusXT LCM (*see* Fig. 3 for a LCM workflow diagram). Both systems are controlled by an integrated computer system that allows accurate and efficient sample cutting and capture.

1. Fix tissue sections directly after cryosectioning with 100 % ethanol for 30 s or use fresh unfixed samples (*see* **Note 4**).
2. Perform LCM microdissection with either a PALM or Arcturus LCM system.
   (a) The PALM laser pressure catapulting (LPC) technique uses a pulsed nitrogen laser (337 nm). The GFP-positive cells in the tissue are visualized using a Zeiss Axiovert fluorescent research microscope. Cells are then microdissected by noncontact precise laser ablation using Robocut software (Microlaser Technologie, Bernried, Germany) and catapulted directly onto PALM adhesive caps (PALM Microlaser Technologies, Bernried, Germany) using the laser pressure catapulting technique of the instrument. PALM adhesive caps contained RNA extraction buffer and were frozen at −80 °C until use.
   (b) The ArcturusXT LCM system combines LCM and UV laser cutting methods. Tissue samples are collected by adherence to a plastic microtube cap. In general, the newer LCM systems have improved capabilities including the ability to visualize whole tissue sections in a grid pattern as shown in Fig. 4. This feature makes it easier to quickly scan the tissue to identify RCMV-GFP+ cells. First, CapSure HS LCM Caps are lowered into place above the tissue sample. Then, an ultraviolet (UV) cut line is drawn around the GFP+ cells using the computer system, and infrared (IR) spots are automatically placed within the region to be cut. The IR laser activates the plastic cap above the site to be captured promoting adherence of the plastic cap to the target cells. Under direction of the computer, the UV laser performs the cutting releasing the cut zone from the remaining tissue. This system allows visualization of tissues after dissection to ensure the sample was extracted properly (Fig. 5). Individual cells can also be acquired with simple adherence by placing individual IR spots over the cells of interest.

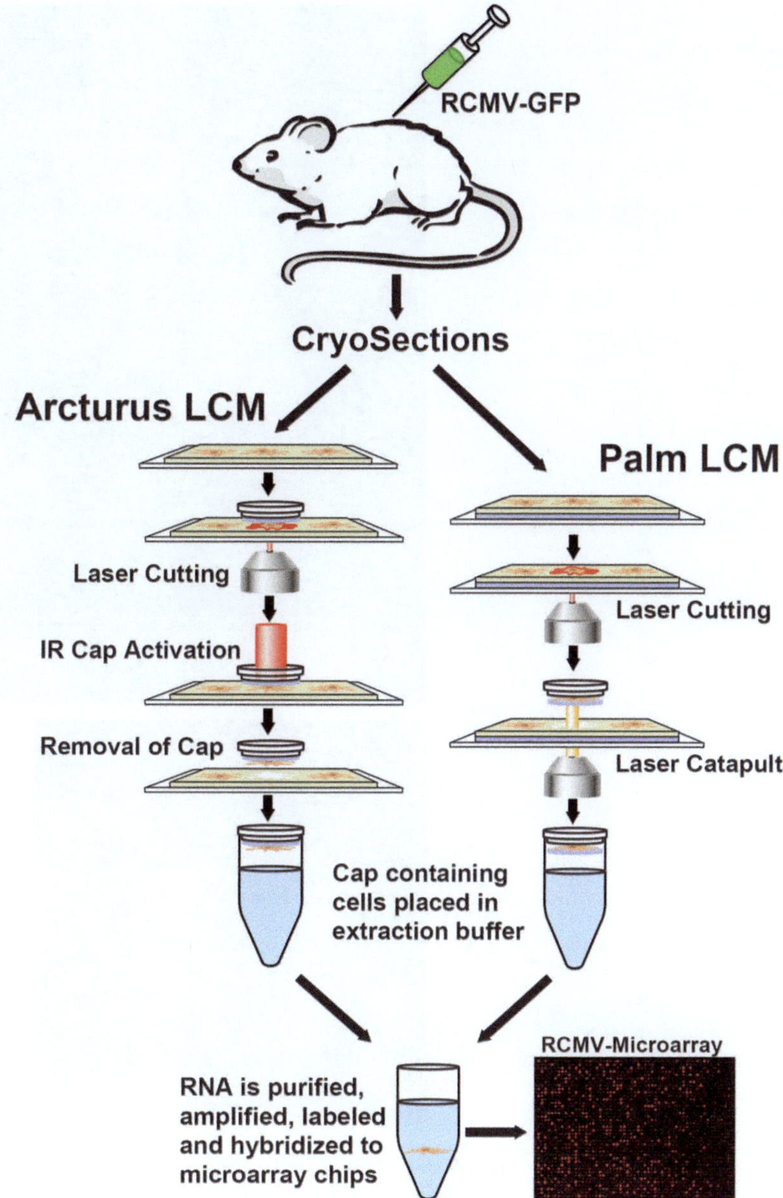


**Fig. 3** LCM workflow diagram. Animals are infected, and at the time of harvest, the tissues are embedded in OCT until cryosectioned. The thin sections are placed on microscope slides as suggested by the LCM system manufacturer. LCM procedures are system specific: *Arcturus LCM system* uses a special plastic tube cap that is lowered just above the sample. A UV laser is used to cut the outline of the infected cell(s), and then an IR laser melts points of the cap to the tissue allowing the cut section to be pulled off of the microscope slide. After microdissection is complete, the cap is placed on a microfuge tube containing RNA extraction buffer. *PALM LCM system* uses a UV laser for cutting around the GFP+ cell(s), and then the cut samples are catapulted by a laser push into a microfuge tube cap preloaded with RNA extraction buffer. RNA is purified, converted to cDNA, and the cDNA is amplified and labeled. The labeled cDNA is hybridized to the microarray chips, and then scanned and analyzed

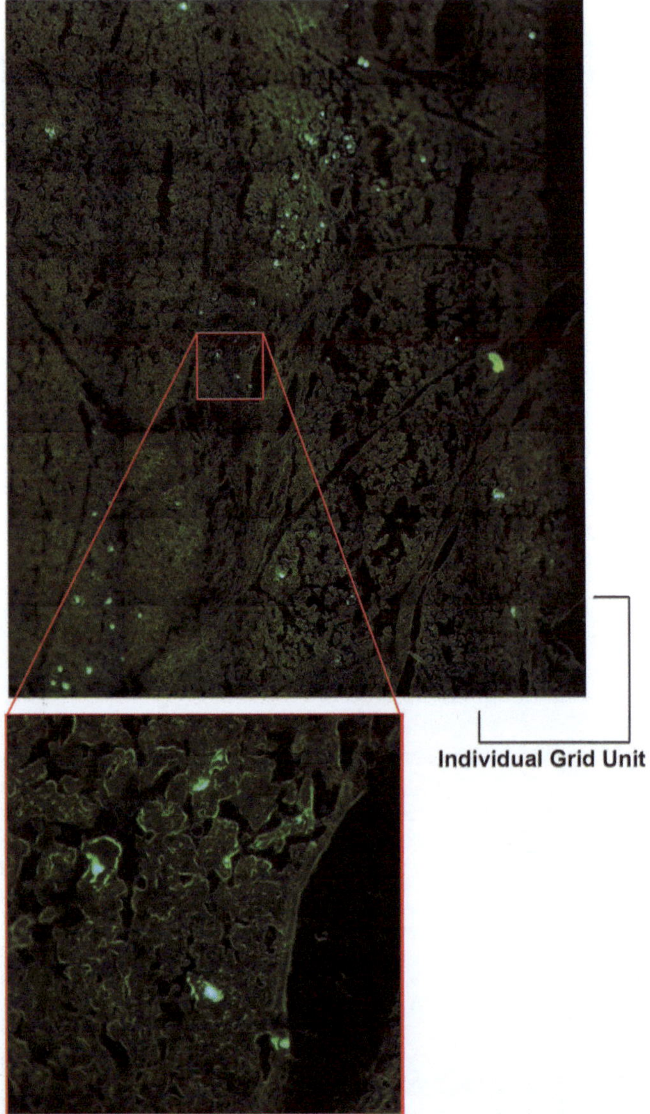


**Fig. 4** Detection of GFP+ cells in salivary glands. In order to identify GFP+ cells in salivary glands, we used the Arcturus LCM system to scan the whole salivary gland tissue sections by fluorescence microscopy. The LCM system performs the scanning in a grid pattern containing multiple miniature scans, and then the system reassembles the complete tissue from the miniscans. This technique allowed us to visually locate GFP+ cells and then mark them for extraction

After plastic adherence and cutting steps are complete for the entire sample, the plastic cap containing the adhered cells is lifted away from the tissue sample. The cap is then removed from the holder and placed on a tube containing RNA extraction buffer.

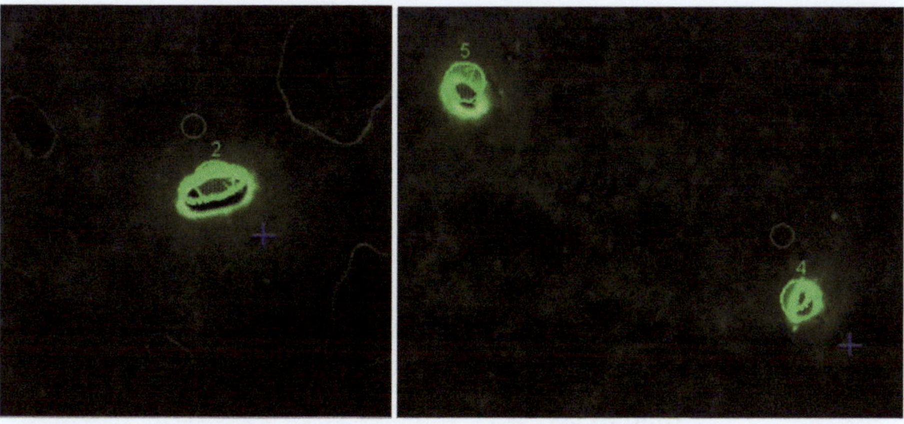


**Fig. 5** Visualization of LCM-extracted samples. Fluorescence microscopy was used to visualize removal of GFP+ cells from salivary gland tissues

3. Scrape a portion of a tissue section directly into extraction buffer without LCM in order to obtain a control sample to determine the effect of the LCM procedure on RNA quality.
4. Store adhesive caps containing LCM sections in RNA extraction buffer (Zymo Research Mini RNA isolation kit, catalog #R1005) (*see* **Note 5**).

### 3.9 RNA Isolation

Total RNA is isolated from laser-captured tissue samples using Zymo Research's Mini RNA isolation kit, and the samples are analyzed for quality using an Agilent 2100 Bioanalyzer.

1. Digest samples in 200 μl of RNA extraction buffer for 20 min on ice. Vortex samples for 10 min.
2. Add 1 volume of 100 % ethanol and incubate on ice for 10 min.
3. Transfer the solution to a Zymo-spin column and centrifuge for 1 min at 10,000 × *g* in a microfuge.
4. Wash the column twice with 200 μl of Wash buffer and centrifuge for 1 min at 10,000 × *g* in a microfuge to remove wash buffer (*see* **Note 6**).
5. Add 10 μl of RNAse-free $dH_2O$ and centrifuge at 10,000 × *g* for 1 min in a microfuge to elute RNA.
6. Analyze RNA quality using the Eukaryotic Total RNA Pico chip and a 2100 Bioanalyzer (Agilent). Total RNA extracted from LCM samples (1 μl) is compared to a control RNA sample, and the quality of the LCM sample is evaluated based on the level and number of peaks present. Representative gel images of isolated RNA samples are shown in Fig. 6.

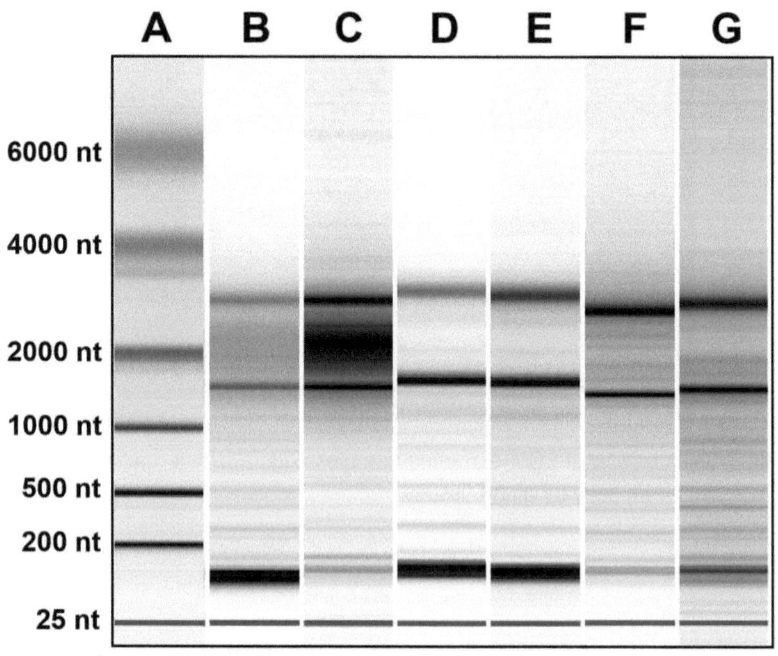


**Fig. 6** Analysis of RNA quality of LCM-captured GFP+ Cells. Shown are representative gel images of isolated total RNA (1 μl) that was analyzed with a 2100 Bioanalyzer (Agilent) using Eukaryotic Total RNA Pico chip. The quality of the LCM sample is evaluated based on the level and number of peaks present in the sample. *Lane A* is the RNA ladder and the band sizes are as labeled. *Lanes B*, *C*, and *D* are RNA harvested from GFP+ cells from spleen. The samples shown in *lanes B* and *C* were not fixed and were on the microscope stage for 1 h and 2 h, respectively. Sample in *lane D* was fixed in 100 % ethanol for 30 s on the microscope slide prior to capture. *Lane D* sample was on the microscope stage for 1 h. *Lanes E*, *F*, and *G* contain samples that were captured from GFP+ cells from native hearts infected for 7 dpi. The sample in *lane E* was acquired by scraping a portion of the tissue section with a sterile razor blade. The sample was immediately placed into RNA extraction buffer. This sample represents the total RNA harvested from the complete tissue. Samples run in *lanes F* and *G* were fixed in 100 % ethanol for 30 s on the slide, and then LCM was used to capture GFP+ cells (time on stage was 1 h). These two samples demonstrate the reproducibility of this technique in extracting quality RNA from GFP+ cells captured by LCM

### 3.10 Microarray Analysis

To perform the microarray analysis, the RNA is first converted to dsDNA and then amplified by conversion to aRNA. The aRNA is converted to cDNA and then labeled with fluorescent dyes. The labeled cDNA is hybridized to the microarray chips.

1. Synthesize first strand cDNA from the entire RNA sample (100–200 ng) isolated above using 1.0-μM oligo dT-T7 primer (GGCCAGTGAATTGTAATACGACTCACTATAGG G(T)$_{24}$), dNTP (0.5 mM), 40-units RNAse inhibitor, 1× first strand buffer (Invitrogen), and 200-units Superscript III reverse transcriptase (Invitrogen).

2. Generate double-stranded cDNA (dsDNA) by the addition of second strand buffer, dNTPs (0.3 mM), 40-units DNA polymerase, 10-units RNAse H, 10-units *E. coli* ligase (Invitrogen), and incubate at 16 °C for 2 h.
3. Add 6 units of T4 DNA polymerase and incubate at 37 °C for 10 min to polish 3′ overhangs.
4. Heat inactivate samples at 70 °C for 10 min.
5. Purify dsDNA by phenol:chloroform:isoamyl alcohol extraction.
6. Concentrate sample with Millipore YM100 columns by centrifugation.
7. Produce amplified RNA (aRNA) using the dsDNA sample as a template and the T7 Megascript kit (Ambion).
8. Purify aRNA with the RNeasy Mini kit (Qiagen).
9. Determine concentration by spectrophotometry.
10. Incubate 5 μg of aRNA with 300-units Superscript III reverse transcriptase in the presence of 9 μg of random hexamers, 0.5-mM dNTPs, 1× first strand buffer (Invitrogen), and 60-units RNAse H inhibitor at 37 °C for 2 h.
11. Hydrolyze aRNA by addition of 0.5N NaOH and Reagent D from the Mirus *Label* IT Cy5 labeling kit and incubate the samples at 65 °C for 30 min.
12. Add 100 μl of neutralizing solution from the Mirus *Label* IT Cy5 labeling kit.
13. Purify the cDNA sample using the Cyscribe GFX Purification kit (Amersham).
14. Florescent labeling is performed with the Mirus *Label* IT Cy5 labeling kit to chemically bind Cy5 to the synthesized cDNA. A total of 1 μg of purified cDNA is added to 1× labeling buffer M, 4 μl of Cy5 labeling reagent and $dH_2O$ to 100 μl. The samples are incubated in the dark at 37 °C for 3 h.
15. Add 0.1 volume of Reagent D to the reaction mixture and incubate for 5 min on ice to stop the labeling reaction.
16. Add 1× neutralization buffer and incubate for 5 min on ice.
17. Purify labeled cDNA using the Cyscribe GFX Purification Kit (Amersham).
18. Resuspend labeled cDNA in 35 μl of Custom Array's Hybridization solution.
19. Incubate the Custom Array slides with prehybridization buffer for 60 min at 50 °C.
20. Remove the prehybridization buffer from the slides, apply the labeled cDNA sample, and hybridize for 18 h at 50 °C.
21. Wash the slide with 6× SSPE, 0.05 % Tween-20 for 5 min at 50 °C.

22. Wash for 1 min at room temperature with 3× SSPE: 0.05 % Tween-20.
23. Wash for 1 min at room temperature with 0.5× SSPE: 0.05 % Tween-20.
24. Wash for 1 min twice with 2× PBS with 0.1 % Tween-20.
25. Wash for 1 min twice with 2× PBS.
26. Scan microarray slides using Bioscience GeneScan Lite laser scanner.
27. Analyze microarray images using Imagene digital processing software and Microarray Imager data analysis software (CustomArray Inc.).
28. Data were subjected to analysis by student's *t*-test. *P* values <0.05 were considered significant. Results from the microarray were compiled and displayed as a heat map (Fig. 7).

## 4 Observations

Laser capture microscopy has proven to be a powerful and valuable tool for examining questions regarding single cell metabolism, especially in whole tissue samples. We have developed LCM techniques to extract individual RCMV-infected (GFP+) cells combined with viral transcriptome analysis via microarray. The RCMV-GFP virus used in this study expresses GFP under the constitutive cellular promoter EF-1α, which enables the expression and accumulation of GFP even during nonproductive replication scenarios such as PBMCs and cells of the bone marrow (Table 1). Importantly, the course of infection and pathogenesis of this virus parallels the reported wild-type RCMV infection patterns with an early involvement of lung and spleen followed by a widespread multiorgan distribution by day 5. This initial infection phase is followed by a long-term persistence phase that usually involves salivary gland tissues. The ability of RCMV-GFP to express even under nonproductive infection conditions enabled us to detect and isolate single cells from several rat tissue types (salivary glands, heart, lung, and spleen) using LCM. We discovered a number of important findings pertaining to in vivo CMV transcription (Fig. 7). First, we have confirmed that CMV transcription is tissue specific, which parallels our previous findings using whole tissue preparations [9]. Second, RCMV gene expression changes in the salivary gland from a profile that is associated with a productive infection at day 7 to one of persistence by day 10 (Fig. 8). Interestingly, viral gene expression in infected cells captured from normal hearts resembles this salivary gland persistence phenotype. Lastly, RCMV gene expression is altered in rats during the inflammatory state produced by allograft rejection, even in tissues not undergoing active rejection (native heart in allograft recipients vs.

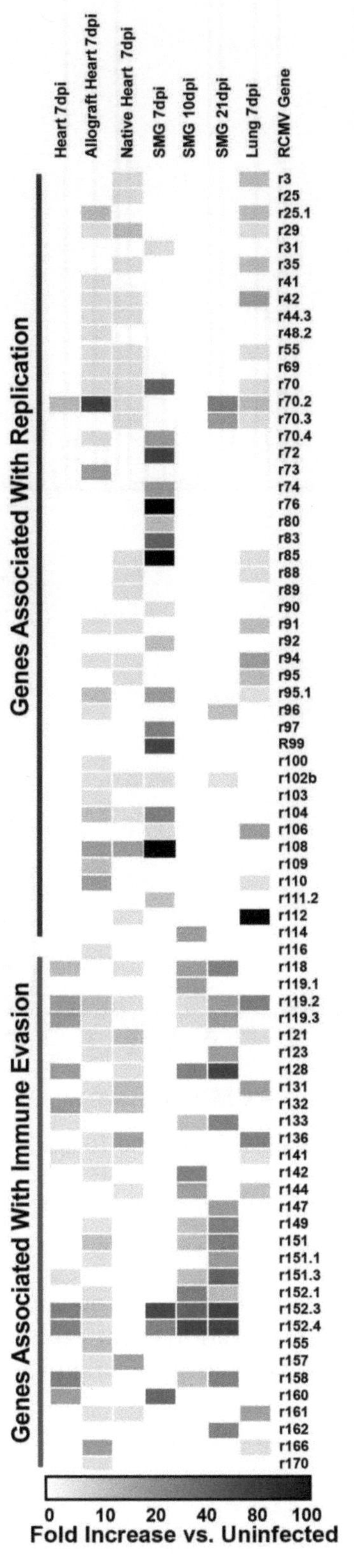


**Fig. 7** Analysis of RCMV-GFP gene expression by LCM/microarray. Shown is a heat map of LCM/microarray analysis of RCMV gene expression in GFP+ cell samples from infected hearts, allograft recipient native and graft hearts, lung (7 dpi), and SMG at 7, 10, and 21 dpi. Shown are the average fold increase values for three biological replicates compared to baseline (uninfected heart tissue)

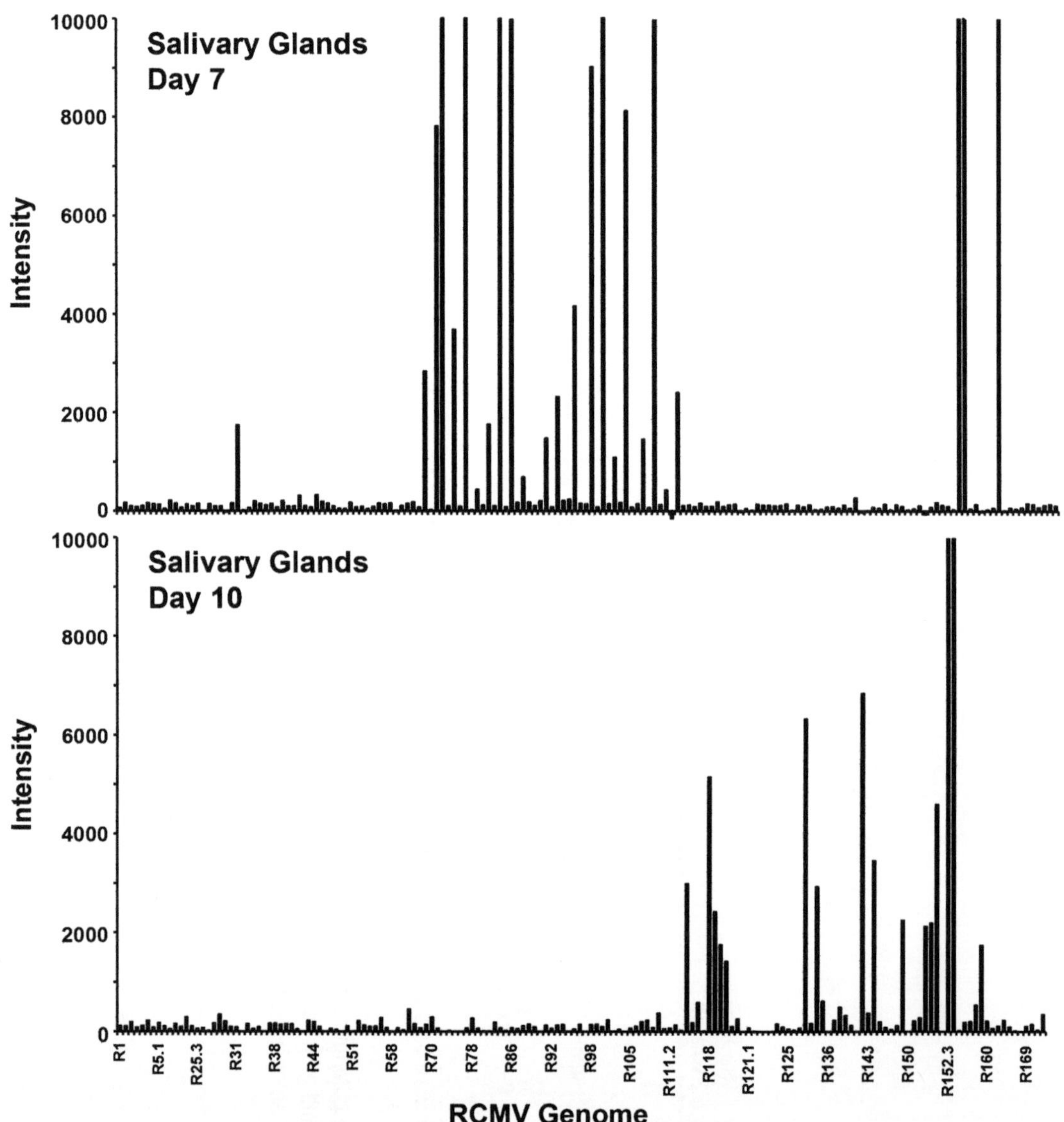


**Fig. 8** RCMV-GFP gene expression in salivary glands at 7 and 10 dpi. Shown are the fluorescence intensities from microarray analysis of GFP+ cells isolated from RCMV-GFP-infected rat salivary glands harvested at 7 and 10 dpi. LCM was used to capture GFP+ cells from three individual infected rats. The RNA was processed for microarray and analyzed on separate microarray chips

native heart in nontransplant controls). This finding suggests a major viral reactivation event occurs during allograft rejection that stimulates CMV activation and promotes viral dissemination.

## 5 Notes

1. When constructing recombinant CMV expressing GFP, the choice of promoter type and placement of the cassette within the genome dramatically affects the level of GFP expression [24].

The use of the EF-1α promoter in our RCMV-GFP construct caused the constitutive expression of the GFP protein and resulted in early and robust levels of expression. The GFP cassette was inserted into the R144–147 region of the RCMV genome, and this may have increased the levels of GFP protein detected due to the availability of the EF-1α promoter to the transcription machinery. We have found that this region of the genome is highly expressed in tissues and during nonproductive infection scenarios. It is possible that other regions of the RCMV genome may be sterically blocked under these same conditions, which would limit accessibility to the promoter [25].

2. Rodents were housed in the Association for Assessment and Accreditation of Laboratory Animals Care (AAALAC)-accredited Portland Veterans Affairs Medical Center animal facility in a specific pathogen-free room, designated for CMV-infected rats, in compliance with guidelines provided by the US Department of Agriculture/Department of Health and Human Services (USDA/HHS).
3. Since we cut our own tissue sections, we always assess tissue section quality by microscopically visualizing GFP fluorescence immediately after thawing tissues on slides. This technique is used to ensure that the subsequent tissue sections are of good quality. We do not counterstain our tissues.
4. In general, GFP fluorescence tends to leak out of the fractured cells obtained during the cryosectioning process. We have employed a number of fixation conditions in order to prevent GFP leakage as well as optimize fluorescence stability and RNA quality and yield. By optimizing these conditions, we determined that 100 % ethanol fixation may stabilize RNA in some tissues, but fixation is not necessary to retrieve high-quality RNA. Freezing of tissue sections after cryosectioning negatively impacts the quality of recovered RNA. LCM directly after cryosectioning gave superior results to storing sections at −80 °C between cryosectioning and LCM. Analysis of the RNA quality by Bioanalyzer is displayed in Fig. 6. For best GFP retention, we advise performing LCM immediately following thawing of the cut tissue sections.
5. Storing adhesive caps containing LCM tissue in Trizol appeared to give only slightly better results compared to RNA extraction buffer. Overall, we obtained the best results when cryosections were taken directly to the LCM and microdissected within 1 h and caps containing tissue were then frozen in Trizol at −80 °C until RNA processing.
6. When purifying RNA or cDNA, we add an additional final centrifugation step prior to elution in order to make sure that the sample is completely free of contaminating ethanol present in the wash buffers.

## Acknowledgements

The work presented in this manuscript was supported by grants from the National Institutes of Health (HL-083194 DNS) and (HL-66238-01 SLO). The Arcturus laser capture microscope used for this study was supported by Award Number S10 RR027503 from the National Center for Research Resources. The content is solely the responsibility of the authors and does not necessarily represent the official views of the National Center for Research Resources or the National Institutes of Health. The Oregon National Primate Research Center Imaging Core is supported by NIH P51 RR000163. We thank Andrew Townsend for his assistance with graphics.

### References

1. Britt W (2008) Manifestations of human cytomegalovirus infection: proposed mechanisms of acute and chronic disease. Curr Top Microbiol Immunol 325:417–470
2. Chee MS, Bankier AT, Beck S, Bohni R, Browne CM, Cerny R, Horsnell T, Hutchison CA III, Kouzarides T, Martignetti JA, Preddie E, Satchwell SC, Tomlinson P, Weston KM, Barrell BG (1990) Analysis of the protein-coding content of the sequence of human cytomegalovirus strain AD169. In: McDougall JK (ed) Cytomegaloviruses. Springer, Berlin, pp 125–171
3. Davison AJ, Dolan A, Akter P, Addison C, Dargan DJ, Alcendor DJ, McGeoch DJ, Hayward GS (2003) The human cytomegalovirus genome revisited: comparison with the chimpanzee cytomegalovirus genome. J Gen Virol 84:17–28
4. Chambers J, Angulo A, Amaratunga D, Guo H, Jiang Y, Wan JS, Bittner A, Frueh K, Jackson MR, Peterson PA, Erlander MG, Ghazal P (1999) DNA microarrays of the complex human cytomegalovirus genome: profiling kinetic class with drug sensitivity of viral gene expression. J Virol 73:5757–5766
5. Goodrum F, Jordan CT, Terhune SS, High K, Shenk T (2004) Differential outcomes of human cytomegalovirus infection in primitive hematopoietic cell subpopulations. Blood 104:687–695
6. Goodrum FD, Jordan CT, High K, Shenk T (2002) Human cytomegalovirus gene expression during infection of primary hematopoietic progenitor cells: a model for latency. Proc Natl Acad Sci U S A 99:16255–16260
7. Goodrum F, Reeves M, Sinclair J, High K, Shenk T (2007) Human cytomegalovirus sequences expressed in latently infected individuals promote a latent infection in vitro. Blood 110:937–945
8. Petrucelli A, Rak M, Grainger L, Goodrum F (2009) Characterization of a novel Golgi apparatus-localized latency determinant encoded by human cytomegalovirus. J Virol 83:5615–5629
9. Streblow DN, van Cleef KW, Kreklywich CN, Meyer C, Smith P, Defilippis V, Grey F, Fruh K, Searles R, Bruggeman C, Vink C, Nelson JA, Orloff SL (2007) Rat cytomegalovirus gene expression in cardiac allograft recipients is tissue specific and does not parallel the profiles detected in vitro. J Virol 81: 3816–3826
10. Bradley AJ, Lurain NS, Ghazal P, Trivedi U, Cunningham C, Baluchova K, Gatherer D, Wilkinson GW, Dargan DJ, Davison AJ (2009) High-throughput sequence analysis of variants of human cytomegalovirus strains Towne and AD169. J Gen Virol 90:2375–2380
11. Gatherer D, Seirafian S, Cunningham C, Holton M, Dargan DJ, Baluchova K, Hector RD, Galbraith J, Herzyk P, Wilkinson GW, Davison AJ (2011) High-resolution human cytomegalovirus transcriptome. Proc Natl Acad Sci U S A 108:19755–19760
12. Almond PS, Matas A, Gillingham K, Dunn DL, Payne WD, Gores P, Gruessner R, Najarian JS (1993) Risk factors for chronic rejection in renal allograft recipients. Transplantation 55:752–756, discussion 6–7
13. Grattan MT, Moreno-Cabral CE, Starnes VA, Oyer PE, Stinson EB, Shumway NE (1989) Cytomegalovirus infection is associated with cardiac allograft rejection and atherosclerosis. JAMA 261:3561–3566

14. Cobbs CS, Harkins L, Samanta M, Gillespie GY, Bharara S, King PH, Nabors LB, Cobbs CG, Britt WJ (2002) Human cytomegalovirus infection and expression in human malignant glioma. Cancer Res 62:3347–3350
15. Melnick JL, Adam E, Debakey ME (1993) Cytomegalovirus and atherosclerosis. Eur Heart J 14(Suppl K):30–38
16. Muhlestein JB, Horne BD, Carlquist JF, Madsen TE, Bair TL, Pearson RR, Anderson JL (2000) Cytomegalovirus seropositivity and C-reactive protein have independent and combined predictive value for mortality in patients with angiographically demonstrated coronary artery disease. Circulation 102:1917–1923
17. Baca Jones CC, Kreklywich CN, Messaoudi I, Vomaske J, McCartney E, Orloff SL, Nelson JA, Streblow DN (2009) Rat cytomegalovirus infection depletes MHC II in bone marrow derived dendritic cells. Virology 388:78–90
18. Minamishima I, Ueda K, Minematsu T, Minamishima Y, Umemoto M, Take H, Kuraya K (1994) Role of breast milk in acquisition of cytomegalovirus infection. Microbiol Immunol 38:549–552
19. Orloff SL, Hwee YK, Kreklywich C, Andoh TF, Hart E, Smith PA, Messaoudi I, Streblow DN (2011) Cytomegalovirus latency promotes cardiac lymphoid neogenesis and accelerated allograft rejection in CMV naive recipients. Am J Transplant 11:45–55
20. Orloff SL, Streblow DN, Soderberg-Naucler C, Yin Q, Kreklywich C, Corless CL, Smith PA, Loomis CB, Mills LK, Cook JW, Bruggeman CA, Nelson JA, Wagner CR (2002) Elimination of donor-specific alloreactivity prevents cytomegalovirus-accelerated chronic rejection in rat small bowel and heart transplants. Transplantation 73:679–688
21. Streblow DN, Kreklywich C, Yin Q, De La Melena VT, Corless CL, Smith PA, Brakebill C, Cook JW, Vink C, Bruggeman CA, Nelson JA, Orloff SL (2003) Cytomegalovirus-mediated upregulation of chemokine expression correlates with the acceleration of chronic rejection in rat heart transplants. J Virol 77: 2182–2194
22. Streblow DN, Kreklywich CN, Andoh T, Moses AV, Dumortier J, Smith PP, Defilippis V, Fruh K, Nelson JA, Orloff SL (2008) The role of angiogenic and wound repair factors during CMV-accelerated transplant vascular sclerosis in rat cardiac transplants. Am J Transplant 8:277–287
23. Streblow DN, Kreklywich CN, Smith P, Soule JL, Meyer C, Yin M, Beisser P, Vink C, Nelson JA, Orloff SL (2005) Rat cytomegalovirus-accelerated transplant vascular sclerosis is reduced with mutation of the chemokine-receptor R33. Am J Transplant 5:436–442
24. Qin JY, Zhang L, Clift KL, Hulur I, Xiang AP, Ren BZ, Lahn BT (2010) Systematic comparison of constitutive promoters and the doxycycline-inducible promoter. PLoS One 5:e10611
25. Nevels M, Nitzsche A, Paulus C (2011) How to control an infectious bead string: nucleosome-based regulation and targeting of herpesvirus chromatin. Rev Med Virol 21: 154–180

# Chapter 14

# Techniques for Characterizing Cytomegalovirus-Encoded miRNAs

**Lauren M. Hook, Igor Landais, Meaghan H. Hancock, and Jay A. Nelson**

## Abstract

microRNAs (miRNAs) are small noncoding RNAs that regulate gene expression at the posttranscriptional level, by binding to sites within the 3′ untranslated regions of messenger RNA (mRNA) transcripts. The discovery of this completely new mechanism of gene regulation necessitated the development of a variety of techniques to further characterize miRNAs, their expression, and function. In this chapter, we will discuss techniques currently used in the miRNA field to express, detect, and inhibit miRNAs as well as methods used to identify their targets.

**Key words** microRNA (miRNA), Northern blot, Stem-loop real-time PCR, RISC immunoprecipitation (RISC-IP), Luciferase assay, Locked nucleic acids (LNA)

## 1 Introduction

The discovery that animal cells and later DNA viruses encode small, noncoding regulatory RNA molecules known as microRNAs (miRNAs) is arguably one of the most important findings in biological research in recent years. Since their initial discovery [1], the miRNA family has expanded exponentially, aided by the development of a variety of identification techniques, from in silico prediction and cross-species identification [2] to small RNA cloning [3] and highly sensitive deep-sequencing approaches [4–7]. Many of these miRNAs are conserved across species and regulate important biological processes including cellular differentiation, DNA repair, metabolism, apoptosis, immunity, aging, and cancer [8].

The vast majority of miRNAs encoded by viruses have been identified in the herpesvirus family, including HSV-1 (16 pre-miRNAs), EBV (25 pre-miRNAs), KSHV (12 pre-miRNAs), HCMV (12 pre-miRNAs), RhCMV (17 pre-miRNAs), MCMV (18 pre-miRNAs), and RCMV (24 pre-miRNAs) [4, 5, 7, 9–15] (http://www.mirbase.org/cgi-bin/mirna_summary.pl?org=hcmv). Virally encoded miRNAs are hypothesized to play a

Andrew D. Yurochko and William E. Miller (eds.), *Human Cytomegaloviruses: Methods and Protocols*, Methods in Molecular Biology, vol. 1119, DOI 10.1007/978-1-62703-788-4_14, © Springer Science+Business Media New York 2014

key role in mediating viral pathogenicity [16]; however, the functions of the majority of the miRNAs are unknown. As an individual miRNA's effect on a given target is usually quite small, the likelihood that multiple miRNAs cooperate to target multiple cellular targets belonging to pathways important for viral pathogenesis is great. After a brief introduction to miRNA biogenesis and mode of action, this chapter will focus on the methods routinely used to express, detect, and inhibit CMV miRNAs as wells as methods used to identify miRNA targets.

The canonical pathway for cellular biogenesis of miRNAs [8, 17] begins with transcription of a single-stranded primary miRNA transcript (pri-miRNA) by an RNA polymerase II, which then folds into one or several hairpin structures. In the nucleus, the hairpin is processed by the endonuclease, Drosha, generating a 60–100 nucleotide (nt) precursor miRNA hairpin species (pre-miRNA) that is exported to the cytoplasm by Exportin 5. There, the pre-miRNA is cleaved by a second endonuclease, Dicer, generating an 18–22 nt double-stranded miRNA duplex. The mature or guide strand is then incorporated into the RNA-induced silencing complex (RISC) that contains argonaute (Ago) proteins. The second strand or passenger strand is usually degraded but in some instances may also be incorporated into RISC, generating two RISC populations from a single double-stranded duplex [8]. Understanding the biogenesis of miRNAs has allowed for the development of several strategies to express exogenous miRNAs in cells, which is detailed below (*see* Subheading 3.1).

After their incorporation into RISC, miRNAs posttranscriptionally regulate cellular gene expression through the miRNA-directed binding of RISC to mRNA targets. This feature is the basis of several related methods aimed at identifying mRNA targets. These methods are based on the affinity purification of ribonucleoprotein (RNP) complexes followed by RNA purification and identification by microarray or deep sequencing. In contrast with the RISC immunoprecipitation (RISC-IP) [18–21] and streptavidin pull-down of biotinylated miRNA mimic strategies [22–24] (detailed below, *see* "Immunoprecipitating" and "Pull-Down" components under the Subheading 3.3.1), other methods such as HITS-CLIP (high-throughput sequencing of RNAs isolated by cross-linking and immunoprecipitation) [25], PAR-CLIP (photoactivatable ribonucleoside-enhanced cross-linking and immunoprecipitation) [26, 27], and iCLIP [28] use UV cross-linking to stabilize the RNP complex. These cross-linking methods have the potential to significantly expand the number of potential CMV mRNA targets.

miRNA target sites are most commonly located in the 3′ untranslated region (UTR) of the mRNA, although sites in the 5′UTR and the coding sequence have also been observed [8, 24]. This last feature has been exploited to develop sensitive luciferase assays (*see*

Subheading 3.2.2) aiming at measuring the posttranscriptional effect of miRNAs on their potential targets.

The fate of the target mRNA depends on whether the miRNA incorporated into RISC is fully or partially complementary with the target mRNA. Extensive complementarity generally leads to Ago-mediated mRNA cleavage, while partial complementarity via the seed sequence, a six to eight residues stretch located between nucleotides 2 and 8 of the miRNA [8], leads to translational repression and/or mRNA decay through mechanisms which are not completely understood. Taking into account the fact that mRNA levels are not always affected by miRNAs, protein-based quantitative methods such as SILAC (stable isotopes labeling by amino acids in cell culture) coupled with mass spectroscopy have also been developed to identify cellular miRNA targets [6, 29]. These methods provide a valuable counterpoint to the RNP affinity purification methods mentioned above for target identification.

## 2 Materials

### 2.1 Transient Expression of miRNAs in Cells

#### 2.1.1 Method 1: Expressing miRNAs Using an Expression Vector (pSIREN)

1. pSIREN-retroQ-ZsGreen (Clontech, Mountain View, CA, USA, *see* **Note 1**).
2. BamHI and EcoRI restriction enzymes and digestion buffers.
3. Purified viral DNA (10–50 ng/μL).
4. Forward and reverse PCR primers designed 50, 200, or 500 base pairs upstream and downstream of the miRNA of interest, respectively (*see* **Note 2**).
5. PCR reagents.
6. Thermocycler.
7. 1 % agarose gel.
8. Ethidium bromide.
9. Agarose gel electrophoresis apparatus and reagents.
10. PCR fragment purification kit.
11. Ligation kit.
12. Competent E. coli bacteria strain (e.g., DH5-α).
13. LB-ampicillin dishes.
14. LB broth + 50 μg/mL ampicillin.
15. Miniprep kit.
16. Maxiprep kit.
17. Opti-MEM I Reduced Serum Media (Invitrogen).
18. Lipofectamine 2000 (Invitrogen, *see* **Note 3**).
19. HEK293 cells (*see* **Note 3**).
20. Fluorescent microscope.

### 2.2 Detecting miRNAs

#### 2.2.1 Method 1: miRNA Northern Blot

##### Urea/Acrylamide Gel Components

1. Urea/acrylamide gel: In a 500 mL glass bottle, mix 150 mL 19:1 acrylamide (30 %), 30 mL 10× TBE (890 mM Tris base, 890 mM boric acid, 20 mM EDTA pH 8.0), and 126 g of urea. Place in 68 °C water bath, stir if necessary, until components are dissolved. Make up to 300 mL using nuclease-free water. Pass through a 0.45 μM Steriflip filter (Millipore) and store at 4 °C, shielded from light.
2. Ammonium persulfate: 10 % solution in water, store at 4 °C.
3. *N*,*N*,*N*′,*N*′-tetramethyl-ethylenediamine (TEMED), store at 4 °C.
4. Running buffer: 1× TBE (diluted from 10× TBE using distilled water).
5. Formamide, store at 4 °C.
6. 6× RNA loading dye: 10 mM Tris–HCl pH 7.6, 0.03 % bromophenol blue, 0.03 % xylene cyanol, 60 % glycerol, 60 mM EDTA. Store at −20 °C.
7. Mini PROTEAN 3 System (Bio Rad) or other mini-gel electrophoresis system.
8. 10 % SDS (*see* **Note 4**).
9. 95 % Ethanol.
10. Ethidium bromide (10 mg/mL solution) (*see* **Note 5**).
11. UV transilluminator with camera.
12. Fluorescent ruler.
13. Small RNA marker.
14. Total RNA samples from cells expressing the miRNA of interest (*see* **Note 6**).

##### Transfer Components

1. GeneScreen Plus membrane (PerkinElmer) or other positively charged nylon membrane.
2. Whatman paper (0.34 mm thickness).
3. Transfer buffer: 1× TBE.
4. Mini Trans-Blot electrophoretic transfer cell.
5. UV Stratalinker 1800.

##### Probe Labeling Components

1. 10× kinase buffer A.
2. Adenosine 5′ triphosphate ($\gamma$-$^{32}$P), 10 μCi/μL (*see* **Note 7**).
3. Polynucleotide kinase, 10 U/μL.
4. 37 °C heat block.
5. G-25 Quick Spin columns.
6. 1× TE buffer: 10 mM Tris–HCl, pH 7.6, 1 mM EDTA, pH 8.0.
7. Tabletop centrifuge.

8. Oligonucleotide probe designed antisense to the miRNA of interest. The sequences of known HCMV miRNAs are readily available online within the miRBase database (http://www.mirbase.org/) [12–15].

Hybridization Components

1. PerfectHyb solution (Sigma).
2. Hybridization tubes.
3. Hybridization oven.
4. 20× SSC: 3 M NaCl, 300 mM sodium citrate.
5. Wash solution 1: 2× SSC, 0.05 % SDS, dissolved in water.
6. Wash solution 2: 0.1× SSC, 0.1 % SDS, dissolved in water.

Detection Components

1. Autoradiography film.
2. Film cassette.
3. Intensifier screen.

*2.2.2 Method 2: Stem-Loop RT-PCR for Detection of miRNAs*

Reverse Transcription Components

1. 200 μM stock of RT primer or 5× stock (Applied Biosystems) or similar.
2. 10× RT buffer (100 mM Tris–HCl, pH 8.3, 500 mM KCl, 5.5 mM $MgCl_2$).
3. 10 μM dNTP stock.
4. Multiscribe enzyme 50 U/μL (Applied Biosystems).
5. RNasin Plus RNase inhibitor, 40 U/μL (Promega).
6. Thermocycler.
7. RNA template (50 ng/μL) (*see* **Note 8**).

Taqman Assay Components

1. 2× Universal PCR Master Mix (Applied Biosystems) or similar.
2. 50 μM forward and reverse primers or 20× Master Mix (Applied Biosystems) or similar.
3. 10 μM probe.
4. Real-time PCR machine.

## 2.3 Identifying mRNA Targets of Viral miRNAs

*2.3.1 Method 1: RNA-Induced Silencing Complex (RISC) Immunoprecipitation (IP)*

Immunoprecipitation and Pull-Down Components

1. DPBS 1×, Dulbecco's Phosphate Buffered Saline. Store at 4 °C.
2. Lysis buffer: 20 mM Tris–HCl, pH 7.5, 2.5 mM $MgCl_2$, 200 mM NaCl, 0.05 % NP-40, 1 mM DTT (added just before use), 1× Protease Inhibitor Cocktail (added just before use). Keep on ice.
3. Complete, Mini, EDTA-free Protease Inhibitor Cocktail Tablets (Roche Applied Science) or similar. Store at 4 °C.
4. Streptavidin-agarose (Invitrogen, Carlsbad, CA, USA), ANTI-c-myc Agarose Conjugate (Sigma, Saint Louis, Missouri, USA), protein A beads, or similar. Store at 4 °C.

5. Antibody directed against one of the RISC complex components (*see* **Note 9**).
6. 10 mg/mL yeast tRNA.
7. 10 mg/mL BSA.
8. RNasin Plus RNase inhibitor 40 U/μL. Store at −20 °C.
9. Scrapers.
10. Tabletop centrifuge capable of refrigeration.
11. 100 mm dishes of cells expressing miRNA(s) of interest (*see* **Note 10**) (*see* Subheading 3.1).

RNA Isolation Components

1. TRIzol reagent (Invitrogen) or similar. Store at 4 °C.
2. Chloroform.
3. Isopropanol.
4. GlycoBlue Coprecipitant (Applied Biosystems) (*see* **Note 11**). Store at −20 °C.
5. 75 % ETOH: Leave an aliquot on ice. Store at 20 °C.
6. Nuclease-free water.

#### *2.3.2 Method 2: Reporter Assays for Identification of Viral miRNA Targets*

Transfection Components

1. HEK293 cells (*see* **Note 3**).
2. Exogenous miRNA source (*see* **Note 12**).
3. Dual-Luciferase reporter vector with test, negative, and positive control 3′UTRs cloned downstream of the reporter: psiCHECK-2 vector (Promega, Madison, WI, USA) (*see* **Note 13**). In the case of the psiCHECK-2 vector, which contains both renilla and firefly luciferase genes, the 3′UTR of the potential target is cloned downstream of the renilla luciferase gene.
4. Black-walled clear-bottom 96-well tissue-culture-treated plate (*see* **Note 14**).
5. Lipofectamine 2000 (Invitrogen, *see* **Note 3**).
6. Opti-MEM I Reduced Serum Media.

Luciferase Assay Components

1. Dual-Luciferase Reporter Assay System.
2. Luminometer (*see* **Note 15**).

### *2.4 Inhibition of HCMV miRNAs Using Locked Nucleic Acids (LNA)*

1. LNA: miRCURY LNA™ microRNA Power Inhibitors (*see* **Note 16**) from Exiqon (www.exiqon.com/mirna-inhibitors). Resuspend 5 nmol in 50 μL or 250 μL RNase-free water to obtain a 100 μM or 20 μM stock, respectively. Aliquot and store at −80 °C. Do not freeze-thaw more than five times.
2. miRNA mimic: *See* Subheading 2.1 for design and ordering information.
3. Luciferase reporter plasmid: *See* Subheading 2.3.2 for design and cloning.

4. Lipofectamine 2000 (Invitrogen, *see* **Note 3**).
5. HEK293 cells.
6. DMEM + 10 % serum + penicillin/streptomycin.
7. Opti-MEM I Reduced Serum Media.

## 3 Methods

### 3.1 Transient Expression of miRNAs in Cells

The simplest method for expressing CMV miRNAs in cells is through infection. This strategy can be very valuable to monitor the effects of physiologically relevant concentrations of miRNAs during the course of CMV infection (e.g., up- or downregulation of target genes, virus growth). However, isolating the effect of a particular miRNA in this setting is complicated by the simultaneous expression of multiple viral genes and miRNAs. To explore the effects of CMV-encoded miRNAs on cellular gene expression in a more defined environment, exogenous miRNA expression is necessary. The choice between the two transfection-based expression methods described below will depend on the objective of the study: if the goal is to express the miRNA at high levels while keeping the cost low, the expression vector strategy is appropriate. The downside of this method is that precisely adjusting the quantity of miRNA expressed in the cell is not possible. This situation may be problematic when comparing the effect of individual or combinations of miRNAs or when performing titration curves. Alternatively, transfection of miRNA mimics allows for the delivery of precise amounts of miRNAs into the cell [30], but the cost can be limiting if large experiments are envisioned. Finally, lentiviral-mediated miRNA expression might be the only possibility in difficult to transfect cells but can require significant optimization (*see* **Note 17**).

#### 3.1.1 Method 1: Expressing miRNAs Using an Expression Vector (pSIREN)

The following protocol explains how to clone and express miRNAs using the pSIREN vector:

##### Cloning into pSIREN Vector

1. Amplify the region surrounding the miRNA of interest using the forward and reverse primers and 10–50 ng purified viral DNA (*see* **Note 18**) using appropriate PCR procedures.
2. Verify the amplification of the predicted product by loading 2 μL of the PCR reaction on a 1 % agarose gel and visualize the DNA with ethidium bromide.
3. Purify the PCR product using a PCR fragment purification kit (*see* **Note 19**).
4. Digest 1 μg pSIREN vector and the purified PCR fragment using the BamHI and EcoRI restriction enzymes following manufacturer's instructions.

5. Purify the digested pSIREN and PCR fragment using the PCR fragment purification kit.
6. Ligate the digested PCR fragment to pSIREN according to the instructions of the ligation kit.
7. Transform the ligation reaction in competent bacteria following standard procedures.
8. Pick colonies and grow small cultures (3–5 mL) in LB-ampicillin media; extract plasmid DNA using a miniprep kit.
9. Verify the presence of an insert of the correct size by colony PCR or BamHI–EcoRI digestion.
10. Confirm positive clones by sequencing using the forward and/or reverse PCR primer.
11. Perform a maxiprep on a validated clone to obtain working quantities of plasmid DNA.

Transfection

1. 24 h before transfection, plate $1 \times 10^5$ HEK293 cells in 1 mL media per well in a 12-well plate or $2 \times 10^6$ cells in 10 mL media per 100 mm dish.
2. For each well of a 12-well plate, add 0.5 μg pSIREN-miRNA DNA to 75 μL Opti-MEM I Reduced Serum Media and 2 μL Lipofectamine 2000 to 75 μL Opti-MEM I Reduced Serum Media. For a 100 mm dish, add 5 μg pSIREN-miRNA DNA to 500 μL Opti-MEM I Reduced Serum Media and 20 μL Lipofectamine 2000 to 500 μL Opti-MEM I Reduced Serum Media.
3. Incubate 5 min at room temperature, combine DNA- and Lipofectamine-containing media, mix gently, and incubate 20 min.
4. Add transfection mix to the cells dropwise.
5. After 24 h, assess plasmid transfection efficiency by evaluating GFP expression using a fluorescent microscope or flow cytometry (*see* **Note 20**).
6. Remove medium and harvest cells using the appropriate protocol (*see* Subheadings 3.2 and 3.3).

*3.1.2 Method 2: Designing and Expressing miRNA Mimics*

Some HCMV miRNA double-stranded RNA mimics are commercially available as duplexes or pre-miRNA hairpins. For miRNAs currently unavailable or to design miRNA duplexes for newly discovered miRNAs, the following protocol can be used [22, 23]:

Design miRNA Mimics

1. Identify the sequence of the mature or guide strand of the miRNA of interest. The sequences of known HCMV miRNAs are readily available online within the miRBase database (http://www.mirbase.org/) [12–15].

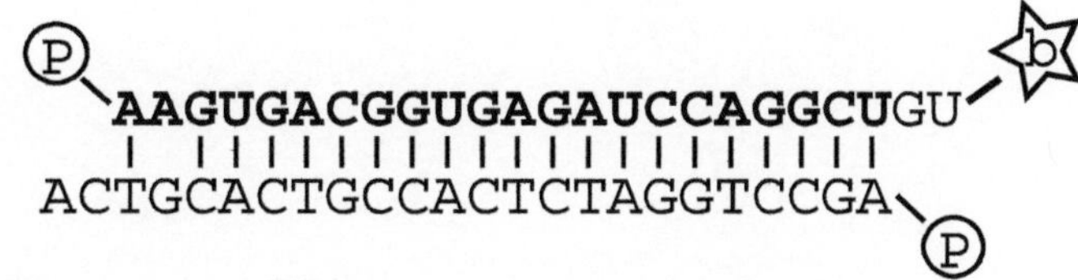


**Fig. 1** Design of custom miRNA mimics. Two nucleotides directly downstream of the mature miRNA (shown here in *bold*) are added by examining the stem-loop structure in mirBase. The sequence of the passenger strand is determined by reverse-complementing the mature strand, followed by the addition of two nucleotides on the 3′ end. A mismatch is then made in the passenger strand at the fourth nucleotide from the 3′ end to facility incorporation of the mature strand into RISC. Phosphates are added to the 5′ end of each strand. A biotin moiety can be added to the 3′ end of the mature strand to facilitate isolation of miRNA containing complexes using affinity purification techniques

2. Include the two nucleotides directly downstream of the mature strand at the 3′ end (*see* Fig. 1). These nucleotides are easily identifiable by examining the miRNA stem-loop structure.
3. Reverse-complement the mature strand sequence to serve as the antisense strand and add the two nucleotides directly downstream as in **step 2**. At this point the sequences of the two strands should be completely complementary except for the two-nucleotide overhangs on the 3′ end of each strand.
4. Make a mutation in the fourth nucleotide from the 3′ end of the antisense strand (*see* **Note 21**).
5. Send RNA sequences to chosen manufacturer (IDT).
6. Request the addition of phosphates to the 5′ end of each strand. If biotinylating the mature strand, request the addition of a biotin moiety at the 3′ end of the mature strand (*see* **Note 22**).
7. Have the two RNA species annealed by the manufacturer. Alternatively the RNAs may be annealed by heating the two RNA species to 95 °C and slowing cooling to room temperature.

#### Transfection

A variety of reagents for transfection of miRNAs and siRNAs are commercially available. Conditions for transfection of both primary cells and established cell lines are typically provided by the manufacturer. Lipofectamine 2000, described in detail in Subheadings 3.1.1 and 3.4, provides a high level of transfection for a variety of cell types and gives the most consistent results when transfecting both plasmids and RNAs together. We typically use 1–2 pmol of miRNA mimic and 1.0 μL of Lipofectamine 2000 to transfect HEK293 cells in a 24-well format. For subsequent use in the RISC-IP, begin optimization using 100–400 pmol of the biotinylated miRNA mimic and 20 μL of Lipofectamine 2000 to transfect HEK293 cells in one 100 mm dish.

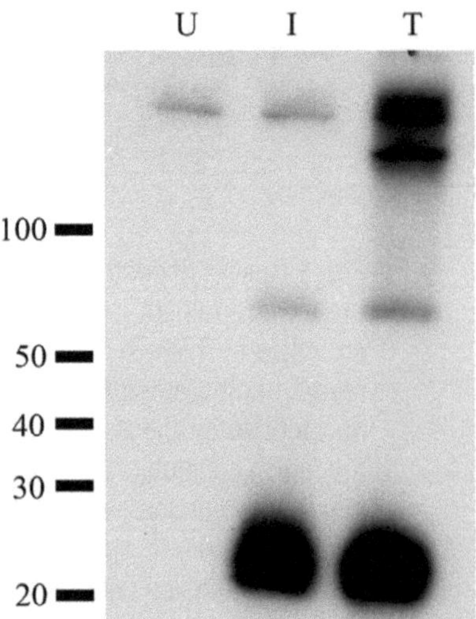


**Fig. 2** miRNA Northern blot analysis of RNA isolated from fibroblasts mock infected (M) or infected with CMV (I) for 72 h or 293 cells transfected (T) with a pSIREN vector for 72 h. Both the pre- (~60 nucleotide) and mature (~22 nucleotide) forms of a miRNA can be detected. The signal at the very *top* of the membrane likely represents the larger sized pri-miRNA transcripts from which the miRNA precursors are derived

### 3.2 Detecting miRNAs

#### 3.2.1 Method 1: miRNA Northern Blot

In this protocol, small RNAs are separated through a urea/acrylamide gel and transferred electrophoretically to a nylon-based membrane. An oligonucleotide probe designed antisense to the miRNA of interest is labeled with $\gamma$-$^{32}$P ATP and hybridized to the membrane [31]. Denaturing urea/acrylamide gels allow for the efficient separation of small, single-stranded RNA species. Therefore, both the pre- and mature forms of a miRNA can be simultaneously visualized (*see* Fig. 2). When compared to the stem-loop RT-PCR assay described elsewhere in this chapter, Northern blotting requires more total RNA and is less sensitive, which may be problematic for detecting miRNAs expressed at low levels. However, this technique is quantitative and technically straightforward and is an important validation step for miRNAs identified by more high-throughput methods.

Carry out all procedures at room temperature, unless otherwise noted.

##### Urea/Acrylamide Gel Electrophoresis

1. Make sure that all glass plates have been cleaned with 10 % SDS and rinsed with water and with 95 % ethanol. Allow the glass plates to dry before casting gel.
2. Prepare a Mini PROTEAN 3 casting tray using 1.5 mm spacers. Mix 10 mL of urea/acrylamide, 12.5 μL of TEMED, and 50 μL of 10 % APS. Pour into casting tray along with appropriate comb and allow gel to set.

3. Prepare gel tank apparatus and fill with 1× TBE. Rinse wells with 1× TBE using a syringe. Pre-run apparatus at 180 V for 1 h.
4. Add 1 volume of formamide and 6× RNA loading dye to 20 μg of concentrated RNA (*see* **Note 6**). Heat to 100 °C for 5 min and snap cool on ice. Load samples, along with 5 μL of small RNA marker, and run gel at 180 V until the bromophenol blue dye front is approximately two thirds of the way down the gel.
5. Carefully remove gel from the Mini PROTEAN 3 glass plates. Stain the gel with 3 μL of ethidium bromide diluted in 50 mL of water for 15 min. An image of the gel can be taken at this stage using an UV transilluminator and a fluorescent gel ruler lined up along the RNA marker lane (*see* **Note 23**).

Electrophoretic Transfer

1. Cut GeneScreen membrane and four pieces of Whatman paper to the correct size for the gel. Pre-wet the membrane by immersing in water and then equilibrate in 1× TBE for 15 min.
2. Pre-wet the Mini Trans-Blot sponges and Whatman paper in 1× TBE.
3. Prepare the transfer as follows: Place a pre-wetted sponge on the left side of the cassette. Add two pieces of pre-wetted Whatman paper. Carefully place the gel on this Whatman paper. Cover the gel with a piece of GeneScreen membrane and two more pieces of Whatman paper. Remove all bubbles from the system by carefully rolling a pipette across the stack. Add the second sponge to the top of the stack and close the cassette. Place the cassette into the electrophoresis unit ensuring the membrane is closest to the positive (red) electrode. Fill the apparatus with 1× TBE and run at the highest voltage possible (15–20 V) for 1 h.
4. Disassemble the transfer system and carefully remove the membrane. An image of the membrane can be taken using an UV transilluminator taking care to note the position of the RNA marker bands on the membrane using a pencil (*see* **Note 23**). Cross-link the RNA and membrane using a UV Stratalinker using the auto cross-link function ($1.2 \times 10^5$ μJ).

Prehybridization

1. Warm the PerfectHyb buffer to 42 °C and swirl to completely dissolve any precipitate. Warm the hybridization oven to 38 °C.
2. Pre-wet the membrane with water and lightly blot to remove excess liquid.
3. Incubate the membrane in 5 mL of PerfectHyb solution in a hybridization tube with continuous rotation at 38 °C for at least 30 min.

Label Probe

1. Add 1 μL of 1 pmol/μL oligonucleotide probe, 2 μL 10× forward buffer, 2 μL γ-$^{32}$P ATP, 1 μL polynucleotide kinase, and 14 μL water to an Eppendorf tube.
2. Mix components well and incubate at 37 °C for 1 h (*see* **Note 24**).
3. About 5–10 min before the end of the probe incubation, prepare the Quick Spin columns as follows: Gently invert the column several times to resuspend the matrix. Remove the cap and break off the bottom tip. Allow the packaged buffer to drain by gravity flow and discard. Place the column in a collection tube and centrifuge at 1,100 × *g* using a tabletop centrifuge for 2 min. Repeat, discarding the eluted buffer.
4. Add 20 μL of TE buffer to the labeling reaction, and mix well.
5. Keeping the column in an upright position in a new collection tube, carefully apply the labeling reaction to the center of the column bed.
6. Centrifuge at 1,100 × *g* for 4 min and collect the eluate.
7. Add an additional 30 μL of TE to the column bed.
8. Centrifuge at 1,100 × *g*, mixing the two eluates.
9. Count 1 μL of the eluate using a scintillation counter. A typical labeling reaction should produce between 70 and 90,000 CPM/μL.

Hybridization and Washes

1. Add all of the labeling reaction to the 5 mL of PerfectHyb in the hybridization tube, making sure not to touch the membrane. Rotate at 38 °C overnight.
2. The next day, rinse the blot in wash solution 1 two times at room temperature. Properly dispose of the radioactive waste (*see* **Note 7**). Wash two times in wash solution 1 at 38 °C for 10 min each.
3. Wash the blot two times in 5 mL wash solution 2 at 38 °C for 10 min each.

Film Development

1. Carefully remove the blot from the hybridization tube and remove excess solution by blotting.
2. Wrap the blot in plastic wrap and expose to autoradiography film in a dark room. An intensifier screen is useful if the miRNA of interest is expressed at low abundance.
3. Store film cassette at −80 °C. The length of time for film exposure will depend on the amount of labeled probe hybridized to the blot. Normally the film can be developed after 24 h of exposure (*see* **Note 25**).

*3.2.2 Protocol 2: Stem-Loop RT-PCR for Detection of miRNAs*

Detection of miRNAs in total RNA preparations can be performed using stem-loop RT-PCR as described in [32]. miRNA species are too short for standard primer pair probe design; thus this technique relies on the hybridization of the last 6 nucleotides of the 3′ end of the miRNA of interest to an RT primer which lengthens the cDNA product to approximately 65 nucleotides. The stem-loop structure prevents binding of the RT primer to the pre- or pri-miRNA due to steric hindrance; therefore no pre- or pri-miRNA should be amplified. miRNA-specific forward primers and probes allow for amplification of the cDNA product with both specificity and sensitivity (*see* **Note 26**). This assay requires only small amounts of input RNA and is easily amenable to high-throughput analyses. Primer and probe sets for most cellular and HCMV miRNAs are now available for purchase from Applied Biosystems. Custom primer and probe sets can also be designed for novel viral miRNAs (*see* **Note 27**).

Anneal the RT Primer (*See* **Note 28**)

1. Mix 25 μL of 200 μM RT primer with 10 μL of 10× RT buffer and 65 μL water in a PCR tube.
2. Cycle at 95 °C for 30 min, 72 °C for 2 min, and 37 °C for 2 min—ramping down 1 °C per second from 72 °C and 25 °C for 2 min—ramping down 1 °C per second from 37 °C.
3. The primer concentration after annealing is 50 μM. This can be frozen at −20 °C (*see* **Note 29**). The working concentration is 5 μM.

Reverse Transcription

1. For each reaction, mix together the following: 0.15 μL 10 μM dNTP mix, 1 μL Multiscribe enzyme, 1.5 μL 10× RT buffer, 0.19 μL RNasin, 0.3 μL 5 μM RT primer, and 9.86 μL water.
2. Aliquot 13 μL per tube. Add up to 100 ng of RNA in 2 μL (*see* **Note 30**).
3. Cycle at 16 °C for 30 min, 42 °C for 30 min, and 85 °C for 5 min. RT mix can be kept at 4 °C for short periods of time (days). Store at −20 °C for longer periods of time.

Taqman Assay

1. Mix together the following for each sample: 10 μL 2× Universal Master Mix, 7.46 μL water, 0.36 μL each of 50 μM forward and reverse primers, 0.32 μL 10 μM probe, and 1.5 μL RT sample (*see* **Note 31**).
2. Cycle at 50 °C for 2 min, 95 °C for 10 min, 95 °C for 15 s, and 60 °C for 1 min. Repeat the last two steps 40 times.

### 3.3 Identifying mRNA Targets of Viral miRNAs

#### 3.3.1 Method 1: RNA-Induced Silencing Complex (RISC) Immunoprecipitation (IP)

One of the more challenging aspects of miRNA biology is determining which mRNA transcripts are regulated by the miRNA. Numerous bioinformatic prediction algorithms have been developed based on complementarity between the miRNA of interest and sites within the 3′UTRs of transcripts; however, these often lead to high levels of false positives. In addition, many of these algorithms rely upon sequence conservation across species and thus cannot accurately predict targets for non-conserved viral miRNAs.

Techniques including microarray and SILAC have been used to identify targeted transcripts and proteins, respectively, that are differentially expressed upon the exogenous addition of the miRNA of interest or antisense inhibitors to the miRNA. However, changes in levels are often small and difficult to detect and distinguishing direct from indirect targets of the miRNA can be difficult. Recently, biochemical approaches, including RISC-IP, PAR-CLIP, and HITS-CLIP, have been developed that directly identify targeted transcripts that have been incorporated into the RISC complex [18, 22–27, 32]. These approaches usually identify transcripts containing seed matches (upwards of 75 %), and experimental validation is exceptionally high compared with bioinformatic approaches [6].

This section of the chapter will outline in detail the RISC-IP procedure. RISC-IP takes advantage of the complex that is formed when a miRNA within the RISC complex interacts with its target. Affinity purification of the RISC complex followed by RNA isolation and microarray analysis of bound transcripts leads to the direct identification of targeted transcripts. Several considerations must be taken into account when designing a RISC-IP screen including the source of miRNA, whether through infection or exogenous expression (described in Subheading 3.1), and the methods of isolating the RISC complex, for example, by using antibodies directed against one of the RISC components (e.g., -Ago1, Ago2, Ago3, or Ago4), cell lines expressing tagged components of the RISC complex (e.g., -c-myc-tagged-Ago2), or miRNAs that have been biotinylated (described in Subheading 3.1.2).

Prepare all solutions using RNase-free DNase-free ultrapure water. It is of utmost importance to perform all manipulations carefully to preserve the interaction between the RNA and protein and to prevent degradation of the RNA.

Carry out all procedures on ice unless otherwise specified.

Preparing the Cell Lysate

1. Determine the number of tubes and tissue culture scrapers needed to collect cell lysates and place on ice. Prepare DPBS 1× and lysis buffer and place on ice until cold. Immediately before lysing cells, remove a small aliquot of lysis buffer (enough for 1 mL/100 mm tissue culture dish) and add RNasin Plus RNase inhibitor to a final concentration of 1 U/μL. Place on ice.
2. Gently remove media from one 100 cm tissue culture dish using an aspirating pipette. Keeping the dish in an upright position, remove the remaining media with a P1000 pipette.
3. Wash cells by slowly adding 10 mL ice-cold DPBS to the side of the dish and gently tip to cover all cells. Remove DPBS wash as in **step 2**.

4. Add 1 mL ice-cold lysis buffer plus RNasin Plus RNase inhibitor to the dish covering the cells and scrape to collect cells and buffer. Work quickly to minimize time spent at room temperature. Collect lysate and transfer to an ice-cold tube and place on ice.
5. Repeat **steps 2–4** to harvest the remaining dishes.
6. Vortex each tube at maximum velocity for 10 s and return to ice for 10 min. Vortex again for 10 s and then centrifuge at 12,000 × *g* for 15 min at 4 °C. While spinning, place additional tubes on ice to chill.
7. Carefully transfer the supernatant to the new chilled tubes and place on ice. Remove 1/20 of the volume to be used as the total RNA sample and transfer to a tube containing 500 μL TRIzol. Vortex and freeze at −80 °C. Removing a small aliquot of cell lysate at this step to serve as the total protein sample is also recommended (*see* **Note 32**).

#### Preparing the Streptavidin-Agarose, ANTI-c-myc Agarose Conjugate, or Protein A Beads

1. Thoroughly mix the streptavidin-agarose, ANTI-c-myc Agarose, or protein A to resuspend the beads and transfer 25 μL/IP sample to a single tube. Centrifuge 12,000 × *g* for 30 s and discard supernatant.
2. Wash the beads by adding 500 μL lysis buffer without RNasin, gently mix, and centrifuge 12,000 × *g* for 30 s and discard supernatant. Repeat wash and resuspend beads in lysis buffer without RNasin (100 μL/IP sample). For each 100 μL, add 10 μL tRNA (10 mg/mL) and 10 μL BSA (10 mg/mL). Rotate 2 h at 4 °C to block. Once the beads are rotating in block, begin the "immunoprecipitating" section below if isolating the RISC complex using an antibody against one of the components.
3. Wash the beads by adding 300 μL lysis buffer, gently mix, and centrifuge 12,000 × *g* for 30 s and discard supernatant. Repeat for a total of two washes, and resuspend in lysis buffer (100 μL/IP sample). Place on ice.

#### Immunoprecipitating

1. If isolating the RISC complex using antibodies directed against one of the components, add the antibody to the cell lysates from **step** 7 in Subheading "Preparing the Cell Lysate" (*see* **Note 9**). If using streptavidin-agarose to purify biotinylated miRNAs or ANTI-myc Agarose to purify myc-tagged Ago2, proceed to Subheading "Pull-Down" below.
2. Rotate 2 h at 4 °C. Proceed to Subheading "Pull-Down."

#### Pull-Down

1. Gently mix the streptavidin-agarose, ANTI-c-myc Agarose Conjugate, or protein A from **step 3** in Subheading "Preparing the Streptavidin-Agarose, ANTI-c-myc Agarose Conjugate,

or Protein A Beads" to resuspend beads, and add 100 μL to the cell lysates.

2. Rotate 2 h at 4 °C.
3. Centrifuge 1 min at 5,000 × *g* 4 °C.
4. Carefully remove supernatant and resuspend in 500 μL lysis buffer. Centrifuge 1 min at 5,000 × *g* 4 °C. Repeat wash three additional times for a total of four washes. After the third wash, transfer suspended beads to a new chilled tube for the final wash.
5. After the final wash, leave the beads in approximately 50 μL lysis buffer. Removing a small aliquot of the immunoprecipitate at this step can serve as a control to ensure efficient immunoprecipitation of RISC complex proteins like Ago (*see* **Note 33**). Add 500 μL of TRIzol to the beads and remaining lysis buffer, mix well, and freeze at −80 °C until ready to process.

RNA Isolation from Pull-Downs

Adapted from TRIzol method for RNA isolation.

1. Add 0.1 mL chloroform to each total and IP sample, shake vigorously for 15 s, and incubate 2–3 min at room temperature.
2. Centrifuge the samples no more than 12,000 × *g* for 15 min at 4 °C.
3. Transfer the aqueous phase to a fresh tube.
4. Precipitate RNA by adding 0.25 mL isopropanol and 1 μL GlycoBlue Coprecipitant (*see* **Note 11**), mix well, and incubate 10 min at room temperature.
5. Centrifuge samples no more than 12,000 × *g* for 15 min at 4 °C.
6. Remove supernatant and wash the pellet once with 0.5 mL of 75 % ETOH. Mix by vortexing and centrifuge no more than 7,500 × *g* for 5 min at 4 °C.
7. Repeat **step 6** once for a total of two washes (*see* **Note 34**).
8. Completely remove the wash, dry briefly, and resuspend IP and total samples in 15 μL and 50 μL Nuclease-free water, respectively.
9. Determine RNA concentration and purity by using an Agilent Bioanalyzer or similar (*see* **Note 35**).

Analysis

1. Determine the success of the affinity purification by using a Western blot analysis to compare the amount of RISC proteins isolated in the purified fraction versus those in the total sample (*see* **Note 36**). Alternatively, the enrichment of the miRNA and a validated mRNA (if known) in the purified fraction compared with the total fraction can be determined by Taqman RT-PCR (*see* Subheading 3.3.2) (*see* **Note 37**).

2. Determine transcript levels in total and IP samples by microarray using an appropriate microarray platform and analysis software (*see* **Note 38**).
3. Determine fold enrichment to identify potential miRNA targets. Cellular mRNA transcripts that are associated with the RISC complex are identified by dividing the amount of mRNA transcripts in the IP fraction by the amount of mRNA in the total RNA fraction in order to account for direct effects of the miRNA expression on transcript levels within the cell. To identify cellular mRNA transcripts specifically enriched within the RISC complex containing exogenously added miRNA and to exclude those transcripts isolated by association with cellular miRNAs (this is particularly important when transfecting in miRNAs and controls using an expression vector and immunoprecipitating or pulling down via a component of the RISC complex), the enrichment profile in miRNA transfected cells is compared to cells transfected with the negative control such that the enrichment value of any given mRNA transcript = $(IP_{miRNA}/Total_{miRNA})/(IP_{Neg}/Tot_{Neg})$ [24]. A similar determination is used when pulling down RISC complexes using biotinylated miRNAs in conjunction with streptavidin-agarose. mRNA transcripts are then ranked according to level of enrichment with the highest enriched transcripts considered potential targets of miRNA.

#### 3.3.2 Method 2: Reporter Assays for Identification of Viral miRNA Targets

Reporter assays have proven invaluable in evaluating gene expression and have been adapted and utilized extensively to identify and characterize potential miRNA targets. Typically, the 3′UTR of a predicted miRNA target is cloned downstream of a reporter gene and is co-transfected into cells along with the miRNA of interest or negative control miRNA. Decreases in reporter activity in cells transfected with the miRNA relative to cells transfected with the negative control indicate miRNA-mediated regulation. A wide variety of reporter constructs and detection systems are commercially available.

Carry out all procedures at room temperature unless otherwise specified.

##### Transfection

1. Plate HEK293 cells at $1–2 \times 10^4$ cells/well in 100 μL media in a black-walled clear-bottom 96-well tissue-culture-treated plate (*see* **Note 39**).
2. 24 h later, prepare (per well) 100 ng luciferase reporter vector plus 0.3–0.4 pmol miRNA in 25 μL Opti-MEM I Reduced Serum Media (*see* **Note 40**)
3. For each well, add 0.5 μL Lipofectamine 2000 to 25 μL Opti-MEM I Reduced Serum Media. A Master Mix is recommended for multiple samples to ensure that a similar amount

of transfection reagent is added to each well. Gently mix by inverting and incubate 5 min at room temperature (*see* **Note 41**).

4. Add 25 μL of the Lipofectamine 2000 Master Mix in **step 3** to the luciferase reporter and miRNA in **step 2**, gently mix, and incubate for 20 min at room temperature.
5. Transfer the 50 μL reaction in **step 4** to each well dropwise and gently swirl plate. Incubate plate 12–24 h at 37 °C, 5 % $CO_2$.

Luciferase Assay

1. Carefully remove media and lyse cells by adding 20 μL 1× Passive Lysis Buffer to each well (*see* **Note 42**). Incubate the plate for 15 min on rocking platform at room temperature.
2. Prepare the necessary quantities of Luciferase Assay Reagent II (LAR II) and Stop & Glo reagents according to the manufacture's recommendations (50 μL/well + 0.5 mL extra) (*see* **Note 43**).
3. Prepare a plate-reading luminometer with two injectors according to the manufacturer's suggestions. Measurements are performed for each well sequentially as follows: inject 50 μL LAR II, delay for a 2 s pre-read, measure firefly luciferase activity for 2 s, inject 50 μL Stop & Glo, delay for a 2 s pre-read, and measure renilla luciferase activity for 2 s (*see* **Note 44**).
4. Calculations: For each sample, calculate the ratio of renilla signal/firefly luciferase signal where renilla is the test reporter and firefly luciferase is the control. Then calculate the average and standard deviation among replicates (*see* **Note 45**). Decreases in reporter activity in cells transfected with the miRNA relative to cells transfected with the negative control indicate miRNA-mediated regulation.

### 3.4 Inhibition of HCMV miRNAs Using Locked Nucleic Acids (LNA)

Several miRNA inhibition strategies based on antisense nucleic acids have been developed including miRNA sponges, decoys, and antagomirs [33]. Each contains sequence complementary to the mature miRNA and prevents the miRNA from binding its intended target. Chemical modifications of these inhibitors, such as the substitution of phosphate bonds with phosphorothioate bonds, methylation of the oxygen at position 2 in the ribose (2′-*O*-methyl), and the addition of an extra bridge between the carbons of the ribose ring (locked nucleic acid, LNA), have greatly increased their stability, affinity, and specificity. Among the chemically modified species, LNAs have recently emerged as the most efficient miRNA inhibitors due to their potency, versatility, and efficacy both in vitro and in vivo [34, 35]. For instance, several clinical trials using LNAs as small as eight nucleotides have demonstrated strong efficacy against hepatitis C virus in primates in vivo [36]. These findings have been confirmed in our lab, with LNAs designed against HCMV miRNAs displaying potent, versatile, and reliable inhibitory activity (data not shown). The following protocol describes how LNAs can be used to inhibit a miRNA mimic in the context of a luciferase assay. LNAs can also be used to antagonize miRNAs derived from the cell or virally expressed miRNAs following infection, though significant optimization is required. Inhibiting the expression of HCMV miRNAs individually or in combination through the use of LNAs is an important tool to evaluate the roles of miRNAs during infection, complementing the more time-consuming generation of miRNA-mutant viruses.

Carry out all procedures at room temperature. The following protocol is designed for HEK293 cells in a 96-well plate format and can be adapted for other cell types and tissue culture formats (*see* **Notes 46–50**).

1. Seed $1 \times 10^4$ HEK293 cells/well in a black-walled 96-well plate (*see* **Note 46**) in 0.1 mL medium.
2. 24 h later, prepare (per well) 0.3 pmol miRNA, 0.6 pmol LNA, and 100 ng luciferase reporter vector in 25 μL Opti-MEM I Reduced Serum Media (*see* **Note 47**).
3. For each well, add 0.25 μL Lipofectamine 2000 to 25 μL Opti-MEM I Reduced Serum Media (*see* **Notes 48** and **49**). A Master Mix is recommended for multiple samples to ensure that a similar amount of transfection reagent is added to each well. Gently mix by inverting and incubate 5 min at room temperature.
4. Combine the reaction mixes from **steps 2** and **3**, mix gently, and incubate for 20 min.
5. Add mix dropwise into wells for a final volume of 150 μL and gently swirl plate to distribute. Incubate plate at 37 °C, 5 % $CO_2$.
6. Perform luciferase assay 12–24 h post-transfection (*see* Subheading 3.2.2) (*see* **Note 50**).

## 4 Notes

1. We use the pSIREN-retroQ-ZsGreen for CMV miRNA expression because (1) the promoter is a RNA polymerase III-dependent promoter not derived from CMV and (2) it constitutively expresses GFP allowing one to easily monitor transfection efficiency.
2. Cloning the viral region surrounding the miRNA locus allows for the expression of the full pri-miRNA and the formation of a native hairpin. Mature miRNAs are considered to be more efficiently processed from a full hairpin than from a truncated one. Since the exact extent of the pri-miRNA is not known in most cases, comparing clones encompassing regions of different sizes around the miRNA locus allows for selection of the region that best expresses the miRNA in one round of cloning.

3. This transfection protocol has been optimized for HEK293 cells. miRNAs can be transfected using other methods (including electroporation) and/or in other cell lines, but each approach must be optimized individually.
4. SDS powder should be handled in a manner to minimize formation of aerosols. Wear proper protective equipment including a facemask when measuring SDS powder.
5. Ethidium bromide is an agent that intercalates into nucleic acids and is therefore believed to be a mutagen. Although the concentrations used in the laboratory are not associated with health risks, ethidium bromide should be handled with care. Wear proper protective equipment when handling ethidium bromide.
6. Due to the small well size of the Mini PROTEAN gel system (wells can hold approximately 40 μL), RNA should be concentrated to at least 1 μg/μL.
7. When using and disposing of radioactive materials, be sure to follow all federal, state, and local regulations. Make sure to wear proper protective equipment.
8. Proper RNA handling techniques should be observed at all times to prevent degradation.
9. RISC-IP has been performed successfully using either commercially available antibodies against Ago2 or serum raised against Ago2 in rabbits. Optimization is required to determine the concentration of antibody required for optimal immunoprecipitation.
10. A pilot experiment should be performed to determine the number of 100 mm plates required to obtain a sufficient amount of RNA for subsequent analysis by microarray.
11. The addition of GlycoBlue during precipitation aids in visualization of the pellet during washing. However, addition will interfere with the quantification of RNA unless the same concentration of GlycoBlue is included in the buffer used to blank the spectrophotometer.
12. Some HCMV miRNA mimics are commercially available for purchase as mature miRNAs or as pre-miRNA hairpins. If unavailable, miRNAs can be cloned into expression vectors as described in Subheading 3.1.1 or can be designed and custom-made by a variety of sources as described in Subheading 3.1.2.
13. A variety of reporter vectors are commercially available. Dual-Luciferase reporters are highly recommended as the effects of the test reporter can be compared to the control reporter. The control reporter provides an internal control that takes into account well-to-well variation due to differences in cell viability, transfection efficiency, pipetting error, and assay efficiency.

14. Performing experiments in a 96-well format is highly recommended. While this protocol is designed for 96-well plates, it can be scaled up for other cell culture formats.

15. For increased accuracy and high-throughput analyses, the use of a plate-reading luminometer equipped with two reagent injectors is highly recommended. If injectors are not available, perform injections manually and read one well at a time (*see* **Note 44**).

16. In addition to negative controls, Exiqon has designed LNAs for all of the HCMV miRNAs deposited in mirBase. LNAs can also be custom-made to target novel miRNAs or ones not present in mirBase.

17. The strategy for cloning a miRNA into a lentiviral vector is very similar to the pSIREN vector cloning strategy described here. Refer to the manufacturer's instructions for cloning details. Make sure to choose a lentiviral vector where miRNA expression is driven by a promoter that is not derived from CMV, since this could affect the level of expression in infected cells.

18. Purify viral DNA by isolating viral particles (e.g., using sorbitol cushion method) followed by DNA extraction (e.g., DNAzol or phenol/chloroform). Alternatively, total DNA from virus-infected cells can be isolated and used as template.

19. If the PCR reaction displays several amplification products, gel-purify the correct band using a DNA gel purification kit.

20. In our hands, transfection efficiency in HEK293 cells is close to 100 % as estimated by GFP expression.

21. The introduction of a mutation in the fourth nucleotide from the 3′ end of the antisense strand results in the preferential incorporation of the mature or guide strand into the RISC complex.

22. miRNAs that are biotinylated can be used in subsequent protocols including RISC-IP to identify miRNA targets. Only the mature strand should be biotinylated.

23. Carefully line up your gel and membrane with the fluorescent ruler when taking images using the UV transilluminator. The small RNA marker generally does appear on the membrane, and the bands should be clearly noted based on the position of the ruler for future reference.

24. This step can be performed during the prehybridization of the membrane.

25. Often, both pre- (~60 nucleotide) and mature (~22 nucleotide) forms of a miRNA can be detected by Northern blotting, so noting the position of the RNA markers is important (*see* Fig. 2). Commonly, radioactive signal can also be detected

at the very top of the membrane. This signal likely represents the larger sized pri-miRNA transcripts from which the miRNA precursors are derived.

26. In these assays, expression of each miRNA is assessed using a unique RT primer. The efficiency of the RT reaction for each miRNA is unknown, and therefore the expression of multiple miRNAs relative to one another cannot be accurately assessed.
27. To design custom primers and probes, it is very important to accurately determine the sequence of the mature miRNA, especially with regard to the 3′ end. The RT primer is designed using the sequence GTCGTATCCAGTGCAGGGTCCG AGGTATTCGCACTGGATACGAC. For each miRNA, the final additional 3′ nucleotides of the RT primer are unique and antisense to the last six nucleotides of the mature miRNA. The forward primers are designed as the first 15 nucleotides of the mature miRNA with 3–5 additional nucleotides at the 5′ end to result in a final annealing temperature of approximately 60 °C. The reverse primer is the same for each reaction (GTGCAGGGTCCGAGGT) and binds within the RT primer sequence. Finally, the probe sequence used with the RT primer outlined above has a sequence of TGGATACGAC followed by six nucleotides designed antisense to the last nucleotides of the mature miRNA (identical to the RT primer). The probe should be synthesized with a 5′ fluorescein (FAM) moiety, and minor groove binder (MGB) nonfluorescent quencher on the 3′ end [32]. Examples of custom-designed primer and probe sets can be found in Hancock et al. [5].
28. If Taqman RT-PCR assays are purchased from Applied Biosystems, RT primers do not need to be annealed.
29. We have found that 50 μM annealed primers are very stable through multiple freeze-thaw cycles.
30. Mock-treated samples are an important control for the RT-PCR reaction. Cross-reactivity of a viral primer and probe set with a cellular miRNA is unlikely but has been observed [5]. In addition, copy number for each miRNA can be estimated by preparing an RT reaction using a known concentration ($10^6$–$10^8$) of HPLC-purified oligonucleotide with the identical sequence to your miRNA of interest (*see* **Note 31**). If generating a standard curve using a miRNA duplex, denature the RNA prior to reverse transcription by incubating at 85 °C for 5 min followed by 60 °C for 6 min.
31. To determine the copy number of each miRNA in your samples, perform a five-step tenfold serial dilution of the RT reaction containing a known number of copies of the oligonucleotide standard (usually $10^6$–$10^9$ copies to be determined for each miRNA depending on their abundance). Prepare a Taqman

reaction for each serial dilution and generate a standard curve by plotting Ct value versus copy number. Determine the copy number of the miRNA in each sample using the standard curve.

32. The total protein sample can be compared with a protein sample taken after affinity purification to confirm the success or troubleshoot the RISC-IP. Western blot analyses for the immunoprecipitated protein or another component of the RISC complex can be performed comparing the total and IP protein samples with the expectation that components of the RISC complex will be pulled down.
33. The IP protein sample can be compared with the total protein sample by Western blot as described above in **Note 32**.
34. Two washes are required to remove any residual phenol:chloroform that will interfere with subsequent steps required for microarray.
35. Determine whether the quality, concentration, and total amount of RNA are sufficient for subsequent microarray analyses. High-quality RNA is required for the labeling reactions. The quantity and purity of each RNA sample is analyzed by applying a small amount for analysis on a bioanalyzer, which determines purity based on the ratio of the 28S to 18S ribosomal subunits. However, since RNA fractions obtained from RISC immunoprecipitation and pull-down based methods are not expected to contain ribosomal RNAs, basing the quality and purity of RNA on this ratio can be misleading.
36. Beads can be suspended directly in protein sample buffer for analysis by Western blot.
37. These steps are helpful in determining the success of the procedure. If the RISC-IP procedure was performed successfully, components of the RISC complex will be detected in the IP sample by Western blot and the miRNA and any known targets will be enriched in the IP sample as detected by stem-loop RT-PCR and real-time PCR respectively. The entire RISC-IP procedure can also be performed on a small scale and used to confirm potential miRNA targets. For example, to confirm potential targets identified by a large RISC-IP screen, the RISC-IP can be repeated on a smaller scale followed by stem-loop RT-PCR, to confirm the incorporation of the miRNA into the RISC complex and subsequent recruitment of the target mRNA.
38. Consulting with your institutional microarray core regarding amount and purity of RNA required, array platform, and post-array normalization as well as overall experimental design is recommended.

39. Cells should be no greater than 30–50 % confluent at the time of transfection. HEK293 will adhere to plastic and remain adherent; however, steps should be taken to minimize washing or replacing media. If using a transfection reagent other than Lipofectamine 2000, consult the manufacturer's guide in respect to cell plating density, amount and type of miRNA to use, and general protocol.
40. The amount of miRNA and vector used may need to be optimized on an individual basis. When transfecting two or more miRNAs simultaneously, the final concentration of miRNA in all samples should be equivalent. A negative control or GFP siRNA can be used to ensure that the same amount of miRNA or siRNA is present within each well. Transfecting more than one miRNA expression vector per well is not recommended.
41. Continue to the following step within 20 min.
42. Prepare 1× Passive Lysis Buffer from the 5× stock provided in the Dual-Luciferase Reporter Assay System by diluting one part 5× stock in four parts distilled water and mixing well. At the time of lysis, cells should be no more than 95 % confluent to achieve optimum lysis.
43. Prepare LAR II by resuspending the LAR II substrate in 10 mL Luciferase Assay Buffer II, aliquoting in small usable quantities, and freezing up to 1 year at −70 °C. We regularly prepare Stop & Glo by combining the vial of Stop & Glo substrate with the bottle of Stop & Glo buffer, mixing well, aliquoting in small usable volumes, and freezing at −70 °C. For small-scale assays, resuspend the Stop & Glo substrate in the Stop & Glo buffer at a ratio of 1:50.
44. The intensity of the firefly and renilla luciferase signal varies depending on the cell line, the amount of reporter plasmid, the transfection efficiency, and other factors. The measurement times may need to be optimized and can range from 1 to 10 s or more. Keep a pre-read delay of 2 s after injection and the same measurement time for firefly and renilla luciferase. If performing injections manually, proceed one well at a time: inject LAR II reagent and read the firefly luciferase signal for 12 s after a 3 s delay and then inject Stop & Glo reagent and read the renilla luciferase signal for 12 s after a 3 s delay.
45. As the effects of miRNAs are often modest, performing assays in triplicate and running multiple experiments is highly recommended.
46. Fibroblasts and THP-1 cells are seeded at $1 \times 10^4$ cells/well in a 96-well format. For THP-1 cells, differentiation and plate attachment can be induced by adding 50 ng/mL phorbol-12-myristate-12-acetate (PMA) to the culture media.

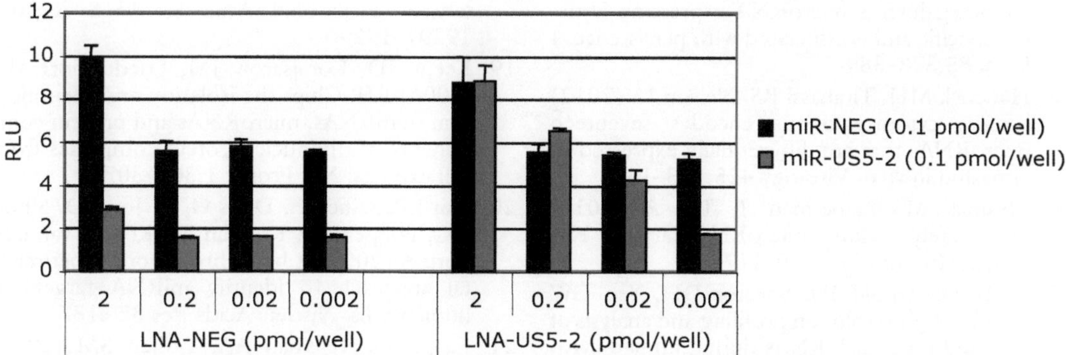


**Fig. 3** LNA-US5-2 inhibits miR-US5-2 targeting of US7 3′UTR. The psiCHECK-2-US7 3′UTR Dual-Luciferase reporter construct (which contains a target site for miR-US5-2 that efficiently decreases luciferase expression) was co-transfected with the indicated combinations of miR-US5-2 and LNA-US5-2 or their corresponding negative controls, miR-NEG and LNA-NEG. *Left panel*: 0.1 pmol/well miR-US5-2 inhibits luciferase expression threefold, while LNA-NEG has no effect. *Right panel*: LNA-US5-2 completely blocks the effect of miR-US5-2 in the range 0.2–2 pmol/well and partially inhibits miR-US5-2 activity at 0.02 pmol/well

47. When co-transfecting an LNA with a miRNA mimic, use a 1:1 to 10:1 LNA:miRNA concentration ratio. Controls consist of combinations of nontargeting miRNAs, negative control (NEG), and LNAs (*see* Fig. 3 for an example of controlled experiment).
48. miRCURY LNA™ microRNA Power Inhibitors enter HEK293 cells even in the absence of transfection reagent and can be added directly to the cell culture before or after the transfection of the other components and at the same concentration as in a regular transfection. LNAs are stable and remain active in the HEK293 cell culture for at least 2 days. It is not clear if these LNAs would exhibit similar properties in other cell types.
49. Luciferase assays can be very sensitive to transfection reagent-mediated cell toxicity. When using other transfection reagents and cell lines, make sure to optimize the quantity of transfection reagent.
50. At the proper LNA–miRNA ratio, the LNA antisense to the miRNA (but not the nontargeting LNA) will block the effect of the miRNA as measured by luciferase assay (*see* Fig. 3).

## References

1. Lee RC, Feinbaum RL, Ambros V (1993) The C. elegans heterochronic gene lin-4 encodes small RNAs with antisense complementarity to lin-14. Cell 75:843–854
2. Grey F, Antoniewicz A, Allen E, Saugstad J, McShea A, Carrington JC, Nelson J (2005) Identification and characterization of human cytomegalovirus-encoded microRNAs. J Virol 79:12095–12099
3. Pfeffer S, Sewer A, Lagos-Quintana M, Sheridan R, Sander C, Grasser FA, van Dyk LF, Ho CK, Shuman S, Chien M, Russo JJ, Ju J, Randall G, Lindenbach BD, Rice CM, Simon V, Ho DD, Zavolan M, Tuschl T (2005) Identification of microRNAs of the herpesvirus family. Nat Methods 2:269–276
4. Meyer C, Grey F, Kreklywich CN, Andoh TF, Tirabassi RS, Orloff SL, Streblow DN (2011)

Cytomegalovirus microRNA expression is tissue specific and is associated with persistence. J Virol 85:378–389

5. Hancock MH, Tirabassi RS, Nelson JA (2012) Rhesus cytomegalovirus encodes seventeen microRNAs that are differentially expressed in vitro and in vivo. Virology 425:133–142
6. Thomas M, Lieberman J, Lal A (2010) Desperately seeking microRNA targets. Nat Struct Mol Biol 17:1169–1174
7. Stark TJ, Arnold JD, Spector DH, Yeo GW (2012) High-resolution profiling and analysis of viral and host small RNAs during human cytomegalovirus infection. J Virol 86:226–235
8. Bartel DP (2009) MicroRNAs: target recognition and regulatory functions. Cell 136: 215–233
9. Dolken L, Perot J, Cognat V, Alioua A, John M, Soutschek J, Ruzsics Z, Koszinowski U, Voinnet O, Pfeffer S (2007) Mouse cytomegalovirus microRNAs dominate the cellular small RNA profile during lytic infection and show features of posttranscriptional regulation. J Virol 81:13771–13782
10. Buck AH, Santoyo-Lopez J, Robertson KA, Kumar DS, Reczko M, Ghazal P (2007) Discrete clusters of virus-encoded microRNAs are associated with complementary strands of the genome and the 7.2-kilobase stable intron in murine cytomegalovirus. J Virol 81:13761–13770
11. Cullen BR (2011) Viruses and microRNAs: RISCy interactions with serious consequences. Genes Dev 25:1881–1894
12. Griffiths-Jones S (2004) The microRNA registry. Nucleic Acids Res 32:D109–D111
13. Griffiths-Jones S, Grocock RJ, van Dongen S, Bateman A, Enright AJ (2006) miRBase: microRNA sequences, targets and gene nomenclature. Nucleic Acids Res 34:D140–D144
14. Griffiths-Jones S, Saini HK, van Dongen S, Enright AJ (2008) miRBase: tools for microRNA genomics. Nucleic Acids Res 36:D154–D158
15. Kozomara A, Griffiths-Jones S (2011) miRBase: integrating microRNA annotation and deep-sequencing data. Nucleic Acids Res 39:D152–D157
16. Umbach JL, Cullen BR (2009) The role of RNAi and microRNAs in animal virus replication and antiviral immunity. Genes Dev 15: 1151–1164
17. Yang JS, Lai EC (2011) Alternative miRNA biogenesis pathways and the interpretation of core miRNA pathway mutants. Mol Cell 43:892–903
18. Karginov FV, Conaco C, Xuan Z, Schmidt BH, Parker JS, Mandel G, Hannon GJ (2007) A biochemical approach to identifying microRNA targets. Proc Natl Acad Sci U S A 104: 19291–19296
19. Keene JD, Komisarow JM, Friedersdorf MB (2006) RIP-Chip: the isolation and identification of mRNAs, microRNAs and protein components of ribonucleoprotein complexes from cell extracts. Nat Protoc 1:302–307
20. Tan LP, Seinen E, Duns G, de Jong D, Sibon OC, Poppema S, Kroesen BJ, Kok K, van den Berg A (2009) A high throughput experimental approach to identify miRNA targets in human cells. Nucleic Acids Res 37:e137
21. Easow G, Teleman AA, Cohen SM (2007) Isolation of microRNA targets by miRNP immunopurification. RNA 13:1198–1204
22. Orom UA, Lund AH (2007) Isolation of microRNA targets using biotinylated synthetic microRNAs. Methods 43:162–165
23. Orom UA, Nielsen FC, Lund AH (2008) MicroRNA-10a binds the 5′UTR of ribosomal protein mRNAs and enhances their translation. Mol Cell 30:460–471
24. Grey F, Tirabassi R, Meyers H, Wu G, McWeeney S, Hook L, Nelson JA (2010) A viral microRNA down-regulates multiple cell cycle genes through mRNA 5′UTRs. PLoS Pathog 6:e1000967
25. Licatalosi DD, Mele A, Fak JJ, Ule J, Kayikci M, Chi SW, Clark TA, Schweitzer AC, Blume JE, Wang X, Darnell JC, Darnell RB (2008) HITS-CLIP yields genome-wide insights into brain alternative RNA processing. Nature 456:464–469
26. Hafner M, Landthaler M, Burger L, Khorshid M, Hausser J, Berninger P, Rothballer A, Ascano M Jr, Jungkamp AC, Munschauer M, Ulrich A, Wardle GS, Dewell S, Zavolan M, Tuschl T (2010) Transcriptome-wide identification of RNA-binding protein and microRNA target sites by PAR-CLIP. Cell 141:129–141
27. Hafner M, Landthaler M, Burger L, Khorshid M, Hausser J, Berninger P, Rothballer A, Ascano M, Jungkamp AC, Munschauer M, Ulrich A, Wardle GS, Dewell S, Zavolan M, Tuschl T (2010) PAR-CliP—a method to identify transcriptome-wide the binding sites of RNA binding proteins. J Vis Exp 41:2034–2038
28. Konig J, Zarnack K, Rot G, Curk T, Kayikci M, Zupan B, Turner DJ, Luscombe NM, Ule J (2011) iCLIP—transcriptome-wide mapping of protein-RNA interactions with individual nucleotide resolution. J Vis Exp 50:2638–2642
29. Vinther J, Hedegaard MM, Gardner PP, Andersen JS, Arctander P (2006) Identification of miRNA targets with stable isotope labeling by amino acids in cell culture. Nucleic Acids Res 34:e107

30. Tirabassi R, Hook L, Landais I, Grey F, Meyers H, Hewitt H, Nelson J (2011) Human cytomegalovirus US7 is regulated synergistically by two virally encoded microRNAs and by two distinct mechanisms. J Virol 85:11938–11944

31. Reinhart BJ, Slack FJ, Basson M, Pasquinelli AE, Bettinger JC, Rougvie AE, Horvitz HR, Ruvkun G (2000) The 21-nucleotide let-7 RNA regulates developmental timing in Caenorhabditis elegans. Nature 403:901–906

32. Chen C, Ridzon DA, Broomer AJ, Zhou Z, Lee DH, Nguyen JT, Barbisin M, Xu NL, Mahuvakar VR, Andersen MR, Lao KQ, Livak KJ, Guegler KJ (2005) Real-time quantification of microRNAs by stem-loop RT-PCR. Nucleic Acids Res 33:e179

33. Stenvang J, Petri A, Lindow M, Obad S, Kauppinen S (2012) Inhibition of microRNA function by antimiR oligonucleotides. Silence 3:1

34. Grunweller A, Hartmann RK (2007) Locked nucleic acid oligonucleotides: the next generation of antisense agents? BioDrugs 21:235–243

35. Veedu RN, Wengel J (2010) Locked nucleic acids: promising nucleic acid analogs for therapeutic applications. Chem Biodivers 7: 536–542

36. Lanford RE, Hildebrandt-Eriksen ES, Petri A, Persson R, Lindow M, Munk M, Kauppinen S, Ørum H (2010) Therapeutic silencing of microRNA-122 in primates with chronic hepatitis C virus infection. Science 327:198–201

# Chapter 15

# What We Have Learned from Animal Models of HCMV

**Pranay Dogra and Tim E. Sparer**

## Abstract

Although human cytomegalovirus (HCMV) primary infection is generally asymptomatic, in immune-compromised patients HCMV increases morbidity and mortality. As a member of the betaherpesvirus family, in vivo studies of HCMV are limited due to its species specificity. CMVs from other species are often used as surrogates to express HCMV genes/proteins or used as models for inferring HCMV protein function in humans. Using innovative experiments, these animal models have answered important questions about CMV's life cycle, dissemination, pathogenesis, immune evasion, and host immune response. This chapter provides CMV biologists with an overview of the insights gained using these animal models. Subsequent chapters will provide details of the specifics of the experimental methods developed for each of the animal models discussed here.

**Key words** Human cytomegalovirus, Mouse cytomegalovirus, Rat cytomegalovirus, Rhesus cytomegalovirus, Pathogenesis, Immune control, Immune evasion, Dissemination, Latency, Reactivation, Vaccine, Animal models for HCMV studies, In vivo

## 1 Introduction

One of the hallmarks of β-herpesviruses is their species specificity. This means that human CMV (HCMV) does not productively infect mouse cells and vice versa. The species barrier is not due to attachment or entry but a combination of factors including blocks in immediate-early gene expression and specificity of antiapoptotic proteins [1–6]. Without the ability to use HCMV in animal models, it necessitates the development and utilization of animal models of HCMV infection. Despite these limitations, animal models have been useful for studying pathogenesis, immune control, immune evasion, dissemination within the host, latency and reactivation, and vaccine/drug development [7–12]. In this chapter, we will discuss the four main animal models used to study HCMV. After a discussion of the advantages and disadvantages of each model, the proceeding sections will focus on the characteristics of

Andrew D. Yurochko and William E. Miller (eds.), *Human Cytomegaloviruses: Methods and Protocols*,
Methods in Molecular Biology, vol. 1119, DOI 10.1007/978-1-62703-788-4_15, © Springer Science+Business Media New York 2014

HCMV infection that have been investigated in the different animal models. The main animal models are:

1. The mouse CMV (MCMV) model
2. The rat CMV (RCMV) model
3. The guinea pig CMV (GPCMV) model
4. The rhesus CMV (RhCMV) model

## 2 Advantages and Disadvantages of the Animal Models

The complexity of host–viral interaction makes an exact mimic of HCMV infection of humans difficult. Nonetheless we have gained tremendous insight into the pathogenesis of HCMV with these animal models. However, it must be kept in mind that these models have their limitations for studying HCMV pathogenesis. Below are some of the major advantages and disadvantages of the different animal models discussed in this chapter.

### 2.1 Mouse Model (Reviewed in Ref. 13)

*Advantages*

- Characteristics of MCMV infection in mice are similar to that of HCMV infection in humans [14–17].
- MCMV contains homologues and/or at least functional homologues of many HCMV genes and gene products. The MCMV genome can be easily manipulated to either delete or exchange genes between HCMV and MCMV [18, 19].
- The mouse has a well-characterized immune system, short gestational periods, and large litter sizes. There are numerous immunologic reagents available including transgenic and knockout mice [19, 20].

*Disadvantages*

- The single major disadvantage of the mouse model is that the placental barrier is refractory to CMV transmission [21], except in a severe combined immune-deficient (SCID) mice [22]. This is most likely due to the three-cell-thick trophoblast layer that separates maternal and fetal circulations [19].

### 2.2 Rat Model

*Advantages*

- Similarity between HCMV and RCMV pathogenesis [12].
- HCMV genetic counterparts in the RCMV genome and the availability of viral mutants [19, 23].
- Availability of immunologic reagents and transgenic animals [19].
- Larger size makes it better suited to surgical manipulation.

*Disadvantages*

- No clear disease phenotype in pups for modeling congenital infection.
- Congenital and placental infections have only recently been described [19].

### 2.3 Guinea Pig Model (Reviewed in Refs. 7, 24)

*Advantages*

- GPCMV can cross the guinea pig placenta, causing infection in utero. This is probably due to single trophoblast layer separating maternal and fetal circulations. This makes the guinea pig well suited for the study of vaccines designed to interrupt transplacental transmission of infection [8, 19].
- The presence of HCMV counterpart genes in the GPCMV genome [25].

*Disadvantages*

- The lack of immunologic reagents for guinea pig studies.
- Lengthy guinea pig gestational periods with relatively small litter size slow down animal studies [19].

### 2.4 Rhesus Model (Reviewed in Ref. 9)

*Advantages*

- High similarity between the pathogenesis of infection of HCMV and RhCMV [9, 26, 27].
- Relatedness of the genomes of RhCMV and HCMV and the availability of viral mutants [19, 28].

*Disadvantages*

- The high cost of the animal maintenance.
- The paucity of RhCMV-seronegative animals because RhCMV infection is ubiquitous in most colonies [19].

## 3 Pathogenesis

Before discussing the animal models used for studying HCMV pathogenesis, it is important to discuss what is known about HCMV disease in humans. HCMV infection causes severe disease in immunocompromised patients including individuals with AIDS, organ transplant patients, cancer [29], and newborns [30]. Infection in these patients can sometimes cause clinical disease including mononucleosis-like syndrome, interstitial pneumonia, gastroenteritis, retinitis, or transplant rejection. Acute rejection and cardiac allograft vascular disease is reduced with suppression of subclinical cytomegalovirus infection [31–33]. HCMV is the leading viral cause of congenital birth defects following infection in utero. Worldwide between 0.5 and 2 % of newborns are infected. The majority of newborns are asymptomatic at birth, but some exhibit outward signs of infection including microcephaly, jaundice, and hepatosplenomegaly [34, 35]. About 10 % of the asymptomatic newborns develop neurological dysfunction, most prominently sensorineural hearing loss (SNHL) that appears when they are older [19, 36, 37]. Animal models provide some insights into these different aspects of HCMV disease.

In immunocompetent individuals, HCMV infection is generally asymptomatic. However, clinical studies point to HCMV's contribution to cardiovascular [38–41] and inflammatory bowel diseases [42–44]. HCMV infection is associated with many types of cancers [45–49]; however, there is no strong evidence of transformation of normal human cells after HCMV infection [47, 48]. In cardiovascular disease, there is an association between HCMV infection and the thickening of the arteriole walls during heart transplantation rejection [50], vasculopathy [51], and arteriolar dysfunction [52].

Some aspects of these HCMV-related diseases can be recapitulated in different animal models. This, in turn, allows the assessment of CMV's contribution to the disease and exploration of potential treatment. Although not all aspects parallel human infection and disease, these animal models are an excellent resource for dissecting particular CMV disease models. Discussed below are the main observations gleaned from animal models of HCMV pathogenesis.

### 3.1 Congenital Infection

Animal model systems that mimic HCMV-induced developmental abnormalities have been used to study the pathological outcomes of CMV congenital disease [7–9, 53–56]. GPCMV and RhCMV, due to their natural ability to cross the placental barrier and cause in utero infection, have been the models of choice for congenital infections [8, 56–59]. Although MCMV and RCMV are very inefficient at crossing the placenta due to the unique features of the trophoblastic layer, MCMV in the SCID model and a new strain of RCMV are capable of crossing the placenta and cause symptoms similar to HCMV congenital disease [10, 22]. Because this is an inefficient process, direct injection of the virus into the central nervous system (CNS) of the fetus, uterus, placenta, brain, or peritoneum of neonatal animals is most commonly used to recapitulate HCMV-induced congenital disease [20, 53–56, 60–64]. One of the drawbacks of these models is assessing whether the infection has led to SNHL or other developmental defects.

Discoveries using these models have led to a better understanding of how HCMV infection leads to developmental defects. Mouse studies have shown that the susceptibility to CMV infection is dependent on gestational age/developmental stage of the embryo. Although embryonic stem cells are resistant to CMV infection initially, as they differentiate they become permissive to

MCMV replication [60, 65]. Once the pup is born, there is a reduction in the susceptibility of the brain during development from neonate to adult. This may be due to a decrease in the number of susceptible cells in the developing brain [66] or increased immune responses, providing protection against infection [67]. CMV targets neural stem cells and the auditory nerve spiral ganglion in the developing brain [27, 56, 68–70]. Several factors that contribute to the development of SNHL have been identified using the murine model of CMV congenital infection, including ultrastructural lesions of the neurons, reduction in the number of spiral ganglion neurons [20], and cytopathic effects of viral replication in the inner ear (cochlea) [19, 27, 56, 69, 71]. The virus spreads to the inner ear most likely via the perilymphal routes [19] where the inflammation induced by CMV viral chemokines may contribute towards pathogenesis [71, 72].

Similarly, the intrauterine model of rhesus CMV infection identified that CMV infection of neuronal stem/progenitor cells before neuronal migration, differentiation, and organization results in more severe outcomes [56]. Infection after these developmental processes are completed results in less severe disease, suggesting the timing of CMV infection during fetal development is one factor in determining disease severity. CMV infection is not limited to the developing CNS. Systemic CMV infection can cause non-CNS diseases like intrauterine growth restriction and renal and hepatic damage to the fetus [27, 56].

Animal models of neonatal CMV infection have provided tremendous insight into CMV neuropathogenesis and the role of immune responses in controlling infection [65, 70, 73–75]. The neonatal mouse model of CMV infection in the presence of maternal antibodies has contributed to current vaccination strategies. The fact that experimental intrauterine infection of rhesus monkeys does not always lead to adverse outcomes implies that other factors limit CMV disease [56].

### 3.2 Vasculopathy and Graft Rejection

The rat and mouse models have been used to investigate the role of CMV infection in cardiovascular disease (CVD). In humans, circumstantial evidence points to the contribution of HCMV infection to the development of arterial restenosis following angioplasty, atherosclerosis, and solid organ transplant vascular sclerosis (TVS) [76, 77]. In the mouse model, MCMV infection accelerates atherosclerosis development in mice with high cholesterol [78–82]. CMV infection has a proinflammatory influence on the microvasculature that increases its susceptibility to both proinflammatory and thrombogenic responses caused by hypercholesterolemia [83]. MCMV and RCMV chemokine receptors M33 and R33, respectively, which are functional homologs of HCMV US28, are required for smooth muscle cell migration to the site of vascular injury. Their accumulation in the vessel intima leads to vessel narrowing and development of CVD [84, 85].

In rat organ transplantation models of chronic rejection (CR), active CMV replication [86, 87] contributes to accelerated graft rejection and increased vasculopathy in allograft vessels [88–92]. Several tissue-specific RCMV genes involved in host modification of inflammatory and tissue repair processes are upregulated in allograft recipients and may contribute to CR [93]. RCMV chemokine receptor R33 also plays an important role in the acceleration of CR [85]. Although prophylactic treatment with ganciclovir in humans delays the time to allograft rejection [94, 95], the rat model shows that recipients of latently infected donor hearts treated with ganciclovir do not prevent CR or TVS. One explanation for this discrepancy is that RCMV induces tertiary lymphoid structure formation and alteration of donor tissue T cell profiles prior to transplantation [96]. Both the mouse and rat models provide unique tools to dissecting the role that CMV plays in CVD and graft rejection.

### 3.3 Retinitis

The mouse model of CMV retinitis is a common model for HCMV retinitis. MCMV readily infects ocular tissue [97–100] and establishes latency in the eye [101]. During the acute phase of the response to CMV infection, there is a rapid expansion and infiltration of $CD8^+$ T cells into the infected retina. This is followed by a contraction phase where viral antigen presentation and $CD8^+$ T cell activation occur in the spleen and the draining lymph nodes but not in the retina or iris [102]. Using this model, TNFα was shown to induce apoptosis of retinal neurons and that bystander cells contribute to the pathogenesis of CMV retinitis [103, 104] while not being T cell dependent [104]. The mouse model also highlights the protective role of $CD8^+$ T cells [105, 106] and NK cells [107] against MCMV-induced necrotizing retinitis. The mouse model, although not perfect, provides a system for exploring not only the mechanism of CMV retinitis but also potential treatment options.

### 3.4 CMV Infection of Immunodeficient Host

HCMV is one of the opportunistic infections in late-stage AIDS patients, leading to pneumonitis, gastroenteritis, and/or retinitis. The animal models that best recapitulate this scenario are mouse retrovirus-induced immunodeficiency syndrome (MAIDS) and either spontaneous simian AIDS (SAIDS) or experimental infection of rhesus monkeys with SIV [108, 109]. In the rhesus model, CMV infects (similar to human infections) the gastrointestinal tract, hepatobiliary system, lungs, and testicles [110, 111]. Much like in humans, reactivation of CMV and the development of disseminated CMV disease are the results of diminished CMV-specific $CD4^+$ T cell and $CD8^+$ T cell immune responses [112, 113]. Interestingly in this model, RhCMV and SIV coinfection suggested that concurrent primary infection with CMV could augment the development of AIDS [9, 114]. In the mouse model, although T cell subsets play a role in MAIDS/MCMV pathology, Dix and colleagues suggested that the type of T cell response (i.e., perforin-mediated cytotoxicity) contributes to the severity of MCMV retinitis [115].

Without an HIV equivalent in the guinea pig model, cyclophosphamide treatment is used as an immune suppressant. The suppression of T and B cell immunity following CMV infection leads to lethal CMV infection in these animals [116]. This once again illustrates the importance of immune cells for maintaining control of CMV infection.

## 4 Immune Control

In the mouse and rhesus models of CMV infection CD8+ T cells, CD4+ T cells, and NK cells are the major cell types responsible for immune control of replication, latency, and reactivation [112, 117–119]. In the rhesus model, the target antigens for cytotoxic T lymphocytes (CTL) responses are the immediate-early proteins 1 and 2, and pp65-2, the homolog of HCMV pp65 [120, 121]. With the identification of the major target antigens, it is now possible to explore vaccination strategies using the major CTL targets as antigens.

Using these animal models, virus-specific antibodies have also been shown to play a crucial role in preventing CMV-induced pathology [122, 123]. The neutralizing antibodies target mainly glycoprotein B (gB). However, in the recently completed human vaccine trials, gB vaccination generated a strong antibody response and with a vaccine efficacy that exceeded predictions albeit with less than 50 % efficacy [124]. In order to test whether inclusion of additional antigens could increase vaccine efficacy, a DNA vaccination/vaccinia virus prime-boost regimen was used to vaccinate rhesus monkeys. They were subsequently challenged with RhCMV, and the amount of viral shedding was measured. Even with these additional antigens, the monkeys were not protected completely from infection and still shed virus [125]. It will be interesting to see if this level of protection is sufficient to protect the developing fetus in RhCMV [56].

For understanding immune control, the mouse model has allowed an in-depth analysis of the immune responses for controlling MCMV infection [126–129]. Besides showing that CTLs and NK cells are important in controlling MCMV infection [130–132], this model system also showed that different immune cells control MCMV infection in different organs. For example, CD4+ T cells expressing IFNγ control MCMV in the liver and IL10 expressing CD4+ T cells are important for clearance in the salivary gland [133–135].

## 5 Immune Evasion

The previous section discussed how animal hosts control CMV infection, but CMV has evolved mechanisms for evading many of these responses. Using animal models of CMV infection, many factors involved in virus immune evasion and their role in CMV survival and damage in their host have been identified. These CMV proteins not only allow the virus to avoid the immune system, but can also activate it to the virus' advantage (reviewed in refs. 136, 137). In the mouse model, the MCMV chemokine homolog MCK-2 (m131-m129) has a potent proinflammatory property and plays a crucial role in dissemination and immune evasion [138–140]. The function of the HCMV viral chemokines, vCXCL-1 and vCXCL-2, in vivo has been inferred from this data, even though they are from a different subclass of chemokines [141, 142]. The constitutively active CMV chemokine receptors utilize the signaling from the chemokine receptors for its advantage. In HCMV there are four chemokine-like receptors: US27, US28, UL33, and UL78 (reviewed in ref. 143). The rodent homologues, M33 and M78 in MCMV and R33 in RCMV, are the counterparts of HCMV UL33 and have a crucial role in immune evasion and/or dissemination [144–146]. Recently the mouse model was used as a surrogate for replacing the function of M33 with the HCMV GPCR homologues US28 and UL33 [147]. The RhCMV genome also encodes six CXC chemokines and five viral GPCRs [28, 148–151], and although they are dispensable for virus growth in vitro, the function of most of these proteins in vivo is unknown [152].

HCMV also encodes proteins with cytokine homology. HCMV encodes a homolog to host IL10 [153–158]. Endogenous IL10 is an immunosuppressive cytokine that downregulates T cell activation. RhCMV also encodes a homolog of rhesus IL10, which possesses potent anti-inflammatory activity that weakens the antibody and cellular immune responses in vivo [159, 160]. MCMV lacks an equivalent IL10 homologue, which reduces its usefulness as a model of HCMV IL10 in vivo.

The MHC class I homologs MCMV m144 and RCMV r144 are the counterparts of UL18 in HCMV. Because of their MHC class I homology, it was speculated that UL18 and its counterparts would therefore be important for preventing NK cell lysis. Using knockout viruses in their respective models, these proteins contribute to virus survival and dissemination [161, 162] (reviewed in ref. 163). In vivo, the MCMV protein m04/gp34, which escorts class I proteins to the cell surface, was shown to prevent NK cell activation [164]. Taken together the mouse system has been valuable for mapping which immune evasion proteins are important for resistance to NK cell lysis.

In human, mouse, and rhesus CMVs, there are several proteins that alter class I expression/antigen processing and presentation in vitro. These proteins are "functional" homologues of proteins in HCMV (i.e., limited sequence homology but similar functions). The HCMV-encoded proteins, gpUS2, gpUS3, gpUS6, and gpUS11, interfere with MHC class I surface expression and antigen presentation [145, 165]. The RhCMV homologues of HCMV gpUS2 and gpUS11 are the functionally related gpRh182 and gpRh189 proteins, while gpRh185 also has many of the functional features of gpUS2, gpUS3, and gpUS11 [166]. It was initially hypothesized that these proteins would diminish CD8+ T cell detection, but in vivo evidence following infection with recombinant viruses lacking some or all of these proteins leads to a similar immune response and equivalent viral titers. Recent evidence from rhesus models points to a role for these MHC homologs in superinfection [167]. Superinfection not only explains how humans can be infected multiple times with the same strain of CMV but also presents a problem for vaccinologists. The data from the rhesus experiments point to the difficulties of effective vaccine development, which will require additional immune responses besides the unprotective CD8+ T cell response [168]. Paradoxically, the MCMV equivalents of the class I immune-modulating proteins (m152/gp40, m04/gp34, and m06/gp48) contribute to an increase in processed and presented antigens leading to a *greater* CD8+ T cell response. This casts into doubt the working hypotheses about these proteins that they function to dampen CD8+ T cell responses [169]. Perhaps this phenomena is the host's countermeasure for the viral counter measure!

CMV infection also induces inflammatory mediators, which seem to have an important role in viral replication. Cyclooxygenase-2 (COX-2), for example, is an enzyme that leads to the generation of inflammatory lipid-derived compounds such as prostaglandin E(2). Although HCMV does not have a COX-2 homolog, it upregulates cellular COX-2 protein expression upon infection. COX-2 and the production of prostaglandin E(2) are necessary for HCMV infection. In fact, COX-2 inhibitors prevent normal HCMV replication [170]. Unlike HCMV, RhCMV encodes a COX-2 homolog, which is critical for viral growth in endothelial cells [171]. Thus, controlling inflammation is not only important for immune responses but also contributes to efficient viral replication. The role of the CMV-induced or CMV-encoded inflammatory mediators in vivo has yet to be determined.

HCMV inhibits apoptotic cellular defenses (reviewed in ref. 172). Using knockout mice and viral deletion mutants, Upton et al. showed the importance of the viral inhibitor of RIP (vIRA), encoded from the M45 locus of MCMV [173]. vIRA inhibits RIP3 activation of necrosis [173]. Although there is no equivalent gene in HCMV, it encodes other inhibitors of apoptosis, which may serve a similar function [174, 175]. Perhaps HCMV must counteract apoptotic pathways in general instead of RIP3/necrosis pathways (reviewed in refs. 176, 177).

## 6 Dissemination Within the Host

Clinical studies have revealed the routes of HCMV person-to-person spread. Vertical transmission of HCMV occurs via transplacental transfer of virus [178–180], intrapartum transmission [181], and breastfeeding from infected mother to child [182–186]. Horizontal transmission includes organ transplantation from an infected donor, exposure to infected secretions (i.e., saliva), contact with infected urine during childhood [187], and sexual activity in adulthood [188]. Inside the host, the infection spreads mainly via leukocytes [189].

MCMV is an excellent experimental model for studying the interaction of CMV with different tissues and cells following infection [190–194]. CMVs can productively infect many different cell types. These include epithelial cells of the salivary glands, kidneys, lung, liver, and intestines [195]. MCMV can infect endothelial cells lining the spleen [196], myocytes, brown fat adipocytes, connective tissue fibrocytes, bone marrow stromal cells, dendritic cells, monocytes, and tissue macrophages, but the B and T cell compartments of lymphoid organs, including the thymus, are not productively infected [197]. The mouse model has demonstrated that circulating leukocytes, predominantly mononuclear cells, disseminate MCMV and that cell-associated viremia is biphasic. First, primary dissemination of MCMV leads to infection of reticuloendothelial organs, the liver and spleen. This is followed by viral amplification and a more intense secondary viremia to organs such as the salivary gland [198]. This system also allowed the identification of the genes that are necessary for virus replication in the different tissues [199]. Recently an MCMV conditional gene expression system was used to quantify viral productivity in specific cell types and determine the role that each one plays in viral dissemination in vivo [200, 201]. Viral factors including chemokines, GPCRs, antiapoptotic genes, and tegument proteins play a role in viral dissemination and full pathogenicity in the mouse [138–140, 145–147, 174, 175, 202].

## 7 Latency and Reactivation

In humans HCMV remains latent in endothelial cells and cells of the myeloid lineage (reviewed in ref. 203). Animal models of latency and reactivation have played a role in our somewhat limited understanding of the maintenance of CMV latency and the signals necessary for reactivation (reviewed in refs. 197, 204). Studies with mouse and guinea pig models of CMV confirmed the role of myeloid lineage cells in virus persistence and the specificity of the $CD8^+$ T cell responses during latent infection [101, 205–209]. Viral and host factors, including novel "unfit" NK cells, have been identified that contribute to persistence and latency in different organs [144, 174, 175, 210]. Also in the murine system, the helper function of $CD4^+$ T cells [211] and antigen presentation on non-hematopoietic and hematopoietic cells [212] play important roles in memory inflation in latently infected hosts.

CMV reactivation in the mouse model includes a kidney transplantation model [213, 221]. Several factors inducing reactivation have been identified [197, 214, 215]. However, immune suppression and cytokine-mediated activation of the productive viral cycle appears to be the most common inducers of recurrence [117, 197, 216]. Transcription of IE1 and the differentiation state of the cell may not be sufficient for virus reactivation [217, 218]. However, allogeneic organ transplantation, tissue implantation, and cell transfer [11] are important factors for inducing the reactivation of MCMV and RCMV [218–220]. The mouse system was also used to understand the source of the reactivated virus in models of organ transplantation. Using a kidney transplantation model in the mouse, Klotman et al. showed that the source of reactivated MCMV in an uninfected recipient comes from the transplanted organ, but if the recipient is latently infected, the majority of the time the reactivated virus comes from the recipient [221]. The rodent models have been very useful for modeling transplantation reactivation and for dissecting mechanisms of reactivation.

## 8 Vaccine and Drug Development and Testing

The animal models of CMV infection have been used to test new candidate vaccines and the potency of existing ones. The gB vaccine that has shown some promise in clinical trials was initially tested in animal models of both congenital infection and immunosuppression [122–124]. All of the models except for the rat have provided some clues to vaccine design. In the guinea pig model, systemic immunity to GPCMV has been shown to protect against hearing loss following congenital infection [75, 222] (reviewed in ref. 24). Antibodies against gB have been shown to be protective against congenital disease, which provided hope that this would also work in humans [223, 224]. Vaccination with gB DNA subunit has shown some capacity to provide protection against congenital CMV infection [225, 226]. More recent vaccination studies with GPCMV matrix protein GP83, a homolog of HCMV pp65, generated protective T cell-mediated immune responses against congenital GPCMV infection and disease [227]. This highlights other possible avenues for protective vaccination. In the rhesus model, DNA vaccination with plasmids encoding gB, pp65-2, and IE-1 has shown promise. The gB-pp65 combined vaccine significantly reduces RhCMV copy numbers in plasma and oral shedding of the virus and has proven to be better than vaccines directed only against gB [125, 228].

Because of its affordability and the availability of reagents for dissecting the immune response, the mouse has been particularly useful for testing potential vaccine strains such as attenuated viruses or viruses that overexpress potential immune stimulators (i.e., for NK cells) [229–234]. Although these animal models of vaccination have not yielded an efficacious vaccine in humans, they have provided the foundational baseline of which proteins, routes, and attenuation genes to target.

The susceptibility of animal CMVs to various antiviral drugs makes them ideal for the identification of potential antiviral compounds (reviewed in ref. 235). MCMV is susceptible to ganciclovir (GCV) [236]. More recently, a number of nucleoside analogues with Z- or E-methylenecyclopropane structures have been evaluated in the mouse model and possess better activity than GCV [237, 238]. Although GPCMV is resistant to GCV [239], it is susceptible to cidofovir and cyclic cidofovir [240]. The guinea pig model has been used to demonstrate the efficacy and safety of cidofovir and cyclic cidofovir [241, 242]. Its administration during pregnancy prevents GPCMV mortality in pups [243]. Other novel non-nucleoside analogues were tested in guinea pigs using the immunosuppressive model of CMV infection [244].

RhCMV has comparable susceptibility to GCV, foscarnet, and benzimidazole nucleosides [245]. The highly conserved sequence of the drug-target proteins in HCMV and RhCMV makes this model ideal to test the efficacy and safety of novel anti-HCMV drugs both in immunocompetent and immunocompromised animals [28, 148, 245].

## 9 Animal Models for HCMV Studies In Vivo

True animal models of HCMV infection are extremely difficult to develop due to its strict species specificity. However, several attempts have been made to develop models of HCMV infection in animals that recapitulate one or more phases of the viral replication cycle. Human cells/tissue fragments implanted into mouse [246] and rats [247] have been shown support viral infection in vivo, but only within the implanted cells.

In SCID mice, human fetal thymus and liver implants were placed under the kidney (SCID-hu mice) [248] or fragments of human fetal retina were placed in the anterior chamber of the eye and supported HCMV growth [238, 249–252]. Using the SCID-hu mouse model, Wang et al. demonstrated that the ULb′ region, encoding 19 open reading frames, present in all virulent strains but deleted from attenuated strains is essential for HCMV replication in vivo [253]. An in vivo model of HCMV retinal infection in athymic rats has also been developed using the same approach [254].

SCID-hu mice do not support systemic infection nor do these mice develop viral latency [248, 252]. In order to study systemic and latent HCMV infection, reactivation, and viral spread within myeloid progenitors, monocytes, and macrophages, a model system in which huCD34+ hematopoietic stem cells (HSCs) are engrafted into NOD/SCID mice has been developed. In this model, granulocyte colony-stimulating factor (G-CSF) leads to reactivation of latent HCMV in monocytes/macrophages that have migrated into organ tissues. The results from this study also suggest that G-CSF-mobilized blood products from seropositive donors pose an elevated risk for HCMV transmission to recipients [255]. These implant-based animal models for HCMV infection have also proved to be valuable for drug testing and vaccine development against HCMV [235].

## 10 Summary

This chapter has provided an overview of CMV infections in the different animal models. Each one has provided important insights into the lifecycle of HCMV and each has its promoters and detractors as models for HCMV infection. In the following chapters, experts in the field will provide details on how to use both animal and cellular systems to address important questions in CMV biology. Importantly, the animal models described in this chapter will ultimately provide the conduit by which discoveries in fundamental virological processes can be examined and extended to an in vivo setting.

### References

1. Lafemina RL, Hayward GS (1988) Differences in cell-type-specific blocks to immediate early gene expression and DNA replication of human, simian and murine cytomegalovirus. J Gen Virol 69(Pt 2):355–374
2. Angulo A et al (1998) Enhancer requirement for murine cytomegalovirus growth and genetic complementation by the human cytomegalovirus enhancer. J Virol 72(11):8502–8509
3. Grzimek NK et al (1999) In vivo replication of recombinant murine cytomegalovirus driven by the paralogous major immediate-early promoter-enhancer of human cytomegalovirus. J Virol 73(6):5043–5055
4. Sandford GR et al (2001) Rat cytomegalovirus major immediate-early enhancer switching results in altered growth characteristics. J Virol 75(11):5076–5083
5. Tang Q, Maul GG (2006) Mouse cytomegalovirus crosses the species barrier with help from a few human cytomegalovirus proteins. J Virol 80(15):7510–7521
6. Lilja AE, Shenk T (2008) Efficient replication of rhesus cytomegalovirus variants in multiple rhesus and human cell types. Proc Natl Acad Sci U S A 105(50):19950–19955
7. Schleiss MR (2006) Nonprimate models of congenital cytomegalovirus (CMV) infection: gaining insight into pathogenesis and prevention of disease in newborns. ILAR J 47(1):65–72
8. Schleiss MR (2002) Animal models of congenital cytomegalovirus infection: an overview of progress in the characterization of guinea pig cytomegalovirus (GPCMV). J Clin Virol 25(Suppl 2):S37–S49
9. Powers C, Fruh K (2008) Rhesus CMV: an emerging animal model for human CMV. Med Microbiol Immunol 197(2):109–115
10. Loh HS et al (2006) Pathogenesis and vertical transmission of a transplacental rat cytomegalovirus. Virol J 3:42
11. Cheung KS, Lang DJ (1977) Transmission and activation of cytomegalovirus with blood transfusion: a mouse model. J Infect Dis 135:841–845
12. Stals FS et al (1990) An animal model for therapeutic intervention studies of CMV infection in the immunocompromised host. Arch Virol 114(1–2):91–107

13. Holtappels R et al (2008) CD8 T-cell-based immunotherapy of cytomegalovirus infection: "proof of concept" provided by the murine model. Med Microbiol Immunol 197(2):125–134
14. Craighead JE, Martin WB, Huber SA (1992) Role of CD4+ (helper) T cells in the pathogenesis of murine cytomegalovirus myocarditis. Lab Invest 66(6):755–761
15. Mutter W et al (1988) Failure in generating hemopoietic stem cells is the primary cause of death from cytomegalovirus disease in the immunocompromised host. J Exp Med 167(5):1645–1658
16. Osborn JE (1986) Cytomegalovirus and other herpesviruses of mice and rats. In: Bhatt PN et al (eds) Viral and mycoplasmal infections of laboratory rodents. Academic, London
17. Shellam GR et al (1985) The genetic background modulates the effect of the beige gene on susceptibility to cytomegalovirus infection in mice. Scand J Immunol 22(2):147–155
18. Rawlinson WD, Farrell HE, Barrell BG (1996) Analysis of the complete DNA sequence of murine cytomegalovirus. J Virol 70(12):8833–8849
19. Cheeran MC, Lokensgard JR, Schleiss MR (2009) Neuropathogenesis of congenital cytomegalovirus infection: disease mechanisms and prospects for intervention. Clin Microbiol Rev 22(1):99–126
20. Juanjuan C et al (2011) Murine model for congenital CMV infection and hearing impairment. Virol J 8:70
21. Medearis DN Jr (1964) Mouse cytomegalovirus infection. 3. Attempts to produce intrauterine infections. Am J Hyg 80:113–120
22. Woolf NK, Jaquish DV, Koehrn FJ (2007) Transplacental murine cytomegalovirus infection in the brain of SCID mice. Virol J 4:26
23. Vink C, Beuken E, Bruggeman CA (2000) Complete DNA sequence of the rat cytomegalovirus genome. J Virol 74(16):7656–7665
24. Schleiss MR (2008) Comparison of vaccine strategies against congenital CMV infection in the guinea pig model. J Clin Virol 41(3):224–230
25. Schleiss MR et al (2008) Analysis of the nucleotide sequence of the guinea pig cytomegalovirus (GPCMV) genome. Virol J 5:139
26. Lockridge KM et al (1999) Pathogenesis of experimental rhesus cytomegalovirus infection. J Virol 73(11):9576–9583
27. Tarantal AF et al (1998) Neuropathogenesis induced by rhesus cytomegalovirus in fetal rhesus monkeys (Macaca mulatta). J Infect Dis 177(2):446–450
28. Hansen SG et al (2003) Complete sequence and genomic analysis of rhesus cytomegalovirus. J Virol 77(12):6620–6636
29. Faber DW et al (1992) Role of HIV and CMV in the pathogenesis of retinitis and retinal vasculopathy in AIDS patients. Invest Ophthalmol Vis Sci 33(8):2345–2353
30. Britt WJ, Alford CA (2001) The human herpesviruses: cytomegalovirus. In: Knipe DM, Howley PM (eds) Fields virology, 4th edn. Lippincott-Raven, Philadelphia, PA, pp 2493–2523
31. Mocarski ES, Tan Courcelle C (2001) Cytomegaloviruses and their replication. In: Knipe DM, Howley PM (eds) Fields virology, 4th edn. Lippincott-Raven, Philadelphia, PA, pp 2629–2673
32. Istas AS et al (1995) Surveillance for congenital cytomegalovirus disease: a report from the National Congenital Cytomegalovirus Disease Registry. Clin Infect Dis 20(3):665–670
33. Potena L et al (2006) Acute rejection and cardiac allograft vascular disease is reduced by suppression of subclinical cytomegalovirus infection. Transplantation 82(3):398–405
34. Demmler GJ (1994) Congenital cytomegalovirus infection. Semin Pediatr Neurol 1(1): 36–42
35. Stagno S et al (1984) Congenital and perinatal cytomegalovirus infections: clinical characteristics and pathogenic factors. Birth Defects Orig Artic Ser 20(1):65–85
36. Fowler KB et al (1999) Newborn hearing screening: will children with hearing loss caused by congenital cytomegalovirus infection be missed? J Pediatr 135(1):60–64
37. Pass RF (2005) Congenital cytomegalovirus infection and hearing loss. Herpes 12(2): 50–55
38. Blum A et al (1998) High anti-cytomegalovirus (CMV) IgG antibody titer is associated with coronary artery disease and may predict post-coronary balloon angioplasty restenosis. Am J Cardiol 81(7):866–868
39. Chiu B et al (1997) Chlamydia pneumoniae, cytomegalovirus, and herpes simplex virus in atherosclerosis of the carotid artery. Circulation 96(7):2144–2148
40. Espinola-Klein C et al (2002) Impact of infectious burden on progression of carotid atherosclerosis. Stroke 33(11):2581–2586
41. Grattan MT et al (1989) Cytomegalovirus infection is associated with cardiac allograft rejection and atherosclerosis. JAMA 261(24):3561–3566
42. Kishore J et al (2004) Infection with cytomegalovirus in patients with inflammatory

bowel disease: prevalence, clinical significance and outcome. J Med Microbiol 53(Pt 11): 1155–1160

43. Papadakis KA et al (2001) Outcome of cytomegalovirus infections in patients with inflammatory bowel disease. Am J Gastroenterol 96(7):2137–2142
44. Soderberg-Naucler C (2008) HCMV microinfections in inflammatory diseases and cancer. J Clin Virol 41(3):218–223
45. Cobbs CS et al (2002) Human cytomegalovirus infection and expression in human malignant glioma. Cancer Res 62(12):3347–3350
46. Cinatl J et al (2004) Molecular mechanisms of the modulatory effects of HCMV infection in tumor cell biology. Trends Mol Med 10(1):19–23
47. Michaelis M, Doerr HW, Cinatl J (2009) The story of human cytomegalovirus and cancer: increasing evidence and open questions. Neoplasia 11(1):1–9
48. Barami K (2010) Oncomodulatory mechanisms of human cytomegalovirus in gliomas. J Clin Neurosci 17(7):819–823
49. Soroceanu L, Cobbs CS (2011) Is HCMV a tumor promoter? Virus Res 157(2):193–203
50. Helantera I et al (2003) The impact of cytomegalovirus infections and acute rejection episodes on the development of vascular changes in 6-month protocol biopsy specimens of cadaveric kidney allograft recipients. Transplantation 75(11):1858–1864
51. Koskinen PK et al (1993) Cytomegalovirus infection accelerates cardiac allograft vasculopathy: correlation between angiographic and endomyocardial biopsy findings in heart transplant patients. Transpl Int 6(6):341–347
52. Petrakopoulou P et al (2004) Cytomegalovirus infection in heart transplant recipients is associated with impaired endothelial function. Circulation 110(11 Suppl 1):II207–II212
53. Li RY, Tsutsui Y (2000) Growth retardation and microcephaly induced in mice by placental infection with murine cytomegalovirus. Teratology 62(2):79–85
54. Tsutsui Y (1995) Developmental disorders of the mouse brain induced by murine cytomegalovirus: animal models for congenital cytomegalovirus infection. Pathol Int 45(2): 91–102
55. Tsutsui Y et al (1993) Microphthalmia and cerebral atrophy induced in mouse embryos by infection with murine cytomegalovirus in midgestation. Am J Pathol 143(3):804–813
56. Barry PA et al (2006) Nonhuman primate models of intrauterine cytomegalovirus infection. ILAR J 47(1):49–64
57. Vogel P et al (1994) Seroepidemiologic studies of cytomegalovirus infection in a breeding population of rhesus macaques. Lab Anim Sci 44(1):25–30
58. Kumar ML, Nankervis GA (1978) Experimental congenital infection with cytomegalovirus: a guinea pig model. J Infect Dis 138(5):650–654
59. London WT et al (1986) Experimental congenital disease with simian cytomegalovirus in rhesus monkeys. Teratology 33(3):323–331
60. Kashiwai A et al (1992) Susceptibility of mouse embryo to murine cytomegalovirus infection in early and mid-gestation stages. Arch Virol 127(1–4):37–48
61. Kosugi I et al (2000) Cytomegalovirus infection of the central nervous system stem cells from mouse embryo: a model for developmental brain disorders induced by cytomegalovirus. Lab Invest 80(9):1373–1383
62. Malm G, Grondahl EH, Lewensohn-Fuchs I (2000) Congenital cytomegalovirus infection: a retrospective diagnosis in a child with pachygyria. Pediatr Neurol 22(5):407–408
63. Perlman JM, Argyle C (1992) Lethal cytomegalovirus infection in preterm infants: clinical, radiological, and neuropathological findings. Ann Neurol 31(1):64–68
64. van den Pol AN, Reuter JD, Santarelli JG (2002) Enhanced cytomegalovirus infection of developing brain independent of the adaptive immune system. J Virol 76(17):8842–8854
65. Matsukage S et al (2006) Mouse embryonic stem cells are not susceptible to cytomegalovirus but acquire susceptibility during differentiation. Birth Defects Res A Clin Mol Teratol 76(2):115–125
66. Kawasaki H et al (2002) The amount of immature glial cells in organotypic brain slices determines the susceptibility to murine cytomegalovirus infection. Lab Invest 82(10): 1347–1358
67. Kosugi I et al (2002) Innate immune responses to cytomegalovirus infection in the developing mouse brain and their evasion by virus-infected neurons. Am J Pathol 161(3): 919–928
68. Mutnal MB et al (2011) Murine cytomegalovirus infection of neural stem cells alters neurogenesis in the developing brain. PLoS One 6(1):e16211
69. Keithley EM, Woolf NK, Harris JP (1989) Development of morphological and physiological changes in the cochlea induced by cytomegalovirus. Laryngoscope 99(4):409–414
70. Tsutsui Y, Kosugi I, Kawasaki H (2005) Neuropathogenesis in cytomegalovirus

infection: indication of the mechanisms using mouse models. Rev Med Virol 15(5): 327–345

71. Schachtele SJ et al (2011) Cytomegalovirus-induced sensorineural hearing loss with persistent cochlear inflammation in neonatal mice. J Neurovirol 17(3):201–211
72. Schraff SA et al (2007) The role of CMV inflammatory genes in hearing loss. Otol Neurotol 28(7):964–969
73. Cheeran MC et al (2004) Intracerebral infection with murine cytomegalovirus induces CXCL10 and is restricted by adoptive transfer of splenocytes. J Neurovirol 10(3):152–162
74. Reuter JD et al (2004) Systemic immune deficiency necessary for cytomegalovirus invasion of the mature brain. J Virol 78(3): 1473–1487
75. Harris JP et al (1984) Immunologic and electrophysiological response to cytomegaloviral inner ear infection in the guinea pig. J Infect Dis 150(4):523–530
76. Melnick JL et al (1983) Cytomegalovirus antigen within human arterial smooth muscle cells. Lancet 2(8351):644–647
77. Speir E et al (1994) Potential role of human cytomegalovirus and p53 interaction in coronary restenosis. Science 265(5170):391–394
78. Burnett MS et al (2001) Atherosclerosis in apoE knockout mice infected with multiple pathogens. J Infect Dis 183(2):226–231
79. Hsich E et al (2001) Cytomegalovirus infection increases development of atherosclerosis in Apolipoprotein-E knockout mice. Atherosclerosis 156(1):23–28
80. Vliegen I et al (2004) Cytomegalovirus infection aggravates atherogenesis in apoE knockout mice by both local and systemic immune activation. Microbes Infect 6(1):17–24
81. Vliegen I et al (2002) MCMV infection increases early T-lymphocyte influx in atherosclerotic lesions in apoE knockout mice. J Clin Virol 25(Suppl 2):S159–S171
82. Vliegen I et al (2004) Murine cytomegalovirus infection directs macrophage differentiation into a pro-inflammatory immune phenotype: implications for atherogenesis. Microbes Infect 6(12):1056–1062
83. Khoretonenko MV et al (2010) Cytomegalovirus infection leads to microvascular dysfunction and exacerbates hypercholesterolemia-induced responses. Am J Pathol 177(4):2134–2144
84. Melnychuk RM et al (2005) Mouse cytomegalovirus M33 is necessary and sufficient in virus-induced vascular smooth muscle cell migration. J Virol 79(16):10788–10795
85. Streblow DN et al (2005) Rat cytomegalovirus-accelerated transplant vascular sclerosis is reduced with mutation of the chemokine-receptor R33. Am J Transplant 5(3):436–442
86. Tikkanen J et al (2001) Cytomegalovirus infection-enhanced chronic rejection in the rat is prevented by antiviral prophylaxis. Transplant Proc 33(1–2):1801
87. Zeng H et al (2005) Mechanistic study of malononitrileamide FK778 in cardiac transplantation and CMV infection in rats. Transplantation 79(1):17–22
88. Orloff SL et al (2002) Elimination of donor-specific alloreactivity prevents cytomegalovirus-accelerated chronic rejection in rat small bowel and heart transplants. Transplantation 73(5):679–688
89. Orloff SL et al (2000) Tolerance induced by bone marrow chimerism prevents transplant vascular sclerosis in a rat model of small bowel transplant chronic rejection. Transplantation 69(7):1295–1303
90. Streblow DN et al (2003) Cytomegalovirus-mediated upregulation of chemokine expression correlates with the acceleration of chronic rejection in rat heart transplants. J Virol 77(3):2182–2194
91. Soule JL et al (2006) Cytomegalovirus accelerates chronic allograft nephropathy in a rat renal transplant model with associated provocative chemokine profiles. Transplant Proc 38(10):3214–3220
92. Streblow DN, Orloff SL, Nelson JA (2007) Acceleration of allograft failure by cytomegalovirus. Curr Opin Immunol 19(5):577–582
93. Streblow DN et al (2007) Rat cytomegalovirus gene expression in cardiac allograft recipients is tissue specific and does not parallel the profiles detected in vitro. J Virol 81(8):3816–3826
94. Merigan TC et al (1992) A controlled trial of ganciclovir to prevent cytomegalovirus disease after heart transplantation. N Engl J Med 326(18):1182–1186
95. Valantine HA et al (1999) Impact of prophylactic immediate posttransplant ganciclovir on development of transplant atherosclerosis: a post hoc analysis of a randomized, placebo-controlled study. Circulation 100(1):61–66
96. Orloff SL et al (2011) Cytomegalovirus latency promotes cardiac lymphoid neogenesis and accelerated allograft rejection in CMV naive recipients. Am J Transplant 11(1):45–55
97. Duan Y, Ji Z, Atherton SS (1994) Dissemination and replication of MCMV after supraciliary inoculation in immunosuppressed BALB/c mice. Invest Ophthalmol Vis Sci 35(3):1124–1131

98. Zhang M et al (2005) Infection of retinal neurons during murine cytomegalovirus retinitis. Invest Ophthalmol Vis Sci 46(6): 2047–2055
99. Zinkernagel MS et al (2010) In vivo imaging of ocular MCMV infection. Invest Ophthalmol Vis Sci 51(1):369–374
100. Zhang M, Xin H, Atherton SS (2005) Murine cytomegalovirus (MCMV) spreads to and replicates in the retina after endotoxin-induced disruption of the blood-retinal barrier of immunosuppressed BALB/c mice. J Neurovirol 11(4):365–375
101. Kercher L, Mitchell BM (2002) Persisting murine cytomegalovirus can reactivate and has unique transcriptional activity in ocular tissue. J Virol 76(18):9165–9175
102. Zinkernagel MS et al (2011) Kinetics of ocular and systemic antigen-specific T-cell responses elicited during murine cytomegalovirus retinitis. Immunol Cell Biol 90(3):330–336
103. Zhang M, Atherton SS (2002) Apoptosis in the retina during MCMV retinitis in immunosuppressed BALB/c mice. J Clin Virol 25(Suppl 2):S137–S147
104. Zhou J, Zhang M, Atherton SS (2007) Tumor necrosis factor-alpha-induced apoptosis in murine cytomegalovirus retinitis. Invest Ophthalmol Vis Sci 48(4):1691–1700
105. Bigger JE et al (1999) Protection against murine cytomegalovirus retinitis by adoptive transfer of virus-specific CD8+ T cells. Invest Ophthalmol Vis Sci 40(11):2608–2613
106. Igietseme JU et al (1991) Mechanisms of protection against herpes simplex virus type 1-induced retinal necrosis by in vitro-activated T lymphocytes. J Virol 65(2):763–768
107. Bigger JE, Thomas CA III, Atherton SS (1998) NK cell modulation of murine cytomegalovirus retinitis. J Immunol 160(12): 5826–5831
108. Henrickson RV et al (1983) Epidemic of acquired immunodeficiency in rhesus monkeys. Lancet 1(8321):388–390
109. Baskin GB (1987) Disseminated cytomegalovirus infection in immunodeficient rhesus monkeys. Am J Pathol 129(2):345–352
110. Kaup F et al (1998) Gastrointestinal pathology in rhesus monkeys with experimental SIV infection. Pathobiology 66(3–4):159–164
111. Kuhn EM et al (1999) Immunohistochemical studies of productive rhesus cytomegalovirus infection in rhesus monkeys (Macaca mulatta) infected with simian immunodeficiency virus. Vet Pathol 36(1):51–56
112. Kaur A et al (2002) Decreased frequency of cytomegalovirus (CMV)-specific CD4+ T lymphocytes in simian immunodeficiency virus-infected rhesus macaques: inverse relationship with CMV viremia. J Virol 76(8): 3646–3658
113. Kaur A et al (2003) Direct relationship between suppression of virus-specific immunity and emergence of cytomegalovirus disease in simian AIDS. J Virol 77(10):5749–5758
114. Sequar G et al (2002) Experimental coinfection of rhesus macaques with rhesus cytomegalovirus and simian immunodeficiency virus: pathogenesis. J Virol 76(15):7661–7671
115. Dix RD, Podack ER, Cousins SW (2003) Loss of the perforin cytotoxic pathway predisposes mice to experimental cytomegalovirus retinitis. J Virol 77(6):3402–3408
116. Aquino-de Jesus MJ, Griffith BP (1989) Cytomegalovirus infection in immunocompromised guinea pigs: a model for testing antiviral agents in vivo. Antiviral Res 12(4):181–193
117. Polic B et al (1998) Hierarchical and redundant lymphocyte subset control precludes cytomegalovirus replication during latent infection. J Exp Med 188(6):1047–1054
118. Kaur A et al (1996) Cytotoxic T-lymphocyte responses to cytomegalovirus in normal and simian immunodeficiency virus-infected rhesus macaques. J Virol 70(11):7725–7733
119. Schlub TE et al (2011) Comparing the kinetics of NK cells, CD4, and CD8 T cells in murine cytomegalovirus infection. J Immunol 187(3):1385–1392
120. Chan KS, Kaur A (2007) Flow cytometric detection of degranulation reveals phenotypic heterogeneity of degranulating CMV-specific CD8+ T lymphocytes in rhesus macaques. J Immunol Methods 325(1–2):20–34
121. Yue Y et al (2006) Characterization and immunological analysis of the rhesus cytomegalovirus homologue (Rh112) of the human cytomegalovirus UL83 lower matrix phosphoprotein (pp65). J Gen Virol 87(Pt 4): 777–787
122. Cekinovic D et al (2008) Passive immunization reduces murine cytomegalovirus-induced brain pathology in newborn mice. J Virol 82(24):12172–12180
123. Yue Y, Zhou SS, Barry PA (2003) Antibody responses to rhesus cytomegalovirus glycoprotein B in naturally infected rhesus macaques. J Gen Virol 84(Pt 12):3371–3379
124. Pass RF et al (2009) Vaccine prevention of maternal cytomegalovirus infection. N Engl J Med 360(12):1191–1199
125. Abel K et al (2011) Vaccine-induced control of viral shedding following rhesus cytomegalovirus

challenge in rhesus macaques. J Virol 85(6):2878–2890

126. Jonjic S et al (1990) Efficacious control of cytomegalovirus infection after long-term depletion of CD8+ T lymphocytes. J Virol 64(11):5457–5464
127. Jonjic S et al (1994) Antibodies are not essential for the resolution of primary cytomegalovirus infection but limit dissemination of recurrent virus. J Exp Med 179(5): 1713–1717
128. Arens R et al (2008) Cutting edge: murine cytomegalovirus induces a polyfunctional CD4 T cell response. J Immunol 180(10):6472–6476
129. Snyder CM et al (2009) CD4+ T cell help has an epitope-dependent impact on CD8+ T cell memory inflation during murine cytomegalovirus infection. J Immunol 183(6):3932–3941
130. Bukowski JF, Woda BA, Welsh RM (1984) Pathogenesis of murine cytomegalovirus infection in natural killer cell-depleted mice. J Virol 52(1):119–128
131. Scalzo AA et al (1992) The effect of the Cmv-1 resistance gene, which is linked to the natural killer cell gene complex, is mediated by natural killer cells. J Immunol 149(2):581–589
132. Dokun AO et al (2001) Specific and nonspecific NK cell activation during virus infection. Nat Immunol 2(10):951–956
133. Humphreys IR et al (2007) Cytomegalovirus exploits IL-10-mediated immune regulation in the salivary glands. J Exp Med 204(5):1217–1225
134. Jonjic S et al (1989) Site-restricted persistent cytomegalovirus infection after selective long-term depletion of CD4+ T lymphocytes. J Exp Med 169(4):1199–1212
135. Lucin P et al (1992) Gamma interferon-dependent clearance of cytomegalovirus infection in salivary glands. J Virol 66(4): 1977–1984
136. Powers C et al (2008) Cytomegalovirus immune evasion. Curr Top Microbiol Immunol 325:333–359
137. Miller-Kittrell M, Sparer TE (2009) Feeling manipulated: cytomegalovirus immune manipulation. Virol J 6:4
138. Fleming P et al (1999) The murine cytomegalovirus chemokine homolog, m131/129, is a determinant of viral pathogenicity. J Virol 73(8):6800–6809
139. Saederup N et al (2001) Murine cytomegalovirus CC chemokine homolog MCK-2 (m131-129) is a determinant of dissemination that increases inflammation at initial sites of infection. J Virol 75(20):9966–9976
140. Noda S et al (2006) Cytomegalovirus MCK-2 controls mobilization and recruitment of myeloid progenitor cells to facilitate dissemination. Blood 107(1):30–38
141. Sparer TE et al (2004) Expression of human CXCR2 in murine neutrophils as a model for assessing cytomegalovirus chemokine vCXCL-1 function in vivo. J Interferon Cytokine Res 24(10):611–620
142. Miller-Kittrell M et al (2007) Functional characterization of chimpanzee cytomegalovirus chemokine, vCXCL-1(CCMV). Virology 364(2):454–465
143. Vischer HF, Leurs R, Smit MJ (2006) HCMV-encoded G-protein-coupled receptors as constitutively active modulators of cellular signaling networks. Trends Pharmacol Sci 27(1):56–63
144. Cardin RD et al (2009) The M33 chemokine receptor homolog of murine cytomegalovirus exhibits a differential tissue-specific role during in vivo replication and latency. J Virol 83(15):7590–7601
145. Davis-Poynter NJ, Degli-Esposti M, Farrell HE (1999) Murine cytomegalovirus homologues of cellular immunomodulatory genes. Intervirology 42(5–6):331–341
146. Beisser PS et al (1998) The R33 G protein-coupled receptor gene of rat cytomegalovirus plays an essential role in the pathogenesis of viral infection. J Virol 72(3):2352–2363
147. Farrell HE et al (2011) Partial functional complementation between human and mouse cytomegalovirus chemokine receptor homologues. J Virol 85(12):6091–6095
148. Rivailler P et al (2006) Genomic sequence of rhesus cytomegalovirus 180.92: insights into the coding potential of rhesus cytomegalovirus. J Virol 80(8):4179–4182
149. Penfold ME et al (2003) Characterization of the rhesus cytomegalovirus US28 locus. J Virol 77(19):10404–10413
150. Oxford KL et al (2008) Protein coding content of the U(L)b′ region of wild-type rhesus cytomegalovirus. Virology 373(1):181–188
151. Lesniewski M et al (2006) Primate cytomegalovirus US12 gene family: a distinct and diverse clade of seven-transmembrane proteins. Virology 354(2):286–298
152. Alcendor DJ et al (2009) Patterns of divergence in the vCXCL and vGPCR gene clusters in primate cytomegalovirus genomes. Virology 395(1):21–32
153. Chang WL et al (2009) Human cytomegalovirus suppresses type I interferon secretion by

plasmacytoid dendritic cells through its interleukin 10 homolog. Virology 390(2): 330–337

154. Chang WL et al (2007) Exposure of myeloid dendritic cells to exogenous or endogenous IL-10 during maturation determines their longevity. J Immunol 178(12):7794–7804
155. Chang WL et al (2004) Human cytomegalovirus-encoded interleukin-10 homolog inhibits maturation of dendritic cells and alters their functionality. J Virol 78(16):8720–8731
156. Raftery MJ et al (2004) Shaping phenotype, function, and survival of dendritic cells by cytomegalovirus-encoded IL-10. J Immunol 173(5):3383–3391
157. Spencer JV et al (2008) Stimulation of B lymphocytes by cmvIL-10 but not LAcmvIL-10. Virology 374(1):164–169
158. Spencer JV et al (2002) Potent immunosuppressive activities of cytomegalovirus-encoded interleukin-10. J Virol 76(3):1285–1292
159. Chang WL, Barry PA (2010) Attenuation of innate immunity by cytomegalovirus IL-10 establishes a long-term deficit of adaptive antiviral immunity. Proc Natl Acad Sci U S A 107(52):22647–22652
160. Lockridge KM et al (2000) Primate cytomegaloviruses encode and express an IL-10-like protein. Virology 268(2):272–280
161. Kloover JS et al (2002) A rat cytomegalovirus strain with a disruption of the r144 MHC class I-like gene is attenuated in the acute phase of infection in neonatal rats. Arch Virol 147(4):813–824
162. Farrell HE et al (1997) Inhibition of natural killer cells by a cytomegalovirus MHC class I homologue in vivo. Nature 386(6624): 510–514
163. Farrell H et al (2000) Cytomegalovirus MHC class I homologues and natural killer cells: an overview. Microbes Infect 2(5):521–532
164. Babic M et al (2010) Cytomegalovirus immunoevasin reveals the physiological role of "missing self" recognition in natural killer cell dependent virus control in vivo. J Exp Med 207(12):2663–2673
165. Beisser PS et al (2000) The r144 major histocompatibility complex class I-like gene of rat cytomegalovirus is dispensable for both acute and long-term infection in the immunocompromised host. J Virol 74(2): 1045–1050
166. Pande NT et al (2005) Rhesus cytomegalovirus contains functional homologues of US2, US3, US6, and US11. J Virol 79(9): 5786–5798
167. Hansen SG et al (2010) Evasion of CD8+ T cells is critical for superinfection by cytomegalovirus. Science 328(5974):102–106
168. Hengel H, Koszinowski UH (2010) Virology. A vaccine monkey wrench? Science 328(5974):51–52
169. Bohm V et al (2008) The immune evasion paradox: immunoevasins of murine cytomegalovirus enhance priming of CD8 T cells by preventing negative feedback regulation. J Virol 82(23):11637–11650
170. Zhu H et al (2002) Inhibition of cyclooxygenase 2 blocks human cytomegalovirus replication. Proc Natl Acad Sci U S A 99(6):3932–3937
171. Rue CA et al (2004) A cyclooxygenase-2 homologue encoded by rhesus cytomegalovirus is a determinant for endothelial cell tropism. J Virol 78(22):12529–12536
172. Mocarski ES, Upton JW, Kaiser WJ (2012) Viral infection and the evolution of caspase 8-regulated apoptotic and necrotic death pathways. Nat Rev Immunol 12(2):79–88
173. Upton JW, Kaiser WJ, Mocarski ES (2010) Virus inhibition of RIP3-dependent necrosis. Cell Host Microbe 7(4):302–313
174. Manzur M et al (2009) Virally mediated inhibition of Bax in leukocytes promotes dissemination of murine cytomegalovirus. Cell Death Differ 16(2):312–320
175. McCormick AL et al (2003) Differential function and expression of the viral inhibitor of caspase 8-induced apoptosis (vICA) and the viral mitochondria-localized inhibitor of apoptosis (vMIA) cell death suppressors conserved in primate and rodent cytomegaloviruses. Virology 316(2): 221–233
176. McCormick AL (2008) Control of apoptosis by human cytomegalovirus. Curr Top Microbiol Immunol 325:281–295
177. Goldmacher VS et al (1999) A cytomegalovirus-encoded mitochondria-localized inhibitor of apoptosis structurally unrelated to Bcl-2. Proc Natl Acad Sci U S A 96(22):12536–12541
178. Schmidt GM et al (1991) A randomized, controlled trial of prophylactic ganciclovir for cytomegalovirus pulmonary infection in recipients of allogeneic bone marrow transplants; The City of Hope-Stanford-Syntex CMV Study Group. N Engl J Med 324(15): 1005–1011
179. Schopfer K, Lauber E, Krech U (1978) Congenital cytomegalovirus infection in newborn infants of mothers infected before pregnancy. Arch Dis Child 53(7):536–539

180. Stagno S et al (1977) Congenital cytomegalovirus infection. N Engl J Med 296(22): 1254–1258
181. Stagno S et al (1982) Maternal cytomegalovirus infection and perinatal transmission. Clin Obstet Gynecol 25(3):563–576
182. Diosi P et al (1967) Cytomegalovirus infection associated with pregnancy. Lancet 2(7525): 1063–1066
183. Dworsky M et al (1983) Cytomegalovirus infection of breast milk and transmission in infancy. Pediatrics 72(3):295–299
184. Hamprecht K et al (2001) Epidemiology of transmission of cytomegalovirus from mother to preterm infant by breastfeeding. Lancet 357(9255):513–518
185. Stagno S et al (1980) Breast milk and the risk of cytomegalovirus infection. N Engl J Med 302(19):1073–1076
186. Vochem M et al (1998) Transmission of cytomegalovirus to preterm infants through breast milk. Pediatr Infect Dis J 17(1):53–58
187. Adler SP (1991) Cytomegalovirus and child day care: risk factors for maternal infection. Pediatr Infect Dis J 10(8):590–594
188. Meyers J (1985) Cytomegalovirus infection after organ allografting. The Herpesviruses 4:201
189. Pass RF (2001) Cytomegalovirus. In: Knipe DM, Howley PM (eds) Fields virology, 4th edn. Lippincott-Raven, Philadelphia, PA
190. Farroway LN et al (2005) Transmission of two Australian strains of murine cytomegalovirus (MCMV) in enclosure populations of house mice (Mus domesticus). Epidemiol Infect 133(4):701–710
191. Cheung KS et al (1981) Murine cytomegalovirus infection: hematological, morphological, and functional study of lymphoid cells. Infect Immun 33(1):239–249
192. Ho M (1991) Cytomegalovirus: biology and infection, 2nd edn. Plenum Medical Books, New York
193. Hudson JB (1979) The murine cytomegalovirus as a model for the study of viral pathogenesis and persistent infections. Arch Virol 62(1):1–29
194. Osborn JE, Shahidi NT (1973) Thrombocytopenia in murine cytomegalovirus infection. J Lab Clin Med 81(1):53–63
195. Klotman ME et al (1990) Detection of mouse cytomegalovirus nucleic acid in latently infected mice by in vitro enzymatic amplification. J Infect Dis 161(2):220–225
196. Mercer JA, Wiley CA, Spector DH (1988) Pathogenesis of murine cytomegalovirus infection: identification of infected cells in the spleen during acute and latent infections. J Virol 62(3):987–997
197. Reddehase MJ, Podlech J, Grzimek NK (2002) Mouse models of cytomegalovirus latency: overview. J Clin Virol 25(Suppl 2): S23–S36
198. Collins TM, Quirk MR, Jordan MC (1994) Biphasic viremia and viral gene expression in leukocytes during acute cytomegalovirus infection of mice. J Virol 68(10): 6305–6311
199. Hanson LK et al (2001) Products of US22 genes M140 and M141 confer efficient replication of murine cytomegalovirus in macrophages and spleen. J Virol 75(14): 6292–6302
200. Sacher T et al (2008) Conditional gene expression systems to study herpesvirus biology in vivo. Med Microbiol Immunol 197(2):269–276
201. Sacher T et al (2011) The role of cell types in cytomegalovirus infection in vivo. Eur J Cell Biol 91(1):70–77
202. McGregor A, Liu F, Schleiss MR (2004) Molecular, biological, and in vivo characterization of the guinea pig cytomegalovirus (CMV) homologs of the human CMV matrix proteins pp71 (UL82) and pp65 (UL83). J Virol 78(18):9872–9889
203. Jarvis MA, Nelson JA (2007) Molecular basis of persistence and latency. In: Campadelli-Fiume G, Arvin A, Mocarski E, Moore PS, Roizman B, Whitley R, Yamanishi K (eds) Human herpesviruses: biology, therapy, and immunoprophylaxis. Cambridge University Press, Cambridge
204. Reddehase MJ et al (2008) Murine model of cytomegalovirus latency and reactivation. Curr Top Microbiol Immunol 325:315–331
205. Koffron AJ et al (1998) Cellular localization of latent murine cytomegalovirus. J Virol 72(1):95–103
206. Podlech J et al (2000) Murine model of interstitial cytomegalovirus pneumonia in syngeneic bone marrow transplantation: persistence of protective pulmonary CD8-T-cell infiltrates after clearance of acute infection. J Virol 74(16):7496–7507
207. Podlech J et al (1998) Reconstitution of CD8 T cells is essential for the prevention of multiple-organ cytomegalovirus histopathology after bone marrow transplantation. J Gen Virol 79(Pt 9):2099–2104
208. Reddehase MJ et al (1985) Interstitial murine cytomegalovirus pneumonia after irradiation: characterization of cells that limit viral replication during established infection of the lungs. J Virol 55(2):264–273

209. Griffith BP et al (1981) Cytomegalovirus-induced mononucleosis in guinea pigs. Infect Immun 32(2):857–863

210. Tessmer MS, Reilly EC, Brossay L (2011) Salivary gland NK cells are phenotypically and functionally unique. PLoS Pathog 7(1):e1001254

211. Walton SM et al (2011) T-cell help permits memory CD8(+) T-cell inflation during cytomegalovirus latency. Eur J Immunol 41(8):2248–2259

212. Seckert CK et al (2011) Antigen-presenting cells of haematopoietic origin prime cytomegalovirus-specific CD8 T-cells but are not sufficient for driving memory inflation during viral latency. J Gen Virol 92(Pt 9):1994–2005

213. Li Z et al (2012) A mouse model of CMV transmission following kidney transplantation. Am J Transplant 12(4):1024–1028

214. Baskar JF, Stanat SC, Huang ES (1985) Congenital defects due to reactivation of latent murine cytomegaloviral infection during pregnancy. J Infect Dis 152(3): 621–624

215. Mayo DR, Rapp F (1980) Leukaemia reactivates mouse cytomegalovirus. J Gen Virol 51(Pt 2):401–404

216. Zhang M et al (2005) Ocular reactivation of MCMV after immunosuppression of latently infected BALB/c mice. Invest Ophthalmol Vis Sci 46(1):252–258

217. Busche A et al (2009) The mouse cytomegalovirus immediate-early 1 gene is not required for establishment of latency or for reactivation in the lungs. J Virol 83(9):4030–4038

218. Hummel M, Abecassis MM (2002) A model for reactivation of CMV from latency. J Clin Virol 25(Suppl 2):S123–S136

219. Hummel M et al (2001) Allogeneic transplantation induces expression of cytomegalovirus immediate-early genes in vivo: a model for reactivation from latency. J Virol 75(10): 4814–4822

220. Kloover JS et al (2002) Persistent rat cytomegalovirus (RCMV) infection of the salivary glands contributes to the anti-RCMV humoral immune response. Virus Res 85(2):163–172

221. Klotman ME, Starnes D, Hamilton JD (1985) The source of murine cytomegalovirus in mice receiving kidney allografts. J Infect Dis 152(6):1192–1196

222. Woolf NK et al (1985) Hearing loss in experimental cytomegalovirus infection of the guinea pig inner ear: prevention by systemic immunity. Ann Otol Rhinol Laryngol 94(4 Pt 1):350–356

223. Harrison CJ et al (1995) Reduced congenital cytomegalovirus (CMV) infection after maternal immunization with a guinea pig CMV glycoprotein before gestational primary CMV infection in the guinea pig model. J Infect Dis 172(5):1212–1220

224. Schleiss MR et al (2004) Protection against congenital cytomegalovirus infection and disease in guinea pigs, conferred by a purified recombinant glycoprotein B vaccine. J Infect Dis 189(8):1374–1381

225. Schleiss MR et al (2000) Immunogenicity evaluation of DNA vaccines that target guinea pig cytomegalovirus proteins glycoprotein B and UL83. Viral Immunol 13(2):155–167

226. Schleiss MR, Bourne N, Bernstein DI (2003) Preconception vaccination with a glycoprotein B (gB) DNA vaccine protects against cytomegalovirus (CMV) transmission in the guinea pig model of congenital CMV infection. J Infect Dis 188(12):1868–1874

227. Schleiss MR et al (2007) Preconceptual administration of an alphavirus replicon UL83 (pp 65 homolog) vaccine induces humoral and cellular immunity and improves pregnancy outcome in the guinea pig model of congenital cytomegalovirus infection. J Infect Dis 195(6):789–798

228. Yue Y et al (2007) Immunogenicity and protective efficacy of DNA vaccines expressing rhesus cytomegalovirus glycoprotein B, phosphoprotein 65-2, and viral interleukin-10 in rhesus macaques. J Virol 81(3):1095–1109

229. Slavuljica I et al (2010) Recombinant mouse cytomegalovirus expressing a ligand for the NKG2D receptor is attenuated and has improved vaccine properties. J Clin Invest 120(12):4532–4545

230. Mohr CA et al (2010) A spread-deficient cytomegalovirus for assessment of first-target cells in vaccination. J Virol 84(15): 7730–7742

231. Morello CS et al (2005) Systemic priming-boosting immunization with a trivalent plasmid DNA and inactivated murine cytomegalovirus (MCMV) vaccine provides long-term protection against viral replication following systemic or mucosal MCMV challenge. J Virol 79(1):159–175

232. Snyder CM et al (2010) Cross-presentation of a spread-defective MCMV is sufficient to prime the majority of virus-specific CD8+ T cells. PLoS One 5(3):e9681

233. MacDonald MR et al (1998) Mucosal and parenteral vaccination against acute and latent murine cytomegalovirus (MCMV) infection by using an attenuated MCMV mutant. J Virol 72(1):442–451

234. Sandford GR, Burns WH (1988) Use of temperature-sensitive mutants of mouse cytomegalovirus as vaccines. J Infect Dis 158(3): 596–601
235. Kern ER (2006) Pivotal role of animal models in the development of new therapies for cytomegalovirus infections. Antiviral Res 71(2–3): 164–171
236. Shanley JD, Morningstar J, Jordan MC (1985) Inhibition of murine cytomegalovirus lung infection and interstitial pneumonitis by acyclovir and 9-(1,3-dihydroxy-2-propoxymethyl)guanine. Antimicrob Agents Chemother 28(2):172–175
237. Rybak RJ et al (1999) Effective treatment of murine cytomegalovirus infections with methylenecyclopropane analogues of nucleosides. Antiviral Res 43(3):175–188
238. Kern ER et al (2004) Oral activity of a methylenecyclopropane analog, cyclopropavir, in animal models for cytomegalovirus infections. Antimicrob Agents Chemother 48(12): 4745–4753
239. Matthews T, Boehme R (1988) Antiviral activity and mechanism of action of ganciclovir. Rev Infect Dis 10(Suppl 3):S490–S494
240. Beadle JR et al (2002) Alkoxyalkyl esters of cidofovir and cyclic cidofovir exhibit multiple-log enhancement of antiviral activity against cytomegalovirus and herpesvirus replication in vitro. Antimicrob Agents Chemother 46(8):2381–2386
241. Bourne N, Bravo FJ, Bernstein DI (2000) Cyclic HPMPC is safe and effective against systemic guinea pig cytomegalovirus infection in immune compromised animals. Antiviral Res 47(2):103–109
242. White DR et al (2006) The effect of cidofovir on cytomegalovirus-induced hearing loss in a Guinea pig model. Arch Otolaryngol Head Neck Surg 132(6):608–615
243. Schleiss MR, Anderson JL, McGregor A (2006) Cyclic cidofovir (cHPMPC) prevents congenital cytomegalovirus infection in a guinea pig model. Virol J 3:9
244. Schleiss MR et al (2005) The non-nucleoside antiviral, BAY 38-4766, protects against cytomegalovirus (CMV) disease and mortality in immunocompromised guinea pigs. Antiviral Res 65(1):35–43
245. North TW et al (2004) Rhesus cytomegalovirus is similar to human cytomegalovirus in susceptibility to benzimidazole nucleosides. Antimicrob Agents Chemother 48(7): 2760–2765
246. Allen LB et al (1992) Novel method for evaluating antiviral drugs against human cytomegalovirus in mice. Antimicrob Agents Chemother 36(1):206–208
247. Gao L et al (2007) An animal model of human cytomegalovirus infection. Transplant Proc 39(10):3438–3443
248. Mocarski ES et al (1993) Human cytomegalovirus in a SCID-hu mouse: thymic epithelial cells are prominent targets of viral replication. Proc Natl Acad Sci U S A 90(1):104–108
249. DiLoreto D Jr et al (1994) Cytomegalovirus infection of human retinal tissue: an in vivo model. Lab Invest 71(1):141–148
250. Bidanset DJ et al (2001) Replication of human cytomegalovirus in severe combined immunodeficient mice implanted with human retinal tissue. J Infect Dis 184(2):192–195
251. Kern ER et al (2001) Predictive efficacy of SCID-hu mouse models for treatment of human cytomegalovirus infections. Antivir Chem Chemother 12(Suppl 1):149–156
252. Bidanset DJ et al (2004) Efficacy of ganciclovir and cidofovir against human cytomegalovirus replication in SCID mice implanted with human retinal tissue. Antiviral Res 63(1): 61–64
253. Wang W et al (2005) Human cytomegalovirus genes in the 15-kilobase region are required for viral replication in implanted human tissues in SCID mice. J Virol 79(4):2115–2123
254. Laycock KA et al (1997) An in vivo model of human cytomegalovirus retinal infection. Am J Ophthalmol 124(2):181–189
255. Smith MS et al (2010) Granulocyte-colony stimulating factor reactivates human cytomegalovirus in a latently infected humanized mouse model. Cell Host Microbe 8(3):284–291

# Chapter 16

# Rodent Models of Congenital Cytomegalovirus Infection

**Djurdjica Cekinovic, Vanda Juranic Lisnic , and Stipan Jonjic**

## Abstract

Human cytomegalovirus (HCMV) is a leading viral cause of congenital infections in the central nervous system (CNS) and may result in severe long-term sequelae. High rates of sequelae following congenital HCMV infection and insufficient antiviral therapy in the perinatal period make the development of an HCMV-specific vaccine a high priority of modern medicine. Due to species specificity of HCMV, animal models are frequently used to study CMV pathogenesis. Studies of murine cytomegalovirus (MCMV) infections of adult mice have served a major role as a model of CMV biology and pathogenesis, while MCMV infection of newborn mice has been successfully used as a model of perinatal CMV infection. Newborn mice infected with MCMV have high levels of viremia during which the virus establishes productive infection in most organs, coupled with a strong inflammatory response. Productive infection in the brain parenchyma during early postnatal period leads to an extensive non-necrotizing multifocal widespread encephalitis characterized by infiltration of components of both innate and adaptive immunity. As a result, impairment in postnatal development of mouse cerebellum leads to long-term motor and sensor disabilities. This chapter summarizes current findings of rodent models of perinatal CMV infection and describes methods for analysis of perinatal MCMV infection in newborn mice.

**Key words** Cytomegalovirus, Brain, Congenital infections

## 1 Introduction

### *1.1 Background on CMV and Congenital Infection*

Human cytomegalovirus (HCMV) is the most frequent viral cause of congenital infections with annual prevalence between 0.1 and 2 % of newborns [1] and a leading cause of birth defects and developmental disabilities [2].

Vertical transmission of HCMV is a result of either transplacental virus infection of the fetus, intrapartum infection of the delivering child or infection of a breastfeeding newborn, however long-term sequelae are mostly linked to transplacental transmission of the virus. Infants with congenital HCMV infection may develop multisystem disease during which all organs can be affected, including the central nervous system (CNS) [3]. HCMV infection of the developing CNS may result in long-term neurological impairment, which may be manifested in numerous sensorineural sequelae,

Andrew D. Yurochko and William E. Miller (eds.), *Human Cytomegaloviruses: Methods and Protocols*,
Methods in Molecular Biology, vol. 1119, DOI 10.1007/978-1-62703-788-4_16, © Springer Science+Business Media New York 2014

the most frequently found being progressive hearing loss [4]. Although symptomatic disease following congenital HCMV infection occurs in only 10–12 % of infected infants [5], most of these children will suffer mild to severe psychomotor and perceptual handicaps [6]. Moreover, a significant percentage of infected infants who are otherwise asymptomatic at childbirth will go on to develop long-term neurological deficits later on in life [4].

The strict species specificity of CMVs precludes studies of HCMV infection in animal models and none of the models developed thus far completely recapitulate the course of congenital HCMV infection [7]. While the rhesus macaque model of perinatal RhCMV infection shows many similarities with congenital HCMV infection, the main obstacle of this model is the paucity of RhCMV-seronegative animals. Several nonprimate models have been used as models of congenital HCMV infection with varying success. The guinea pig model of congenital GPCMV infection remains unique among rodent models as transplacental transmission of the virus can be detected [8]. Additionally, congenitally infected guinea pigs show systemic viremia with an effect on the CNS [8, 9] and infection of the cochlea can result in development of sensorineural hearing loss (SNHL) [7]. Practical limitations of this model are the need for high doses of the virus for infection of dams with consequent significant fetal loss and small litters (average of three pups per pregnant animal with pregnancy lasting between 65 and 70 days). Therefore experiments in the guinea pig model of congenital GPCMV infection can require high numbers of animals.

While the rat CMV (RCMV) model is frequently used for studying CMV-associated vascular diseases [10], early studies failed to demonstrate congenital infection, as virus could not be recovered from rat embryos from RCMV-infected mothers [11]. However, Loh and colleagues have reported a new strain of RCMV isolated from placental tissue that is able to infect rat fetuses [12, 13]. It remains to be determined if this new strain of RCMV will be useful to model aspects of congenital infection.

The mouse model is the most frequently used model for studying the pathogenesis of CMV infection mainly due to a well defined MCMV genome and proteome, availability of numerous viral deletion mutants and ease of use of mice as experimental animals. However one significant disadvantage of this model is the inability of MCMV to cross the placental barrier and infect mouse embryos. With the aim of overcoming this obstacle, we and others have developed alternative routes of infection of either mouse embryos or newborn mice. Keeping in mind that in congenital HCMV disease, CNS involvement mainly determines the outcome of the infection, most investigators have aimed their models to study infections of the developing mouse brain. MCMV infection of the developing CNS can be established by direct virus inoculation into cerebral hemispheres or lateral ventricles of either mouse embryos or newborn mice [14, 15], intraplacental inoculation of the virus [16], intraperitoneal infection of newborn mice [17], or infection of suckling mice by milk extracted from MCMV-infected dams [18].

The model of perinatal infection of newborn mice developed in our laboratory is based on intraperitoneal inoculation of the virus with no additional pretreatment necessary [17]. This model is in contrast to previously described murine models, which require either cryoanesthesia of pregnant mouse dams or newborn mice and ultrasound-guided inoculation of virus into cerebral ventricles or treatment of dams with TNFα to induce fetal infection following intraplacental inoculation of the virus [15, 16]. In essence, intraperitoneal inoculation of the virus results in systemic infection in which virus replication can be detected in parenchymal tissues, including the brain, as well as in blood cells and plasma (Fig. 1). Moreover, in contrast to intracerebral inoculation of the MCMV which results in localized infection corresponding to virus injection site in the brain parenchyma [19, 20], our model most closely recapitulates the presumed route of CMV entry into the developing CNS via a systemic viremia [21]. Taken together, these properties make the MCMV model an attractive alternative to other animal models, including the primate model of CMV disease.

### 1.2 Developmental Abnormalities in the Brains of MCMV-Infected Newborn Mice

Newborn mice infected with MCMV present with developmental abnormalities of the cerebellum, a finding similar to those described in studies utilizing either cranial ultrasound or magnetic resonance imaging in fetuses and infants congenitally infected with HCMV [17, 22]. The characteristic phenotype of the developing cerebellum in these neonatally infected newborn mice includes reduced cerebellar foliation, decreased cerebellar area and increased thickness of the external granular layer (EGL) [17] (Fig. 1). These morphological impairments result from reduced proliferation of granule neurons in the EGL, delayed migration of postmitotic neurons from EGL into deeper parts of the cerebellar cortex and impaired morphology of Purkinje cells in the cerebella of MCMV-infected newborn mice.

The exact mechanism of observed MCMV induced malformations is not yet fully understood. Impaired responsiveness to brain-derived neurotropins (BDNF) due to decreased expression of the BDNF specific receptor TrkB and/or a strong inflammatory response in the CNS mediated by expression of a number of pro-inflammatory cytokines and chemokines may play important roles. Despite the fact that developmental abnormalities were temporally correlated with virus replication in the CNS, they were symmetric and global in nature, and there was no obvious co-localization of virus with the described abnormalities [17]. This focal nature of the cerebellum maldevelopment argues for an indirect effect of virus infection as a likely mechanism of the disease observed in these animals, rather than direct virus replication in cerebellar neurons.

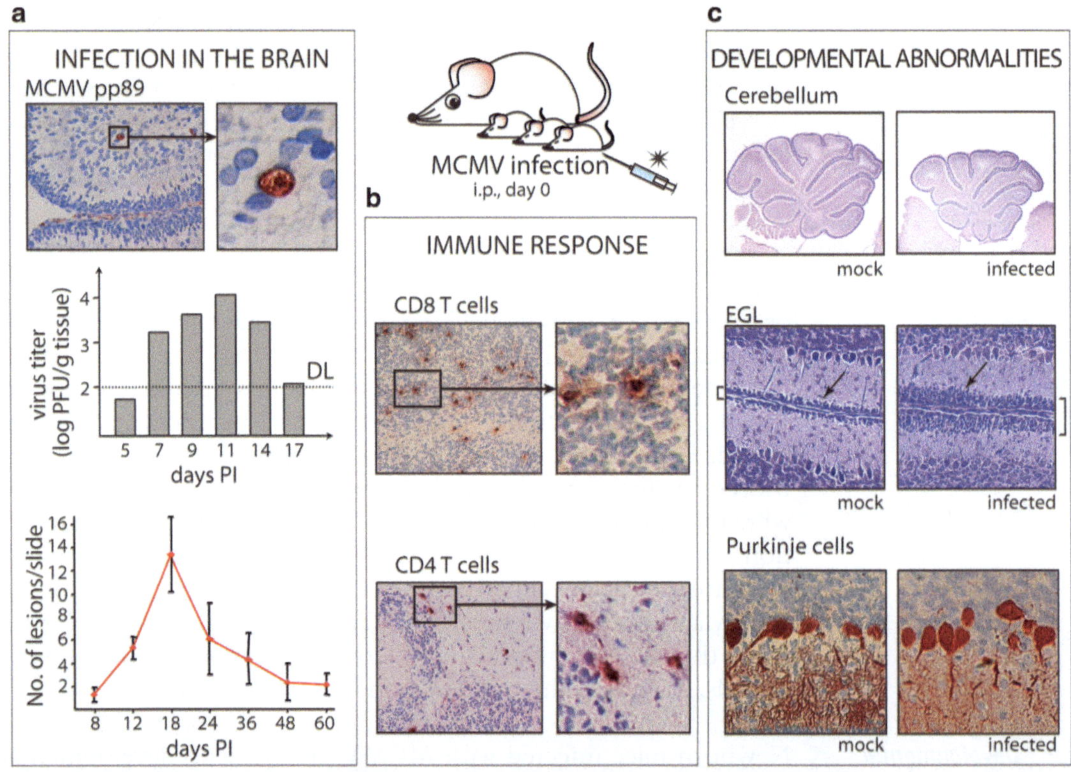


**Fig. 1** MCMV infection of newborn mice
Mouse offsprings are infected with low doses of MCMV by injecting the virus intraperitoneally on day of the delivery (within 24 h after birth) (*graphics*). Intraperitoneal inoculation of the virus results in systemic infection during which the virus enters the CNS. MCMV-infected cells are observed in the brain parenchyma (MCMV pp89$^+$ cells). Productive infection in the CNS is detectable within days 7–17 p.i. (**a**, *middle*). Virus presence in the brain induces development of encephalitis characterized with numerous pathohistological lesions that reside in the brain parenchyma for months after the productive infection is abolished (**a**, *bottom*). Immune response in the MCMV-infected newborn brain is predominately comprised from T lymphocytes among which CD8$^+$ T cells highly outnumber CD4$^+$ T cells (**b**). MCMV-infected newborn mice present with developmental abnormalities of the cerebellum which size is significantly lower than the one of uninfected controls (**c**, *top*). Reduced cerebellar area in MCMV-infected mice is coupled with increased thickness of the external granular layer (EGL) of the cerebellum (**c**, *middle*) and impaired morphology of Purkinje cells (**c**, *down*) in cerebella as compared to control animals. Some parts of the figure are taken from and modified with permission from the Journal of Immunology. J Immunol. 2008 Aug 1;181(3):2111–23

### *1.3 Immune Response in MCMV-Infected Newborn Mice*

Systemic MCMV infection in perinatally infected newborn mice affects nearly all tissues and organ systems and induces a robust inflammatory response in which inflammatory lesions are comprised of both polymorphonuclear and mononuclear cell infiltrates [23]. Numerous cytomegalic cells and cells with eosinophilic inclusions coupled with focal hypoxic necrosis can be observed in specific organs; most frequently in the liver. Sera of these mice contain elevated levels of TNFα that do not correlate temporally with the peak of virus replication in various organs, suggesting that tissue damage in peripheral organs of MCMV-infected newborn mice is cytokine mediated. Although elevated expression of TNFα was also observed in the brains of infected newborn mice, no sign of necrosis or tissue damage was detected [17]. This observation is likely attributable to efficient control of the immune response within the CNS; components of both the innate and adaptive immune response can be detected in the brain (Fig. 1). NK cells can be isolated from the CNS of infected newborn mice by day 7 post infection (unpublished data) and, although NK cells represent the minor immune cell population in infected brain, a robust inflammatory response characterized by upregulation of proinflammatory genes associated with interferon production (IRF-1, IRF-7, USP18, and LRG-47) and chemokine secretion (TNFα, CxCL3, CCL5, CCL21, and CXCL10) [17] coincides with the appearance of NK cells in the brain parenchyma. This suggests that NK cells could serve as an important source of cytokines in MCMV-infected newborn brain. MCMV-infected newborn mice, depleted of NK cells in the early postnatal period showed significantly higher viral burden in the CNS following intracranial (i.c.) inoculation of the virus [19]. Studies performed on newborn mice infected with an attenuated MCMV that expresses a ligand for the activating NK cell receptor NKG2D Rae-1γ show reduced virus replication in various organs, including brain compared to wild type virus [24], arguing for an active role of NK cells in the control of MCMV infection in newborn mice.

In the second week post-infection monocytes are also recruited to the CNS, where they then differentiate into macrophages. Activated microglia cells and infiltrating macrophages can be found in close proximity to virus-infected cells within encephalitic lesions, finding that argues for an active role of these cells in control of MCMV infection in the newborn mouse brain [25].

Although NK cells and monocytes are recruited, the predominant immune cells in MCMV-infected newborn brain are $CD8^+$ T lymphocytes (Fig. 1). Starting from the second week post-infection, activation of T cells and influx into infected CNS is readily observed, suggesting that $CD8^+$ T cells may play a role in controlling MCMV infection in the newborn brain [25]. Infiltration of these cells into the infected brain correlates with a significant reduction of virus replication in the CNS. Moreover, when neonatal mice were depleted of $CD8^+$ T lymphocytes there was a significant increase in viral genome copy numbers in the CNS and these mice succumbed to infection by postnatal day 15. A significant proportion of $CD8^+$ T cells isolated from infected brain are specific for several virus-encoded peptides, with pp89-specific $CD8^+$ T cells being the most predominant. Furthermore, when stimulated ex vivo with MCMV pp89-encoded peptide, these cells produce significant amount of IFN-γ, confirming an active antiviral activity of $CD8^+$ T lymphocytes in the newborn mouse brain [25].

In vivo significance of T lymphocytes isolated from the brain of MCMV-infected newborn mice at later stages post-infection was confirmed by adoptive transfer of brain-isolated mononuclear cells into γ-irradiated, immunocompromised animals. Mice that received CNS-isolated CD8+ T cells had lower viral burden in liver, lungs and spleen, as compared to control mice [25]. This finding closely resembles the effector memory CD8+ T cells accumulating in latently infected lungs of adult mice [26, 27], arguing in favor of a fully functional state for these cells.

### 1.4 Role of Antibodies in the Control of MCMV Infection in Newborn Mice

In adult mice, virus-specific antibodies (Abs) appear to be dispensable for the resolution of acute MCMV infections, but are essential for the prevention of virus spread following reactivation [28, 29]. A pivotal role for virus-specific antibodies has been confirmed in congenital HCMV infection where pre-conceptual immunity reduces the rate of virus transmission from mother to child but fails to provide complete protection of infection to the fetus [30, 31]. This incomplete protection from virus transmission is considered to be the consequence of reinfection with a new serotype(s) of HCMV or reactivation of an endogenous virus [32, 33]. In a model of perinatal MCMV infection, antibodies have a proven role in the control and prevention of infection in the developing CNS [34]. When administered to infected newborn mice early after infection, antibodies provided protection and limited virus replication in the brain parenchyma, and when administered at later time points post-infection, even at the time of peak virus infection in the brain, the antibodies significantly reduced virus titers in the CNS [34]. As a consequence, mice treated with antibodies had less pronounced inflammatory lesions in the brain parenchyma and improved postnatal cerebellar development as compared to MCMV-infected, untreated animals [34]. In addition, we have recently shown that vaccination of female mice before pregnancy with the highly attenuated virus carrying the ligand for NK cell activating receptor NKG2D induces an antibody response in dams that is sufficient to limit MCMV infection in MCMV-infected offspring [24]. Taken together, these data strongly argue for a protective role of maternal antibodies in perinatal MCMV infection.

## 2 Materials

Standard equipment and material for cell culture, virus preparation, infection and sacrifice of animals, and tissue harvesting are necessary (hoods, incubators, centrifuges, microtomes, cryostat, pipettes, tubes, dishes, small surgical instruments, etc.). Additional major materials are listed below:

1. Viruses: MCMV Smith strain (ATCC, VR-1399, MCMV Δm152Rae1γ).
2. Mice: BALB/c, C57BL/6.
3. Cell lines: mouse embryonic fibroblasts (MEF).

4. Minimal essential medium (MEM).
5. Dulbecco's Modified Eagle Medium (DMEM).
6. RPMI-1640.
7. Fetal calf serum (FCS) and newborn bovine serum (NBS).
8. Stainless steel wire-mesh.
9. Cell strainer (BD Biosciences).
10. Methyl cellulose medium (viscous medium): To 320 ml distilled water, add 8.8 g of 0.16 M methyl cellulose (4,000 centipoise, Sigma). Stir with a large magnetic stirrer. Leave the medium for 4 days at 4 °C until methyl cellulose has completely dissolved. Autoclave for 30 min at 121 °C and then supplement the solution with 40 ml of 10× MEM, 40 ml FCS, $4 \times 10^4$ U Penicillin, 40 mg Streptomycin, 4 ml 1 M HEPES, and 0.88 g of $Na_2CO_3$.
11. Phosphate-buffered saline (PBS) (10× stock): Dissolve 2.0 KCl, 2.0 g of $KH_2PO_4$, and 80 g of NaCl in 700 ml of dd$H_2O$. Dissolve 11.4 g of $Na_2HPO_4$ anhydrous in 200 ml of dd$H_2O$ with stirring and heating. Combine both solutions. Adjust pH to 7.4 with 5 M NaOH. Adjust volume with dd$H_2O$ to 1 l.
12. TBS: 1 M Tris–HCl, pH 7.4,1 M NaCl.
13. Trypsin for cell culture: 0.25 % Trypsin plus 1 mM EDTA.
14. 15 % sucrose/VSB buffer: 50 mM Tris–HCl, 12 mM KCl, 5 mM $Na_2$EDTA, adjust pH to 7.8 with HCl.
15. FACS (fluorescence activated cell sorting) medium: PBS, 10 mM EDTA, 20 mM HEPES, pH 7.2, 2 % FCS, 0.1 % $NaN_3$.
16. Erythrocyte lysing buffer (Tris–NH4Cl): 90 ml of 0.16 M $NH_4Cl$, 10 ml of 0.17 M Tris–HCl, pH 7.65. Adjust final pH to 7.2 with HCl.
17. Ca-stock solution (500 mM calcium solution): 1.095 g $CaCl_2 \times 6H_2O$ in 10 ml pure RPMI.
18. Ca-RPMI: 100 ml 10 % FCS supplemented RPMI, 1 ml 100× Ca-stock solution.
19. Collagenase D: Dilute the stock solution to 1 mg/ml in Ca-RPMI.
20. 3 % Acetone.
21. Paraformaldehyde.
22. Xylol.
23. NP 40 detergent: dissolve 500 μl of NP40 in distilled water.
24. Percoll (Amersham): 100 % Percoll: dissolve 9 ml of Percoll with 1 ml 10× PBS (*see* **Note 11**).
25. Cresyl violet (0.1 g of cresyl violet in 100 ml water and 300 ml glacial acetic acid).

26. FITC-conjugated rat IgG2a.
27. FITC-conjugated rat anti-mouse CD8.
28. PE-conjugated rat anti-mouse CD4.
29. Mouse anti-MCMV pp89 mAb (CROMA 101).
30. Biotin-conjugated goat anti-mouse IgG (BD Pharmingen).
31. Rat anti-mouse CD8 (YTS 191.2.1) ATCC.
32. Rat anti-mouse CD4 (YTS 169.4.2) ATCC.
33. Biotin-conjugated goat anti-rat IgG (Caltag).
34. Mouse anti-mouse calbindin (Sigma).
    Tetramers (Produced by NIH tetramer core facility, Bethesda, SAD, http://www.niaid.nih.gov/reposit/tetramer/overview.html).
    (a) H-2L$^d$-IE1$_{168-176}$ (168-YPHFMPRNL-176): MHC class I molecule coupled with MCMV peptide IE1/m123pp89.
    (b) H-2D$^d$-m164$_{257-265}$ (257-AGPPRYSRI-265): MHC class I molecule coupled with MCMV peptide m164.
    (c) H-2D$^d$-m04$_{243-251}$ (243-YGPSLYRRG-COOH-251): MHC class I molecule coupled with MCMV peptide m04.
35. Ethylcarbazole (Sigma-Aldrich): 4 ml deionized water, two drops of acetate buffer, one drop of AEC chromogen (ethylcarbazole), one drop 3 % hydrogen peroxide.
36. DAB (diaminobenzidine) (DAKO): 1 ml substrate buffer, one drop of DAB chromogen.
37. Propidium iodide (Sigma-Aldrich): dilute 10 μl in 990 μl of FACS medium.
38. Entellan mounting medium.

## 3 Methods

### 3.1 Analysis of the MCMV Pathogenesis in Newborn Mice

The methods described below outline the procedures to assess MCMV pathogenesis primarily in the brain of newborn mice. However, most techniques described can also be applied to other organs and to adult mice. Where this does not apply (e.g., isolation of mononuclear cells from blood) we have provided separate protocols.

Following intraperitoneal (i.p.) inoculation of the virus, systemic infection is established, during which practically all organs can be affected, and the virus can be isolated from both blood cells and plasma [17, 34]. Parallel with the infection, strong inflammatory responses have also been observed in peripheral organs of MCMV-infected newborn mice and are comprised of both polymorphonuclear and mononuclear cell infiltrates. Focal hypoxic necrosis, numerous cytomegalic cells and cells with eosinophilic inclusions can also be observed in specific organs (most commonly, the liver) [23]. Productive MCMV infection in the CNS can be detected between days 7 and 19 post infection with the peak of virus replication between days 9 and 14 post infection [17]. Immunohistochemical analysis of the brain tissue from infected newborn mice reveals no preference for the site of infection and development of widespread encephalitis. Both neurons and glial cells are permissive for MCMV infection and while lytic infection in the brain terminates within three weeks post-infection, pathohistological lesions remain in the CNS for several months. The preservation of neuronal lesions argues for either virus persistence in the brain that constantly primes the immune response or consequent development of immunopathology in this immunologically privileged organ.

The pathohistological lesions in virus infected newborn mice closely resemble the ones described in autopsied cases of encephalitis in congenitally infected infants, stillbirths, and fetuses [35], and are comprised of infiltrating monocytes and activated microglia cells spread throughout the brain parenchyma and also within specific sites of infection [17, 25, 34].

#### *3.1.1 Production of Tissue-Derived Virus and Preparation of Virus Stocks*

Tissue culture derived MCMV is produced in permissive murine cell lines such as primary embryonic fibroblasts (MEFs) or immortalized cell lines (NIH/3 T3, BALB/3 T3, B12). Many immortalized cell lines divide much faster than primary ones and continue to divide even after infection with MCMV. This fact should be taken into account when seeding the cells for virus production. Additionally, many immortalized cell lines require a medium supplemented with 10 % FCS in contrast with primary cell lines that require only 3 % of FCS. The protocol outlined below is adapted for primary MEFs grown in DMEM supplemented with 3 % FCS (*see* **Notes 1** and **2**).

All steps should be carried out in a laminar flow hood.

1. Grow MEFs in 3 % FCS supplemented DMEM until they reach ~70–85 % confluence. Use low passage cells (usually third passage is used).
2. Prepare 0.01 PFU/cell virus in warm DMEM supplemented with 3 % DMEM in a volume sufficient to cover the whole surface of the dish to prevent cells from drying.
3. Remove the medium from cells using vacuum pump or a pipette and overlay them with previously prepared virus diluted in 3 % DMEM. Incubate for 4–6 h at 37 °C in a $CO_2$ incubator. Occasionally gently rock the dish to mix the medium above the cells.
4. Add 15–20 ml medium, leave in incubator for 3–5 days, monitoring for plaque formation (*see* **Note 3**).

5. Collect the cells and supernatant using cell-scraper. Centrifuge at 6,400 × *g* for 20 min to separate cell debris from cell-free virus. Collect the supernatant.
6. Pellet the virus from the supernatant in an ultracentrifuge at 26,000 × *g* at 4 °C for 2–3 h.
7. Decant the supernatant leaving 200 μl of medium overlaying the pellet. Store on ice at 4 °C overnight.
8. On the next day, resuspend the pellet. In a new set of ultracentrifuge tubes, place 18 ml of 15 % sucrose/VSB. Carefully overlay resuspended virus from the first set of tubes. Centrifuge at 72,000 × *g* at 4 °C for 2 h.
9. Remove the supernatant leaving 500 μl of sucrose. Leave on ice at 4 °C overnight to facilitate easier resuspension of the pellet.
10. Thoroughly resuspend the pellet in 25 % sucrose/VSB. Pool all resuspended pellets. If suspensions appear too thick, they can be diluted in 15 % sucrose/VSB. Aliquot and store at −80 °C. Additionally, in order to break up microscopic clumps, virus can be sonicated prior to aliquoting.

For infection of newborn mice, virus stocks should be further diluted in DMEM to lower titers. Recommended titers of the stocks are between $10^5$ and $10^6$ PFU/ml.

#### 3.1.2 Infection of Mice

The most often used route of infection of adult mice is intravenous (i.v.) into the tail vein which quickly results in systemic viremia. Alternative routes are footpad (f.p.) and intraperitoneal infection. Footpad infection represents a more natural way of infection mimicking viral transfer in rodents. The most frequently used dose of tissue-cultured virus is 2–5 × $10^5$ PFU/mouse in 100–500 μl of PBS or pure medium (DMEM or RPMI) for adult mice, while for newborns the maximal volume is 50 μl. When using the i.v. route of infection, mice should be heated under an infrared lamp prior to the infection as the heat will cause tail vein to expand making it easier to see and inject.

*Infection of newborn mice*

1. Inject newborn mice intraperitoneally (i.p.) using insulin syringe and insert it under the skin in upper thoracic region.
2. Slide the needle up to the peritoneal cavity. Be careful not to injure the liver or urinary bladder. After inoculation wait for 2–3 s before removing the needle to avoid virus solution leaking out.

#### 3.1.3 Determination of Virus Titers Using Standard Plaque-Forming Assay

As in tissue culture virus production, primary MEFs are cells of choice for determination of virus titers however other permissive cell lines can be used. The protocol outlined here uses primary

MEFs grown in DMEM with 3 % FCS. All steps should be carried on ice and in a laminar flow hood.

1. The day before titration, seed MEFs in 48-well plates at 200,000 cells/well.
2. Thaw mouse organs or virus stocks on ice.
3. Prepare organ homogenates in 2 ml medium by passing them through stainless steel wire-mesh or cell strainer.
4. Make serial tenfold dilutions of each organ homogenate (1:100; 1:1,000; 1:10,000; 1:100,000).
5. Remove the medium from MEF cells and overlay them with 100 μl of organ homogenate in duplicate (on 2 wells). Incubate for 30 min at 37 °C in $CO_2$ incubator. This step allows the virus to attach to the cells.
6. Centrifuge the plates at 800 × *g* for 30 min (*see* **Note 4**).
7. Remove the cells from the centrifuge and incubate at 37 °C in a $CO_2$ incubator for 30 min.
8. Overlay with 500–1,000 μl of methyl cellulose medium. Methyl cellulose is a very viscous medium which prevents horizontal spread of the virus. Because of its viscosity it cannot be added with a pipette. Sterile beakers or graduated cylinders should be used instead.
9. Incubate the plates in $CO_2$ incubator at 37 °C for 3–4 days until visible plaques are formed.
10. Count virus plaques using an inverted microscope. Plaques are clearly visible without any staining. Count only those wells where individual plaques can be observed (no more than 100 plaques per well is a good rule of thumb). Calculate the number of PFUs per organ/aliquot/gram of tissue taking the dilution factor into account.

#### 3.1.4 Immunohistochemical Detection of MCMV-Infected and Immune Cells in Tissues

Immunohistochemistry is performed either on frozen or paraffin embedded sections (*see* **Note 5**). Blocking serum, primary and secondary antibodies, enzyme–avidin complex, and chromogenic substrate are added with a pipette directly onto the slides in total volume of 200 μl per slide.

For every experimental procedure described below intracardiac perfusion with PBS should be performed prior to sacrificing the animals. Paraffin embedded sections are prepared from paraffin blocks that are made by rinsing tissue in growing series of alcohol, xylol, and paraffin using Histokinet (Leica). Paraffin-embedded brain sections are cut to thickness from 2 to 8 μm on microtome. Frozen sections are cut to thickness of 8 μm from snap frozen brain tissue which was immersed into liquid nitrogen immediately after harvesting the brain.

Paraffin Embedded Sections

1. Before immunohistochemical procedures, slides are deparaffinized by rinsing slides in xylol and then decreasing concentrations of alcohol in the following order: Xylol 3× 8 min, 100 % ethanol 2× 4 min, then 90, 80, 70, and 50 % ethanol for 4 min.
2. Wash the slides in deionized water for 1 min.
3. For detection of intracellular antigens rinse slides in trypsin (pH 7.2) for 15 min at 37 °C (heat trypsin 10 min in advance).
4. Wash in deionized water 3× 1 min.
5. Block tissue peroxidase in 30 ml PBS + 30 ml methanol + 1 ml $H_2O_2$ for 30 min. Blocking solution is made right before use.
6. Wash in deionized water 3× 1 min.
7. Rinse the slides in NP40 detergent for 10 min.
8. Wash in deionized water 3× 5 min.
9. Perform antigen retrival if needed (*see* **Note 6**).
10. Add blocking serum and incubate for 25 min at room temperature. Blocking serum must be derived from animal species which are the source of secondary antibodies you will use (e.g., if your secondary antibody is derived from rat, use rat serum to block) (*see* **Note 7**).
11. Add primary antibody (e.g., mouse anti-MCMV pp89) and incubate for 60 min at 37 °C.
12. Wash in PBS 3× 1 min.
13. Add biotinylated secondary antibody (e.g., goat anti-mouse IgG) and incubate for 30 min/RT.
14. Wash in PBS 3× 1 min.
15. Add enzyme–avidin complex (e.g., Streptavidin–POD, Streptavidin–AP) and incubate for 30 min at RT.
16. Wash in PBS 3× 1 min.
17. Add chromogenic substrate (AEC/DAB) and incubate for 5–10 min.
18. Wash in deionized water.
19. Counterstain with hematoxylin.

Frozen Sections

1. Dry sections for 15–20 min at RT.
2. Rinse slides for 3–4 min in cold PBS.
3. Fix slides in cold acetone (4 °C) for 5 min or 2.5 % PFA for 20 min (*see* **Note 8**).
4. Wash in PBS 3× 1 min.
5. Block tissue peroxidase by rinsing slides in 50 ml of PBS + 0.12 ml of $H_2O_2$ for 30 min. Blocking solution is made right before use.

6. Wash in PBS 3 × 1 min.
7. Add blocking serum and incubate for 25 min at RT. Use serum derived from animal species that is the source of secondary antibodies.
8. Add primary antibodies (specific for either $CD4^+$ or $CD8^+$ or NK lymphocytes), incubate for 60 min at RT.
9. Wash in PBS 3× 1 min.
10. Add biotinylated secondary antibody (goat anti-rat igG) and incubate for 30 min at RT.
11. Wash in PBS 3× 1 min.
12. Add SA-POD and incubate for 30 min at RT.
13. Wash in PBS 3× 1 min.
14. Add chromogen (AEC/DAB) and incubate for 5–10 min.
15. Wash in deionized water.
16. Counterstain with hematoxylin.

### 3.2 Assessment of Developmental Abnormalities in Brains of MCMV-Infected Newborn Mice and Consequent Neurological Impairments

Pathohistological lesions and morphometric analyses are performed on cresyl violet-stained paraffin-embedded brain sections. For morphometrical analyses we recommend either IPLab or CellB software.

#### 3.2.1 Histomorphometrical Analyses of the Brain

##### Cresyl Violet Staining

Cresyl violet staining is used on paraffin embedded sections exclusively. It specifically stains neurofilaments. Cresyl violet solution can be made in advance and stored at 4 °C. Prior to use it needs to be filtered through filter-paper.

1. Deparaffinize the sections in xylol 2× 10 min, then hydrate the slides in 100 % ethanol 2× 5 min, 95 % ethanol 3 min and 75 % ethanol 3 min.
2. Wash the slides by quickly immersing in water 2–3 times (tap water can be used).
3. Repeat with deionized water.
4. Stain the slides in 0.1 % cresyl violet 5 min on room temperature (*see* **Note 9**).
5. Quick wash in deionized water.
6. Rinse slides in 90 % ethanol for 15 min to wash off excess dye.
7. Dehydrate the slides in 100 % ethanol 2× 5 min.

8. Dehydrate the slides in xylol 2× 5 min.
9. Mount the slides in entellan.

Measurement of Cerebellar Area and the External Granular Layer Thickness

1. Measure cerebellar area on serial cresyl violet-stained sections on the microscope, under the magnification 2×.
2. Determine the thickness of the external granular layer by marking six points along the primary cerebellar fissure on serial cresyl violet-stained sections on the microscope, under 20× magnification.
3. Count the number of Purkinje cells along 500 mm of both sides of primary cerebellar fissure on serial cresyl violet-stained sections on the microscope, under 20× magnification.

*3.2.2 Balance Beam Test*

1. Use a narrow beam, 90 cm long and 2 cm in diameter.
2. Position a beam 10 cm above the ground and place a mouse in the middle of a beam.
3. Measure the distance crossed in 2 min.
4. Repeat the process (*see* **Note 10**).

### 3.3 Assessment of Inflammatory Response in MCMV-Infected Newborn Brain

Following virus entry into newborn mouse CNS, the initial innate immune response is mediated by resident glia cells. Beside activated astrocytes and microglia cells, infiltrating macrophages and NK cells are the other components of the innate immune system that can be isolated from MCMV-infected brain in the early post infection period. The primary activity of infiltrating immune cells is the elimination of virus from infected cells while limiting the damage which may have long-term detrimental consequences to the host. In the first week post infection, numerous soluble mediators are activated in the brains of infected newborn mice, and they are largely directed towards activation of adaptive immunity [17]. It is important to point out that NK cells can be detected in infected brain within the first week post infection. This raises the possibility that, although NK cells represent the smallest population of immune cells in MCMV-infected newborn brain, they are an important source of proinflammatory cytokines.

*3.3.1 Isolation of Mononuclear Cells from Tissue*

1. Infect animals as described above.
2. Place the brain in a petri dish in ~10 ml of cold medium (RPMI or DMEM) on ice.
3. Cut the organ into small pieces using sterile scissors or a scalpel and then pass through metal wire-mesh to obtain single cell suspension.
4. Centrifuge at 200 ×*g* for 5 min.
5. Resuspend the cell pellet in cold medium and add Percoll to desired density. The total volume of medium and Percoll should be 10 ml (*see* **Note 11**).

6. Slowly overlay the suspension onto 10 ml of diluted Percoll (*see* **Note 11**).
7. Centrifuge at 700 × *g* for 30 min at 4 °C without brake.
8. Remove buffy middle layer and transfer into a new tube.
9. Add the medium to bring volume to 10 ml and pellet the cells at 200 × *g* for 10 min at 4 °C to remove Percoll (*see* **Note 12**).
10. Resuspend pellet in desired volume of medium or PBS.

#### *3.3.2 Isolation of Mononuclear Cells from Spleen*

1. Prepare single cell suspension of splenocytes by passing the spleen through wire mesh or cell strainer. To facilitate passing through cell strainer you may cut the spleen into smaller pieces with scissors.
2. Pellet the cells by centrifugation at 500 × *g* for 5 min at room temperature.
3. Lyse erythrocytes by resuspending cell pellet in 5 ml of erythrocyte lysis buffer and incubating on ice for 5 min. Stop the lysis by addition of 10 ml medium. Pellet the cells at 200 × *g* for 5 min at room temperature.
4. Resuspend in 10 ml medium.

#### *3.3.3 Isolation of Mononuclear Cells from Blood*

1. Prepare tubes for blood collection by adding 60 μl of 0.5 M EDTA to prevent clotting. 150–300 μl of blood can be obtained from one mouse by cardiac puncture procedure or from orbital vein from adult mice, while less blood can be obtained from the tail vein.
2. Dilute the blood with double volume of PBS and overlay over 3–5 ml of Lymphoprep.
3. Spin at 800 × *g* for 10 min with no brake.
4. Collect middle layer (buffy coat) with pipette and transfer to new tube.
5. Add medium/PBS to total volume of 10 ml to wash Lymphoprep.
6. Spin at 500 × *g* for 5 min at room temperature.
7. Lyse erythrocytes with erythrocyte lysis buffer, if needed as described above.

#### *3.3.4 Isolation of Mononuclear Cells from Lungs*

1. Remove lungs, place in one well of a 6-well plate and cut them to small pieces with scissors.
2. Add 3 ml collagenase D (1 mg/ml) and incubate at 37 °C for 60 min.
3. Pass the tissue through wire mesh or cell strainer and wash with 5 ml RPMI.
4. Pellet the cells by centrifugation at 500 × *g* for 5 min.

5. Resuspend in 4 ml of RPMI Overlay over 30 % Percoll in a new falcon tube.
6. Spin at 800 × *g* for 30 min with no brake.
7. Remove supernatant carefully and resuspend in 5 ml of RPMI.
8. Spin at 500 × *g* for 5 min.
9. Lyse erythrocytes by addition of 2 ml of erythrocyte lysis buffer if needed as described above.
10. Stop the lysis by addition of 5 ml of RPMI.
11. Pellet the cells at 500 × *g* for 5 min.
12. Resuspend the pellet in 1 ml of RPMI.

#### *3.3.5 Mononuclear Cell Adoptive Transfer (Via Cell Sorting)*

Adoptive cell transfer into MCMV-infected immunodeficient mice is a frequently used method to analyze the functional capacity of these cells to control MCMV infection in vivo. Interestingly, adoptive transfer of brain-isolated $CD8^+$ cells from MCMV-infected newborn mice reduced subsequent virus replication in organs of adult immunocompromised mice [25].

Blood and spleen are the most accessible sources of mononuclear cells, however, any organ can be used. Cell adoptive transfer can be done either by cell sorting or by transferring total number of cells from cell sample normalized to the desired number of target cell subpopulation. In both approaches, care should be taken to make sure that donor and acceptor mice are syngeneic to avoid graft-versus-host disease. Prior to adoptive transfer donor mice may be treated with monoclonal antibodies to deplete unwanted cell subpopulations that could influence the results of the experiment. Acceptor mice may also be treated with depleting antibodies or γ-irradiation. Benefits of cell sorting are clear and defined cell population with no interfering cell subpopulations (e.g., CD8 T cells in NK cell adoptive transfer). However direct labeling of cells with monoclonal antibodies can cause their activation. Additionally, transfer of cells marked with antibodies can lead to rejection of transferred cells via antibody-mediated cellular cytotoxicity (ADCC). Stripping of bound antibodies post sorting from the surface of the cells is not very efficient and leads to loss of viable cells. ADCC and activation of cells can be avoided by utilizing negative sorting—i.e., by labeling all the unwanted cell subpopulations. However, the purity of such sorts is much lower in comparison to directly labeled and sorted cells since committed cell precursors may not express markers used for sorting out a particular cell subpopulation. Additionally, negative cell sorting requires significant amount of monoclonal antibodies. Finally, cell sorting presents a stress for the cells, which can influence their expression patterns and behavior post-transfer.

Adoptive transfer of all cells normalized to target cell numbers is simple and fast method with very little manipulation of transferred cells. However, since all cells are transferred and in many cases that includes unwanted cell populations that can have a significant impact on the outcome of the experiment. Nowadays there exist many inbred mouse strains lacking different cell type, which can be used as donors in adoptive transfer experiments (e.g., SCID mice as NK cell donors to avoid contamination with B and T cells).

#### Adoptive Cell Transfer by Cell Sorting

1. Remove the organ–source of transferred cells from the mouse aseptically in laminar flow cabinet.
2. Obtain single cell suspension by cutting the organ to smaller pieces with surgical scissors and then passing it through stainless-steel wire mesh or cell strainer in 10 ml of RPMI.
3. Pellet the cells by centrifuging at 500 × *g* for 5 min.
4. Lyse erythrocytes by addition of 5 ml sterile lysing buffer and incubate on ice for 5 min.
5. Stop the lysis by addition of 10 ml sterile RPMI.
6. Pellet the cells by centrifuging at 500 × *g* for 5 min.
7. Resuspend in 10 ml of sterile RPMI and count the cells.
8. Stain the cells with desired antibodies making sure that the antibodies and/or FACS medium do not contain any azide (*see* **Note 13**).
9. Prepare acceptor tubes for cell sorting and fill them with 1–2 ml of RPMI supplied with 10–20 % of fetal calf serum (*see* **Note 14**).
10. Sort the cells into acceptor tubes.
11. Retrieve an aliquot of sorted cells, add 100 μl of 1:100 diluted propidium iodide solution in FACS medium and check purity of sort and viability on FACS analyzer or sorter.
12. Pellet the cells by centrifugation at 800 × *g* for 10 min and resuspend in 150–500 μl of pure RPMI.
13. Inject the sorted cells into tail vein.

#### Adoptive Cell Transfer of All Cells Normalized to Target Cell Numbers

1. Start from **step 7** of the previous Subheading "Adoptive Cell Transfer by Cell Sorting".
2. Collect an aliquot of cells (typically 1–2 million of cells is enough) for flow cytometric analysis.
3. Stain with antibodies to determine percentage of live target cells. Use propidium iodide to determine viability and exclude dead cells.
4. Calculate number of all cells needed to transfer selected number of target cells.

   Example: if you want to transfer 5,000 CD8 T cells from spleen which contains 10 % of live CD8 T cells in order to transfer that number of CD8 T cells you need 5,000/0.10 = 50,000 total splenocytes.

5. Resuspend the number of total cells in maximal total volume of 500 μl of RPMI and inject the cells into tail vein.

### 3.4 Assessment of Antiviral Antibody (Ab)-Mediated Protection of MCMV Infection of Developing Brain

A reduction in the level of virus replication to undetectable levels in the brains of newborn mice, which received Abs prior to virus spread to the CNS, suggested that antibodies could prevent virus spread to the CNS by either direct neutralization of cell-free virus or prevention of cell-associated virus infection by mechanisms such as Ab dependent cellular cytotoxicity (ADCC). Moreover, virus clearance from brains of infected newborn mice that received Abs during peak virus replication in the brain implies that Abs could also have antiviral effect within the CNS itself in order to limit virus spread and/or replication. Abs have been detected in brain parenchyma of newborn mice, surrounding neural cells, and blood vessels [34]. Reduced virus titers in brains of newborn mice (Fig. 2.), which received Abs at the time of peak virus replication, are in accordance with results of MCMV-specific Ab-mediated control of virus spread in the liver even when administered during an already established infection in the liver parenchyma [36]. Studies performed in the guinea pig model suggested that antiviral Abs prevented hearing loss in infected pups. Ab-treated animals exhibited preserved cochlear morphology and insignificant hearing loss [37]. However, a possible protective effect of Abs in preventing the development of hearing loss in MCMV-infected newborn mice has yet to be proven.

#### 3.4.1 Preparation of Hyperimmune Serum

1. Infect mice with $2 \times 10^5$ PFU of w.t. MCMV via i.v. injection.
2. Sacrifice mice 3–4 months post infection.
3. Use insulin syringe and perform cardiac function to aspirate the blood.
4. Leave the blood at room temperature for 10 min to allow it to clot.
5. Centrifuge the blood at 500 × *g* for 5 min to separate the cells from serum.
6. Pool the serum from several animals and store it at −20 °C.

#### 3.4.2 *In Vivo* Administration of Immune Serum or Monoclonal Antibodies to Newborn Mice

1. Infect newborn mice within the first 24 h after birth with 200–500 PFU of w.t. MCMV (*see* Subheading 3.1.2).
2. Inject 100 μl of immune serum intraperitoneally using insulin syringe at 5 or 9 days post infection by inserting the syringe subcutaneously in upper thoracic region. Slide the needle up to peritoneal cavity carefully not to injure liver or urinary bladder.

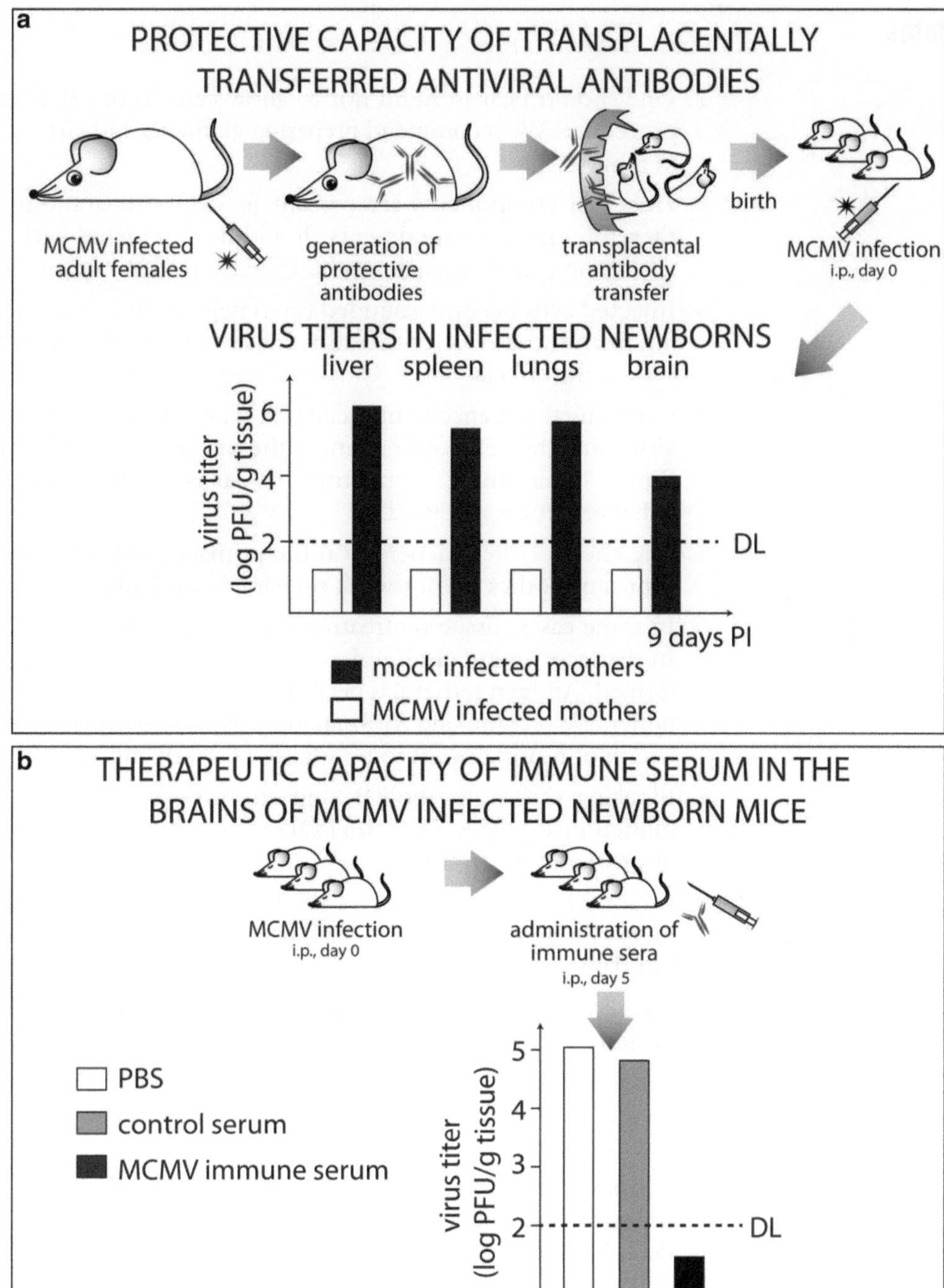


**Fig. 2** Antibody-mediated protection of MCMV infection in newborn mice Transplacentaly transferred antibodies from immunized dams provide efficient protection from MCMV infection to newborns since no productive infection was observed in organs of MCMV-infected newborn mice delivered from mothers infected with MCMV prior to pregnancy (**a**). Passively transferred virus-specific antibodies administered to MCMV-infected newborn mice prior to virus replication in the CNS reduce productive infection in the brain to undetectable levels (**b**). Some parts of the figure are copyright© American Society for Microbiology, J Virol (2008), 82(24): 12172–12180

## 4 Notes

1. One should bear in mind not to allow cells to dry during the procedure. We recommend preparing all media and virus stocks in advance.
2. Host cell components are present in virus preparations, and therefore, in vivo experiments should be conducted with virus grown on cells syngeneic to the animals used in experiments.
3. Infected cells become rounded (cytomegalic) and start detaching from the surface of the dish. Ideally all cells should exhibit cytopathic effects.
4. Centrifugal enhancement facilitates synchronous entry of the virus into the cells and enhances the infection by 10–25-fold. When determining virus titer in virus stocks centrifugal enhancement is not used.
5. The choice largely depends on the primary antibodies used as some antibodies do not work on paraffin embedded samples.
6. In some cases, tissue pretreatment may mask antigen epitopes. In that case antigen retrieval as additional step needs to be performed. Antigen retrieval is performed by boiling slides in citrate buffer. We recommend 3 × 5min in microwave oven however the length of heating needs to be adjusted to each antibody.
7. Blocking serum, antibodies, and enzyme–avidin complex are diluted in 1 % BSA-TBS. SA-POD should be prepared 15 min prior to administration.
8. Fixating in acetone preserves the tissue structure better but is more damaging to the antigens, while PFA conserves antigens but damages the tissue.
9. For slides thicker than 20 μm cresyl violet solution should be heated to 37–50 °C.
10. If the mouse falls from the beam, it should be excluded from further analysis and additional beam test should be performed 5 min later.
11. For isolation of mononuclear cell from brain, the upper suspension should contain 30 % Percoll, while the lower 70 % Percoll. For liver, upper solution contains 40 % Percoll and lower 80 % Percoll. 100 % Percoll is prepared by mixing 9 ml of Percoll and 1 ml of 10× PBS. 70 % Percoll is made by mixing 7 ml of 100 % Percoll and 3 ml of the medium.
12. If perfusion or density gradient centrifugation was not efficient and coagulated blood is still visible after the step 8, add 1 ml of erythrocyte lysis buffer onto pellet and incubate in ice for 1–2 min. Stop the lysis by addition of 10 ml medium (Note: this step will lower your yield).

13. Sodium azide is an irreversible inhibitor of cellular metabolism.
14. The cells are very sensitive to drying. This is effectively prevented by filling the receptor tubes with the medium and rolling them in palms in order to cover whole tube with a sheath of medium. The high percentage of fetal calf serum is needed since cells are sorted each in a small drop so the medium gets diluted.

## References


1. Britt W (2010) Cytomegalovirus. In: Remington JS et al (eds) Infectious diseases of the fetus and newborn infant. Elsevier, Philadelphia, pp 706–756
2. Cannon MJ (2009) Congenital cytomegalovirus (CMV) epidemiology and awareness. J Clin Virol 46(Suppl 4):S6–S10
3. Becroft DMO (1981) Prenatal cytomegalovirus infection: epidemiology, pathology and pathogenesis. In: Rosenberg HS, Bernstein J (eds) Perspective in pediatric pathology. Masson, New York, pp 203–241
4. Morton CC, Nance WE (2006) Newborn hearing screening – a silent revolution. N Engl J Med 354(20):2151–2164
5. Kenneson A, Cannon MJ (2007) Review and meta-analysis of the epidemiology of congenital cytomegalovirus (CMV) infection. Rev Med Virol 17(4):253–276
6. Ross SA, Boppana SB (2005) Congenital cytomegalovirus infection: outcome and diagnosis. Semin Pediatr Infect Dis 16(1):44–49
7. Cheeran MC, Lokensgard JR, Schleiss MR (2009) Neuropathogenesis of congenital cytomegalovirus infection: disease mechanisms and prospects for intervention. Clin Microbiol Rev 22(1):99–126, Table of Contents
8. Griffith BP, Lucia HL, Hsiung GD (1982) Brain and visceral involvement during congenital cytomegalovirus infection of guinea pigs. Pediatr Res 16(6):455–459
9. Keithley EM, Woolf NK, Harris JP (1989) Development of morphological and physiological changes in the cochlea induced by cytomegalovirus. Laryngoscope 99(4):409–414
10. Streblow DN et al (2008) Mechanisms of cytomegalovirus-accelerated vascular disease: induction of paracrine factors that promote angiogenesis and wound healing. Curr Top Microbiol Immunol 325:397–415
11. Priscott PK, Tyrrell DA (1982) The isolation and partial characterisation of a cytomegalovirus from the brown rat, *Rattus norvegicus*. Arch Virol 73(2):145–160
12. Loh HS et al (2003) Characterization of a novel rat cytomegalovirus (RCMV) infecting placenta-uterus of *Rattus rattus diardii*. Arch Virol 148(12):2353–2367
13. Loh HS et al (2006) Pathogenesis and vertical transmission of a transplacental rat cytomegalovirus. Virol J 3:42
14. Li L et al (2008) Induction of cytomegalovirus-infected labyrinthitis in newborn mice by lipopolysaccharide: a model for hearing loss in congenital CMV infection. Lab Invest 88(7):722–730
15. Ishiwata M et al (2006) Differential expression of the immediate-early 2 and 3 proteins in developing mouse brains infected with murine cytomegalovirus. Arch Virol 151(11):2181–2196
16. Tsutsui Y (2009) Effects of cytomegalovirus infection on embryogenesis and brain development. Congenit Anom (Kyoto) 49(2):47–55
17. Koontz T et al (2008) Altered development of the brain after focal herpesvirus infection of the central nervous system. J Exp Med 205(2): 423–435
18. Wu CA et al (2011) Transmission of murine cytomegalovirus in breast milk: a model of natural infection in neonates. J Virol 85(10): 5115–5124
19. Kosugi I et al (2002) Innate immune responses to cytomegalovirus infection in the developing mouse brain and their evasion by virus-infected neurons. Am J Pathol 161(3):919–928
20. van Den Pol AN et al (1999) Cytomegalovirus cell tropism, replication, and gene transfer in brain. J Neurosci 19(24):10948–10965
21. Stagno S, Britt WJ (2006) Cytomegalovirus. In: Remington JS, Klein JO (eds) Diseases of the fetus and newborn infant. Saunders, Philadelphia, PA
22. de Vries LS et al (2004) The spectrum of cranial ultrasound and magnetic resonance imaging abnormalities in congenital cytomegalovirus infection. Neuropediatrics 35(2):113–119
23. Trgovcich J et al (1998) Pathogenesis of murine cytomegalovirus infection in neonatal

mice. In: Scholz M et al (eds) CMV-related immunopathology monographs in virology. Karger, Basel, Switzerland

24. Slavuljica I et al (2010) Recombinant mouse cytomegalovirus expressing a ligand for the NKG2D receptor is attenuated and has improved vaccine properties. J Clin Invest 120(12):4532–4545
25. Bantug GR et al (2008) CD8+ T lymphocytes control murine cytomegalovirus replication in the central nervous system of newborn animals. J Immunol 181(3):2111–2123
26. Holtappels R et al (2000) Enrichment of immediate-early 1 (m123/pp 89) peptide-specific CD8 T cells in a pulmonary CD62L(lo) memory-effector cell pool during latent murine cytomegalovirus infection of the lungs. J Virol 74(24):11495–11503
27. Podlech J et al (2000) Murine model of interstitial cytomegalovirus pneumonia in syngeneic bone marrow transplantation: persistence of protective pulmonary CD8-T-cell infiltrates after clearance of acute infection. J Virol 74(16):7496–7507
28. Jonjic S et al (1994) Antibodies are not essential for the resolution of primary cytomegalovirus infection but limit dissemination of recurrent virus. J Exp Med 179(5):1713–1717
29. Polic B et al (1998) Hierarchical and redundant lymphocyte subset control precludes cytomegalovirus replication during latent infection. J Exp Med 188(6):1047–1054
30. Fowler KB et al (1992) The outcome of congenital cytomegalovirus infection in relation to maternal antibody status. N Engl J Med 326(10):663–667
31. Adler SP et al (1995) Immunity induced by primary human cytomegalovirus infection protects against secondary infection among women of childbearing age. J Infect Dis 171(1):26–32
32. Ross SA et al (2011) Mixed infection and strain diversity in congenital cytomegalovirus infection. J Infect Dis 204(7):1003–1007
33. Boppana SB et al (2001) Intrauterine transmission of cytomegalovirus to infants of women with preconceptional immunity. N Engl J Med 344(18):1366–1371
34. Cekinovic D et al (2008) Passive immunization reduces murine cytomegalovirus-induced brain pathology in newborn mice. J Virol 82(24): 12172–12180
35. Becroft DM (1981) Prenatal cytomegalovirus infection: epidemiology, pathology and pathogenesis. Perspect Pediatr Pathol 6:203–241
36. Wirtz N et al (2008) Polyclonal cytomegalovirus-specific antibodies not only prevent virus dissemination from the portal of entry but also inhibit focal virus spread within target tissues. Med Microbiol Immunol 197(2):151–158
37. Harris JP et al (1984) Immunologic and electrophysiological response to cytomegaloviral inner ear infection in the guinea pig. J Infect Dis 150(4):523–530

# Chapter 17

# Recent Approaches and Strategies in the Generation of Antihuman Cytomegalovirus Vaccines

**Suresh B. Boppana and William J. Britt**

## Abstract

The development of prophylactic and to lesser extent therapeutic vaccines for the prevention of disease associated with human cytomegalovirus (HCMV) infections has received considerable attention from biomedical researchers and pharmaceutical companies over the previous 15 years, even though attempts to produce such vaccines have been described in the literature for over 40 years. Studies of the natural history of congenital HCMV infection and infection in allograft recipients have suggested that prophylaxis of disease associated with HCMV infection could be possible, particularly in hosts at risk for more severe disease secondary to the lack of preexisting immunity. Provided a substantial understanding of immune response to HCMV together with several animal models that faithfully recapitulate aspects of human infection and immunity, investigators seem well positioned to design and test candidate vaccines. Yet more recent studies of the role of a maternal immunity in the natural history of congenital HCMV infection, including the recognition that reinfection of previously immune women by genetically distinct strains of HCMV occur in populations with a high seroprevalence, have raised several questions about the nature of protective immunity in maternal populations. This finding coupled with observations that have documented a significant incidence of damaging congenital infections in offspring of women with immunity to HCMV prior to conception has suggested that vaccine development based on conventional paradigms of adaptive immunity to viral infections may be of limited value in the prevention of damaging congenital HCMV infections. Perhaps a more achievable goal will be prophylactic vaccines to modify HCMV associated disease in allograft transplant recipients. Although recent descriptions of the results from vaccine trials have been heralded as evidence of an emerging success in the quest for a HCMV vaccine, careful analyses of these studies have also revealed that major hurdles remain to be addressed by current strategies.

**Key words** Congenital CMV infection, Maternal infection, Maternal reinfection, Adaptive immunity, Animal models of CMV immunity, Immune evasion, Vaccine development, Vaccine testing

## 1 Epidemiology of HCMV Infections

Human cytomegalovirus (HCMV) infections have been recognized in every human population that has been studied [1, 2]. HCMV acquisition in a population is characterized by an age-dependent rise in seroprevalence, and correlates most closely with socioeconomic level and race [3–7]. About 50 % of women of childbearing age

Andrew D. Yurochko and William E. Miller (eds.), *Human Cytomegaloviruses: Methods and Protocols*, Methods in Molecular Biology, vol. 1119, DOI 10.1007/978-1-62703-788-4_17, © Springer Science+Business Media New York 2014

**Table 1**
**Sources and routes of transmission of HCMV infection**

| Mode of exposure and transmission | |
|---|---|
| Community acquired | |
| Age | |
| Perinatal | Intrauterine fetal infection (congenital); intrapartum exposure to virus; breast milk acquired; mother-to-infant transmission |
| Infancy and childhood | Exposure to saliva and other body fluids; child-to-child transmission |
| Adolescence and adulthood | Exposure to young children; sexual transmission; possible occupational exposures |
| Hospital acquired | |
| Source | |
| Blood products | Blood products from seropositive donors; multiple transfusions; white blood cell containing blood products |
| Allograft recipients | Allograft from seropositive donors |

Reproduced with permission from Boppana SB, Fowler KB (2006) Persistence in the population: epidemiology and transmission, In: Arvin A, Campadelli-Fiume G, Mocarski E, Moore PS, Roizman B, Whitley R, Yamanishi K (eds), Human Herpesviruses

are seronegative in industrialized countries [7, 8]. In this population, HCMV acquisition occurs at a rate of 1–7 %/year, and usually follows frequent and prolonged contact with young children (less than 3 years of age) [1, 8–11]. By comparison, in resource-poor communities in industrialized countries and in developing countries, HCMV is usually acquired very early in life owing to breast milk transmission and crowded living conditions, and far fewer adult women are seronegative [1, 12–15].

### 1.1 Transmission of HCMV

Although the exact mode of HCMV acquisition is unknown, it is presumed to be through direct contact with body fluids from an infected person. Breast feeding, group care of children, overcrowded living conditions, and sexual activity have all been associated with high rates of HCMV infection. Sources of the virus include oropharyngeal, urine, cervical and vaginal secretions, semen, breast milk, blood products, and allografts (Table 1) [16–19]. Presumably, exposure to saliva and other body fluids containing infectious virus is a primary mode of spread because infected infants typically excrete significant amounts of HCMV for months to years following infection. Even older children and adults shed virus for prolonged periods (>6 months) following primary HCMV infection. In addition, a significant proportion of seropositive individuals continue to shed virus intermittently. An important determinant of the frequency of congenital and perinatal HCMV infection is the seroprevalence rate in women of childbearing

**Table 2**
**Rates of maternal HCMV seroprevalence and congenital HCMV infection in various populations**

| Location | Maternal HCMV seroprevalence (%) | Prevalence of congenital HCMV infection (%) | Reference |
|---|---|---|---|
| Aarhus-Viborg, Denmark | 52 | 0.4 | [262] |
| Abidjan, Ivory Coast | 100 | 1.4 | [22] |
| Birmingham, USA | | | [10] |
| Low-income group | 77 | 1.25 | |
| Middle-income group | 36 | 0.53 | |
| Hamilton, ON, Canada | 44 | 0.42 | [263] |
| London, UK | 56 | 0.3 | [264] |
| Seoul, South Korea | 96 | 1.2 | [28] |
| New Delhi, India | 99 | 2.1 | [25] |
| Ribeirão Preto, Brazil | 96 | 1.1 | [24] |
| Sukuta, The Gambia | 96 | 5.4 | [29] |

age (Table 2). Studies from the USA and Europe have shown that the seropositivity rates in young women range from less than 50–85 % [1, 20, 21]. In contrast, the vast majority of women of childbearing age in less well-developed regions are HCMV antibody positive [4, 22–25].

### 1.2 Vertical Transmission

Perinatal HCMV acquisition, including congenital infection, contributes significantly to the spread of HCMV because infected infants excrete large amounts of virus for prolonged periods. An additional and less well-appreciated mode of virus spread is through breast milk. It is estimated that almost all breast-fed infants born to persistently infected mothers will be exposed to HCMV as a result of breast feeding [16, 26, 27]. Rates of congenital CMV infection are higher in developing countries and for low-income groups in developed countries [4, 22, 24, 25, 28, 29]. The prevalence of congenital HCMV infection is directly related to the maternal seroprevalence rates (Table 2). Studies of risk factors for congenital HCMV infection showed that young maternal age, non-white race, single marital status, and history of sexually transmitted diseases have been associated with increased rates of congenital CMV infection [30–32].

Similar to congenital infections, infants infected through breast feeding will excrete virus for prolonged periods, making them ideal vectors for the spread of virus. Children continue to acquire HCMV infection throughout childhood, and the rate of infection continues to increase during adolescence and early adulthood secondary to sexual exposure. Significant titers of infectious HCMV can be found in semen and cervical secretions, suggesting that exposure to body fluids could result in the transmission of HCMV [33–36]. The natural history of HCMV infection in adolescents and adults has been shown to parallel sexually transmitted diseases (STDs) [37, 38]. Homosexual men and women attending STD clinics have an increased incidence of HCMV infection [35]. Thus, HCMV should be considered an STD in adults that can effectively spread virus through a sexually active population (Table 1).

### 1.3 Congenital HCMV Infection Following Maternal Non-primary Infections

It was recognized shortly after the description of the cytomegalic inclusion disease in infants that preexisting maternal immunity to HCMV does not prevent transmission to the fetus (non-primary maternal infections), suggesting that the natural history of congenital HCMV was unique among other perinatal infections [39–41]. The rate of congenital infection in a given population is directly related to the HCMV seroprevalence rates in women of childbearing age, such that populations with high seroprevalence have higher rates of congenital HCMV infections (Table 2) [4, 24, 25, 29]. It was initially thought that this increased rate of congenital HCMV infection in highly seroprevalent maternal populations resulted from an increased likelihood of infection during pregnancy of non-immune women and delivery of an infected infant. However, the increased rates of congenital infection have been reported from equatorial Africa, Brazil, and India with near-universal seroimmunity among women of childbearing age. This finding differs dramatically to other causes of congenital viral infections such as rubella [42]. In the case of rubella, the incidence of congenital rubella syndrome fell dramatically when the rate of maternal seroimmunity was above 80–85 %, suggesting that a sufficient pool of susceptible seronegative women was necessary to result in maternal infection and transmission to the developing fetus [43]. In contrast, the rates of congenital HCMV infection increase as the rates of maternal seroimmunity increase past 90 %. The significance of this finding was not appreciated until recently because it was thought that congenital infections following non-primary maternal infections are rarely associated with disease and/or sequelae. Since many early studies included patients referred to single centers because of symptomatic infection, the natural history of this infection was not derived from screened populations that would eliminate much of this bias [44, 45]. However, natural history studies in Sweden by Ahlfors et al. in the early 1980s demonstrated symptomatic infections and hearing loss in infants born to women with non-primary infections [39, 41]. Subsequent studies in screened populations have demonstrated that congenitally infected infants born to

**Table 3**
**HCMV-related hearing loss according to type of maternal infection**

| Maternal HCMV infection ($n = 85$) | | | |
|---|---|---|---|
| Hearing status | Primary ($n = 3$) | Non-primary ($n = 40$) | Indeterminate or samples not available (95 % CI) |
| Moderate to severe unilateral HL | 0 | 4 | 1 |
| Moderate to profound bilateral HL | 1 | 2 | 2 |
| Normal hearing | 2 | 34 | 39 |

Reproduced with permission from Yamamoto AY et al. (2011) Congenital cytomegalovirus infection as a cause of sensorineural hearing loss in a highly immune population. Pediatr Infect Dis J 30:1043–1046

women following non-primary infections account for the majority of HCMV infected newborns [24, 34, 46–48]. Finally, recent studies from populations in which maternal seroimmunity exceeds 96 % and almost all infected infants are born to women with non-primary infection have shown that the prevalence of symptomatic infection and the development of long-term sequelae such as sensorineural hearing loss in infected infants are very similar to those in reported studies of infants infected following primary maternal infections (Table 3) [24, 47]. Therefore, for a vaccine to be effective in preventing or reducing the burden of congenital HCMV infection, the candidate vaccine has to perform better at preventing HCMV reinfection than naturally acquired immunity.

Non-primary maternal infections were initially believed to be secondary to recurrences of preexisting maternal infection, presumably following reactivation of latent infections. Using restriction fragment length polymorphisms of viral DNA, early studies argued for identity between viral isolates, thus suggesting that non-primary maternal infection could follow reactivation of existing infections and ultimately lead to the infection of the fetus [49]. However, a careful review of the findings from these studies suggests that the identity of viral isolates could be questioned. These studies were carried out with labeled viral DNA and required extensive passage in vitro to adapt the viral isolates for growth on human fibroblasts, a process that would put significant pressure on virus populations that may have been present in the initial inoculum. Thus, it is uncertain whether definitive information was indeed derived from these studies. More recent studies utilizing direct sequencing, including next-generation sequencing of viruses isolated from mothers and infants, have demonstrated remarkable genetic diversity and the presence of distinct virus strains in specimens from different compartments from the same child [50, 51]. However, the implications of infection with multiple viral

**Table 4**
**Infection with multiple HCMV strains in mothers according to serologic responses to two polymorphic determinants on gH and gB**

| Variable | Mothers of infected infants *n* (%) (*n*=40) | Mothers of uninfected infants *n* (%) (*n*=109) | *P* value |
|---|---|---|---|
| Antibody reactivity against ≥2 HCMV strains at first prenatal visit | 14 (35.0 %) | 17 (15.6 %) | 0.009 |
| Seroconversion to new HCMV strain during pregnancy | 7 (17.5 %) | 5 (4.6 %) | 0.02 |
| Infection with ≥ 2 CMV strains before and/or during pregnancy | 21 (52.5 %) | 22 (20.2 %) | <0.0001 |

Reproduced with permission from Yamamoto AY et al. (2010) Human cytomegalovirus reinfection is associated with intrauterine transmission in a highly cytomegalovirus-immune maternal population. Am J Obstet Gynecol 202:295

strains in the pathogenesis and outcome in children with congenital CMV infection are not yet known, and remain an important area for future investigation. Defining the sources of virus in infants infected following non-primary maternal infections represents a challenge. Two sources can be imagined. The first is recurrence of an existing infection and the second is from a reinfection. Reinfection of previously immune individuals has been demonstrated in young children attending daycare, individuals with sexually acquired HCMV, transplant recipients, and women undergoing non-primary infection [52–56]. Reinfection appears to be a relatively common occurrence in populations with significant exposure to HCMV [53, 54, 57]. However, the importance of reinfection in the natural history of non-primary maternal infections and subsequent congenital infections has not been adequately defined. Our early studies clearly demonstrated that maternal reinfections can result in a damaging congenital infection, but the parameters of this infection in previously immune women remain poorly defined (Table 4) [58]. However, symptomatic congenital HCMV infections and the occurrence of hearing loss in infants infected following a maternal reinfection are findings that are seemingly at odds with the concept of protective maternal immune responses [46, 47, 56].

## 2 HCMV-Specific Immunity

Invasive HCMV infection and clinically apparent disease are seen most consistently in individuals with deficits in adaptive immune responses. Although developmental immaturity of the adaptive immune system has been argued to explain the susceptibility of fetus and newborn infant to HCMV infection, there is little direct evidence to support this claim. Prolonged periods of virus replication at high levels and virus shedding are consistently observed in infants with congenital HCMV infection. Several early reports suggested that infants with congenital CMV infection lacked HCMV-specific CD4+ T lymphocyte responses; these studies used methodologies that likely underestimated the CD4+ responses in infants [3, 59, 60]. Recent studies demonstrated the presence of HCMV-specific CD4+ T lymphocyte responses in infants with congenital HCMV infection albeit at a significantly lower level compared to adults [61, 62]. Moreover, findings from a study of congenitally infected infants demonstrated the presence of HCMV-specific CD8+ T lymphocyte responses in preterm infants infected in utero with HCMV [63]. Therefore, the role of HCMV-specific CD4+ and CD8+ T lymphocyte responses in the control of virus replication in congenitally infected infants remains unclear.

### 2.1 Maternal and Congenital Infections

While HCMV infection and risk of transmission to the fetus are closely linked to immunity, the temporal appearance and quality of the humoral and cell-mediated responses against HCMV during primary and non-primary maternal infection remains incompletely understood. The importance of immune responses in protecting against intrauterine transmission of HCMV is demonstrated by the differences in intrauterine transmission in women with seroimmunity prior to pregnancy (somewhere between 1 and 6 %) as compared to women undergoing primary infection during pregnancy (30 %) [6, 40]. However, the number of women undergoing primary HCMV infections during pregnancy is considerably smaller than the number of those with non-primary infections, especially in highly seropositive populations. Therefore, when the rates of congenital HCMV infections in the USA are examined, it was estimated that greater than two-thirds of congenitally infected infants are born to women with preexisting HCMV seroimmunity [4, 64]. In populations with near-universal seroimmunity, it is likely that most infants with congenital HCMV infection are born to women with non-primary infection [24, 40, 47].

Differential levels of neutralizing antibody titers directed against glycoprotein B at the time of delivery in mothers with infected infants (transmitters) and those with uninfected infants (non-transmitters) following primary maternal infection suggest that the humoral arm of the immune system modulates intrauterine transmission of HCMV [65]. The gB protein is relatively well conserved among different virus strains, and is considered the major target of the neutralizing antibody response to CMV [66]. Recent data suggests that the pentameric complex comprising gH, gL, UL128, UL130, and UL131 is an important antigenic complex for neutralizing antibody responses [67]. The potential beneficial effects of administering hyperimmune globulin (HIG) in women with primary infection during pregnancy to prevent intrauterine transmission and improve outcome in infected infants has also been suggested from non-randomized studies without control groups [68, 69]. However, we are still awaiting the results of ongoing randomized controlled trials evaluating the role of HCMV HIG (NCT01376778) for the prevention of intrauterine transmission. High titers of neutralizing antibodies are thought to protect against transmission by blocking receptor-mediated transcytosis of HCMV in the placenta [70] and by reducing viral replication [71]. The temporal appearance of humoral responses against the pentameric glycoprotein complex in both primary and non-primary infections in pregnant women has not been sufficiently studied to determine whether these antibodies can be used as a potential marker for risk of transmission and/or congenital HCMV disease.

There is increasing evidence in transplant recipients that high levels of virus replication and disease are associated with suboptimal quality of T lymphocyte responses against HCMV [72, 73]. In addition, in healthy adolescents, it has been shown that plasma CMV DNAemia was still evident despite the detection of a strong neutralizing antibody response within 6–8 weeks following primary infection, although lymphoproliferative responses were weak [74]. Therefore, cellular immunity is indispensable during the acute phase of infection, as well as for the control of chronic infection and the prevention of reinfection [75–78]. Although a range of CMV proteins are targeted by the CD4 and CD8 immune system [79], major targets include the pp65 tegument protein and the IE1 antigen (and to a lesser extent gB) [80]. In pregnant women experiencing primary infection, the evolution of the lymphoproliferative response has been shown to be relatively slow until a memory T-cell response develops [81]. The cytokine profile of these T-cells is dominated by interferon gamma producers with relatively little IL-2 production. In mothers who experienced primary infection and who transmitted the virus to their fetus, CMV CD4+ T-cell responses appear to be delayed and of lower frequency and lower CMV CD45RA+ cells in mothers who transmitted CMV to their fetuses [81, 82]. In seropositive pregnant women, it has been shown that naive CD8+ T-cells were reduced by 50 % with the CD45RA effector population showing a more highly differentiated state (CD27 and CD28 low) while the CD45RA+ revertant memory cell population was expanded and was mainly composed of CMV-specific cells [83].

### 2.2 Transplant Recipients

Studies in immunosuppressed transplant patients have demonstrated that deficits in adaptive immune responses, particularly HCMV-specific T lymphocyte responses, represent an independent risk factor for the development of invasive HCMV infection and end-organ disease [84–93]. Adoptive transfer of in vitro propagated HCMV-specific CD8+ T lymphocytes provided significant protection against HCMV infection and disease in allograft recipients [94, 95]. Similarly, in HIV-infected patients, the loss of HCMV-specific T lymphocyte responses was shown to be a significant risk factor in the development of HCMV end-organ disease [96–98]. Limited data also suggests that the loss of CD4+ T lymphocytes in HIV-infected patients was also associated with impaired anti-HCMV antibody responses and this decrease in anti-HCMV antibody activity may play a role in HCMV-associated disease progression [99–101]. In solid organ transplant recipients, the transplantation of an organ from a seropositive donor into a seronegative donor (D+ R–) is associated with the highest risk of clinically significant infection, presumably because HCMV-specific adaptive immune responses are slow to develop in these patients because of the potent immunosuppression which targets adaptive immune functions. This observation provided further evidence for the critical role of adaptive immunity in the control of HCMV replication, particularly in HCMV infections characterized by high levels of virus replication.

In patients with compromised adaptive immune responses, HCMV infections can result in a variety of manifestations of end-organ damage. These include pneumonitis, hepatitis, colitis, retinitis, encephalitis, and bone marrow dysfunction. Disease is assumed to be secondary to the cytopathic effect of lytic virus replication, an assumption based on finding of viral inclusions and necrotic damage in infected organs. Treatment of patients with disease and active HCMV replication with antivirals such as ganciclovir result in decreased levels of virus and improvement in clinical symptoms and laboratory abnormalities. Together, these observations strongly argue for a role of lytic virus gene expression and disease. In addition, these findings also provide convincing evidence of the role of the adaptive immune response in the control of virus replication in the normal host. However, it is also of interest that the clinical severity of disease often appears to exceed the degree of organ involvement, even in patients with profound depression of adaptive immune responses and HCMV disease associated with high levels of virus replication [102–106]. This observation suggests that residual host adaptive and/or innate responses contribute to disease in these patients. Future studies that dissect these responses could provide insight into the disease syndromes that are associated with low levels of virus replication and restricted viral gene expression.

## 3 Animal Models of Protective Immunity

### 3.1 Introduction

The host restriction of HCMV has limited the development of animal models of infection with HCMV and required investigators to utilize models in small animals and nonhuman primates which employ CMVs from the respective species. Although there are well known differences in the biology of these models compared to humans that prevent precise recapitulation of HCMV infections, a wealth of information has been gathered from studies in these animal models. Model systems have been developed to study HCMV replication in (1) immunocompetent animals to further the understanding of the relationships between normal host immune responses and control of viral replication, (2) immunocompromised animals generated by treatments with immunosuppressive agents or radiation, or co-infected with simian immunodeficiency virus (SIV) as models to characterize key immune responses and control of virus replication and disease, and (3) models of congenital CMV infection that include both intrauterine infection of developing embryos and infection in the immediate newborn period to study the pathogenesis of congenital HCMV infection [107, 108]. As could be predicted, each of these models has obvious limitations and often may not precisely recapitulate human infection and disease associated with HCMV. However, findings from almost every model have increased our existing knowledge about HCMV infection and have led to the development of more informative and focused studies in humans. Model systems that have been developed to investigate each of these categories of HCMV associated disease will be discussed below.

### 3.2 Animal Models Using Immunocompetent Animals to Investigate Responses Associated with Control of Virus Replication and Shedding

Much of the fundamental understanding of the immune response to HCMV has been provided by studies in the murine model of HCMV infections. Murine CMV (MCMV) is closely related to HCMV and utilizes similar replication and immune evasion strategies. Early studies in normal mice infected with MCMV described the role of NK cells, antiviral antibodies, and CD4/CD8+ T-lymphocytes in the control of virus replication [109–118]. Almost every area of CMV biology has been studied utilizing the murine model. Key findings in the genetics of NK cell responses, the role of B-cells and antiviral antibodies in limiting virus dissemination, specific contributions of both CD4+ and CD8+ T lymphocytes in the control of virus replication and dissemination, and perhaps most importantly, the definition and mechanism of action of the multitude of viral genes that modulate host responses to infection. The functions of many viral immune evasion genes have been described and range from inhibiting intrinsic cellular responses such as apoptosis in infected to evasion functions that limit the efficacy of adaptive immune responses [119–131]. Studies in mice and nonhuman primates have defined particularly important roles for these viral functions in modifying the early events of an infection and thus, almost certainly in determining the long-term parameters of infection in an individual host [132–138]. Representative evasion functions from some of these studies carried out in mice and known HCMV orthologs are listed in Table 5.

**Table 5**
**Representative MCMV immune evasion genes and proposed functions**

| Viral gene | Proposed function(s) | HCMV ortholog | Reference |
|---|---|---|---|
| m143/m143 | Inhibit PKR response | TRS1/IRS1 | [265] |
| m38.5/m41.5 | Inhibit apoptosis | UL37 | [266–269] |
| m45 | Inhibits RIP3-mediated programmed necrosis | | [130, 270] |
| m145,m138,m152,m155 | Inhibit NK cell activating antigen expression | UL141, UL40 | [123, 124, 128, 271, 272] |
| m36 | Inhibits caspase dependent and independent apoptosis in macrophages | UL36 | [269, 273] |
| m27 | IFN production (stat-2 inhibition) | | [121] |
| m157 | NK cell responses | UL18; UL142 | [146, 179, 182, 189, 240, 274] |
| m152 | Inhibits antigen presentation | US 3,US6 (TAP) | [125, 128, 275] |
| m06 | Degradation of MHC class I | | [276, 277] |
| m04 | Inhibits MHC class I recognition | | [126, 278] |
| m06 | Degrades MHC class II | US2 | [279, 280] |

Protective responses in immunocompetent murine models have detailed the normal immune response to MCMV and further defined functional aspects of the immune response, often in experiments using transfer of specific cell populations to immunocompromised animals. These experiments have demonstrated the activity of specific responses in the control of virus replication and clearance of virus infected cells. Such studies have defined the protective activity of NK cells, CD8+ cytotoxic T lymphocytes, cytokine secretion, and by antiviral antibodies [110–112, 114, 115, 139–145]. A hierarchy of responses has been detailed and points to the paradigm that all components of the innate and adaptive immune response contribute to the overall resistance to invasive MCMV infection, albeit at different times during infection. This paradigm is almost certain to extend to humans infected with HCMV. Long-term immunological memory is thought to be confined to the adaptive immune responses but more recent studies have suggested that memory functions are present in NK cell responses to MCMV [146, 147]. Induction of antiviral responses thought to be protective, including CD8+ CTL and virus neutralizing antibody responses, have been accomplished using both replicating and inactivated virus, defective viruses with only single cycle infectivity, recombinant viral proteins expressed in heterologous vectors, and recombinant proteins [141, 148–154]. A novel approach using a recombinant MCMV expressing a NK cell ligand or ligands for activating NK cell receptors demonstrated the induction of protective adapative immunity in animals inoculated with the replicating virus in the face of a dramatic reduction in the viral load [155]. Because immunocompetent mice are resistant to even large inoculums of MCMV, protective responses have been quantified by reduction in virus burden, accelerated clearance of replicating virus, and decreased levels of latent virus in salivary glands. In each of these measures, components of the innate immune response (NK cells, γδ T cells) and the adaptive response (antiviral antibodies, CD8+ CTL, and CD4+ T lymphocytes) have all been shown to control virus replication. Targets of CD8+ CTL have been defined in several mice strains and immunization with a CTL epitope has been shown to elicit protective responses [150]. Several virion glycoproteins have been shown to be targets of antiviral antibodies and immunization with these viral glycoprotein or passive transfer of antisera or mAbs directed at these glycoproteins results in virus clearance [140, 144, 148, 156]. In contrast to the clearance of replicating virus by both innate and adaptive immune responses, latent virus cannot be cleared by similar responses. Similar to the situation in normal humans, reinfection of normal mice with unrelated strains of MCMV has been demonstrated [157, 158]. Studies that could determine the quantity and quality of immunity necessary to prevent or limit reinfection of previously immune mice with another strain of MCMV have not been reported.

Studies of immunity in nonhuman primate models have been limited to studies in the Rhesus macaque. Extensive studies in immunocompetent macaques have been carried out utilizing various routes of virus inoculation [159–162]. Resolution of viremia and clearance of infectious virus from the urine and/or saliva have used as end points for control of virus replication. Adaptive immune responses, including both anti-CD8+ and CD4+ T lymphocyte responses and virus neutralizing antibodies, have been correlated with virus clearance [161–165]. As discussed above, immunization of macaques with virion envelope glycoproteins can elicit virus-neutralizing antibodies that can limit viremia and in some cases when combined with other RhCMV antigens, can limit excretion of RhCMV at mucosal surfaces [166]. DNA immunization with antigens that elicit both virus neutralizing antibodies and CD8+ CTL responses not only accelerated virus clearance from the blood but also decreased virus excretion in the saliva of immunized monkeys [167]. It is of interest, however, that immunization of rhesus macaques with a modified vaccinia Ankara recombinant virus (MVA) expressing the UL18-131/gH/gL pentameric complex could elicit virus-neutralizing antibodies and induce immunity in monkeys that accelerated virus clearance from the blood but this level of immunity did not alter virus excretion from the oral mucosa [168]. Thus, a case has been made for the role of both virus specific CD8+

CTL responses and virus-neutralizing antibodies in the control of RhCMV replication in normal monkeys. Similar studies detailing the role of NK cells and other effector mechanisms such as γδ T cells have not been reported. Lastly, reinfection with RhCMV apparently occurs with regularity in Rhesus macaques with existing immunity. This was most clearly demonstrated by studies utilizing the RhCMV as a replicating vaccine platform to serve as a vector for SIV antigens into Rhesus macaques [169]. These authors demonstrated that inoculation of previously immune monkeys with limiting amounts of the same strain of RhCMV expressing a heterologous antigen derived from SIV resulted in reinfection of these previously immune animals and that the reinfected animals excreted this chimeric RhCMV in their urine for prolonged periods of time [169]. Further studies in this same model system have demonstrated that efficient reinfection of rhesus macaques requires immune evasion function of RhCMV that modulate expression of class I MHC molecules important for recognition of virus infected cells by CD8+ CTL [75]. Findings from this study argued that one biological function of the immune evasion functions encoded by CMVs could be to limit T lymphocyte recognition of virus infected cells and therefore block early virus clearance, leading to reinfection of an immune animal [75].

### 3.3 Models Utilizing Immune-Compromised Animals to Characterize Immune Responses Responsible for Control of Virus Replication and Disease

Small animal models including mice, rats, and to a much lesser extent, guinea pigs have been developed to investigate immune responses that are associated with the control of virus replication following immunosuppression. Although both the rat models and guinea pig models could be envisioned to offer specific advantages for such studies, lack of immunologic reagents and genetically engineered animals has limited interest in these models as evidenced by the overwhelmingly larger number of studies carried out in mice. Immunosuppression has been achieved with cytotoxic and immunosuppressive drugs, radiation, and depletion of specific components of the immune system either through genetically engineered mice or by antibody treatment, including anti-CD8 and anti-CD4 specific antibodies. Early studies in rodent models that involved ablation of the immune system followed by reconstitution with specific cell populations demonstrated the importance of immune lymphocytes in control of virus replication [170–172]. Similarly, studies using antiviral antibodies also indicated that antibodies were biologically active and could affect control of virus replication, although the mechanisms of virus clearance have not been definitively established [111, 140, 144, 156]. Natural killer cell activity has also been shown to have a very important role in virus clearance and resistance to MCMV-induced disease. In fact, early studies that mapped genetic resistance to MCMV infection identified a gene designated CMV-1 that provided resistance to MCMV infection in mice with a C57 black genetic background [173, 174]. Subsequent studies identified

CMV-1 as Ly49H, an activating receptor for NK cells that was engaged by the product of the m157 open reading frame of MCMV [129, 146, 174–182]. The Ly49H activating receptor represents only one of group of such receptors associated with NK cell activation in mice; functionally equivalent NK cell activating receptors have been described in humans [183–189]. The difficulty in assigning a single functional response as the major protective responses in immunocompromised small animal models is perhaps secondary to the complexity of host immune responses to MCMV and the interplay between the host and the virus. Clearly, transfer of MCMV specific CD8+ T lymphocytes to irradiated and infected recipients can limit virus replication and virus neutralizing antibodies can limit disease in immunosuppressed animals. In early studies, CD4+ T lymphocytes from immune animals could limit virus replication in recipient mice depleted of CD8+ T lymphocytes [110]. Interestingly, this study also argued that CD4+ T lymphocytes from immune animals could not directly mediate protective effector function in CD8+ T lymphocyte depleted animals arguing that their activity required some other cell populations [110]. In the most frequently used model of latent virus infection, virus specific CD4+ T lymphocytes have been shown to be required to clear latent infection from the salivary gland by non-cytolytic mechanisms, including secretion of γ-interferon [142, 190, 191]. Remarkably, other investigators have demonstrated that passively transferred B lymphocytes from immune donors can protect immunocompromised MCMV infected mice that lack T lymphocytes secondary to a genetic lesion (Rag−/−) even after depletion of NK cell activity [140]. These data argue strongly that B lymphocytes alone, presumably through the production of antiviral antibodies, can affect protective activity [140]. More recent studies from this same group of investigators have demonstrated the potent protective activities of antiviral antibodies in this same model [110]. Together, these and other findings have provided a very strong case for the cooperative effects of several different arms of the immune system in the control of MCMV replication in murine models using immunocompromised animals. Thus, to argue that only one response will be completely protective in normal animals appears to be naïve, and data derived from immunocompetent animals must be viewed in context of the background of the totality of immune responses generated in those animals following challenge with replicating virus.

Under specific experimental conditions, individual components of the adaptive and innate immune systems have been shown to limit virus replication and in some cases provide protective immunity in murine models of lethal CMV infections. Perhaps the most carefully studied model has involved lethally irradiated mice that have been reconstituted with syngeneic bone marrow [170, 192, 193]. Infection of these animals with MCMV leads to disseminated infection, multi-organ disease, and death [170, 192, 193].

Investigators using this model have provided a detailed description of the importance of different components of the adaptive immune response in resistance to uncontrolled virus replication and lethal infection in these mice. Several studies using this model have demonstrated that infusion of MCMV specific CD8+ CTL can effect virus clearance and resistance to lethal infection, in particular the lethal pneumonitis associated with MCMV infection in these immunosuppressed animals [143, 172, 192]. Furthermore, MCMV specific CD8+ T cell lines have been shown to control MCMV replication in the absence of other cellular effector functions in this model and thus have provided convincing evidence of the importance of this singular adaptive immune response in the control of virus replication in vivo [143, 170]. Interestingly, this response appears to persist in the lungs and has been suggested to be required for the maintenance of latency in animals that have been reconstituted by syngeneic bone marrow [194].

Studies in immunosuppressed nonhuman primates have also provided a better understanding of the parameters of protective immunity. As was discussed above, these investigations have been carried out exclusively in the rhesus macaque and have been limited almost entirely to studies in animals with deficits in adaptive immunity induced by infection with the simian immunodeficiency virus (SIV). Severe immunodeficiency in Rhesus macaques inoculated with pathogenic SIV results in loss of CD4+ T lymphocyte responses as well as virus specific CD8+ T lymphocyte and antiviral antibody responses that have been associated with dissemination of RhCMV and end-organ disease [165, 195, 196]. The primary immunodeficiency in these animals appears to follow that of HIV/AIDS and results from loss of CD4+ T lymphocytes, but it is uncertain whether these cells represent a component of the effector responses that control RhCMV. Investigators have more recently utilized antibodies to deplete CD4+ and CD8+ T lymphocytes in rhesus macaques and animals depleted of functional CD4+ or CD8+ T lymphocytes can be used to address the role of these components of the adaptive immune system in resistance and clearance of RhCMV infection.

### 3.4 Animal Models of Congenital HCMV Infection That Include Both Intrauterine Infection of Developing Embryos and Infection in the Immediate Newborn Period to Study the Pathogenesis of Congenital HCMV Infection

Since the early descriptions of the natural history of congenital HCMV, investigators have been searching for relevant animal models to study the pathogenesis of this infection. Following isolation of murine CMV, it was quickly determined that transplacental transmission of MCMV does not occur in normal animals. Recently, isolated reports have suggested that transplacental transmission of MCMV can be accomplished in severely immunocompromised pregnant mice, although the results of these studies have not been reproduced in other laboratories [197, 198]. Yet findings from such studies are potentially distant from the pathogenesis of congenital HCMV infection and the physiology of human pregnancy to make them of uncertain value to model this complex human infection. In contrast, early studies in the guinea pig provided evidence of transplacental transmission in this small animal model of HCMV infections [107, 199–201]. This model has been further developed and will be described in more detail below. Finally, primate models of congenital HCMV infection have been surprisingly difficult to develop. Congenital RhCMV infection is not well described and initially it was believed that this was secondary to the near universal infection of breeding colonies of Rhesus macaques. More recently, two findings argue that RhCMV infection in colonies of macaques may infrequently lead to transplacental transmission of this virus. The first is that even in colonies of monkeys that have been fostered to limit exposure to RhCMV, models of congenital HCMV infection have not been reported. However, since colonies of non-infected monkeys have only become available relatively recently, additional experimental protocols could lead to the development of a suitable model of congenital HCMV infection. The second finding suggesting that the natural history of RhCMV infection during pregnancy could be different than in humans, is the lack of congenitally infected newborn monkeys in colonies where reinfection of previously infected monkeys with RhCMV have been shown to take place. Although speculative at this time, this finding would argue that similar to humans, reinfection of previously infected monkeys is commonplace, yet in contrast to humans, congenital infections do not or uncommonly arise from such reinfections.

The guinea pig model of congenital HCMV infection represents the most widely studied model of transplacental infection and has been used to define parameters of maternal immunity that limit infection in some cases and disease in the offspring. In guinea pig dams without prior immunity to gpCMV, induction of immunity to gpCMV or viral envelop and tegument proteins encoded by gpCMV by various strategies can modify the natural history of transplacental virus transmission and disease in the offspring [202–204]. In addition, passive transfer of antiviral antibodies either as polyvalent immune serum or mono-specific polyclonal immune serum can limit transmission and disease in the offspring of gpCMV infected dams [205, 206]. Although the induction of virus neutralizing antibodies and adaptive cellular immune responses has been shown to be responsible for such protective activity, the mechanism(s) of protection remain undefined. An important caveat to studies in the guinea pig model of congenital HCMV infection is that most studies that have successfully used this model have noted that in order to reproducibly achieve transplacental transmission; the pregnant dam must be infected with an inoculum sufficient to cause clinical symptoms. Furthermore, such high inoculums lead to significant loss of newborn pups in utero and in the immediate newborn period, often on the order of 20 %, and the surviving pups consistently exhibit runting. Induction of protective responses in the pregnant guinea pig limits the severity of the disease in the pregnant dam and improves the outcome of the pups. These later findings have raised some debate about the specific mechanisms of protective immunity in this model and suggest that much of the protective effects may involve preservation of placental function in the pregnant guinea and to a lesser extent, induction of protective immunity to limit infection/disease in the offspring. Finally, the latter aspect of the model has also raised some questions about the interpretation of the use of the model to study CNS abnormalities in the offspring of infected dams, including hearing loss [207]. Initially, there was considerable excitement about the application of the guinea pig model of congenital HCMV infection to investigate mechanisms of hearing loss in human infection, but thus far most studies have more descriptive findings than mechanistic details of hearing loss [208]. However, there remains a core group of laboratories with interest in this model.

As discussed above, definitive evidence of transplacental transmission of MCMV in normal mice has not been reported. More recently a potentially informative model using peripheral inoculation of newborn mice with minimal amounts of MCMV has been shown to recapitulate many aspects of congenital HCMV infection, including CNS disease and hearing loss [209]. A major advantage of this model over existing murine models in which CNS infection is accomplished by direct intracranial inoculation, this model relies on intraperitoneal infection with virus replication and hematogenous dissemination [210]. This model has been used to begin explore mechanisms of neuroinvasion and neuropathogenesis and because newborn are similar in terms of neurodevelopment to a second trimester human fetus, this animal model could prove informative. This model has been used to investigate the role of antiviral antibodies and protection from CNS infection and prevention of CNS damage, i.e., altered neurodevelopment [144]. In this report, antiviral antibodies, including virus neutralizing monoclonal, limited infection and disease phenotype [144]. It is of interest to note that in one of these studies a recombinant virus expressing an NK cell activating ligand was used to vaccinate mice prior to pregnancy to model the efficacy of an immunologically attenuated replication competent virus in the induction of maternal antiviral antibodies that following transplacental transmission to the offspring of the pregnancy, prevented CNS disease maternal antibodies and subsequent prevention of CNS disease in the offspring [155]. Lastly, the pathogenesis of HCMV infection of the developing CNS is far from understood and recent findings in a murine model of this infection suggests that host-derived inflammation could contribute to the pathogenesis of this infection [211]. If these data reflect a mechanism of disease in the developing human fetus, the requirements for vaccine-induced immunity could include prevention of virus entry into the CNS, a potentially challenging target for any vaccine.

## 4 Approaches for Vaccine-Induced Protective Immunity

### 4.1 Introduction

The induction of protective immunity to HCMV remains the goal of current vaccine development programs and most current candidate vaccines have been developed as prophylactic vaccines and not as therapeutic vaccines. In addition, studies of the natural history of HCMV infections have suggested that at best, HCMV vaccines could limit the severity of infection with HCMV but were unlikely to prevent acquisition of the virus [40, 44, 212]. Results from previous studies using attenuated replicating viral vaccines have been consistent with observations from natural history studies [213]. In contrast to this paradigm, a recent clinical trial suggested that prevention of infection was accomplished following vaccination with an adjuvanted glycoprotein vaccine [214]. The results of this trial have stirred debate, and many investigators have questioned the protection afforded by this candidate vaccine because the efficacy of the vaccine appeared to be very short-lived [214]. Furthermore, although a primary end point of prevention of infection achieved by the vaccine did reach statistical significance, the group size was small and statistical significance was dependent on limited differences between vaccine and non-vaccine groups, both reflected by the large confidence interval in the statistical analysis [214]. Prior to this study, there was little evidence that any candidate vaccine could induce sufficient immunity to prevent infection and accumulating data from studies in human, primate, and murine models continues to argue that immunity produced following natural infection with CMVs is insufficient to prevent infection with a new viral strain [158, 169, 215]. Thus, it is likely that the majority of candidate vaccines will be developed with goals of preventing disease that follows infection with HCMV.

### 4.2 Generation of Protective Immunity by Induction of Adaptive Immunity to HCMV

The evidence from animal models and natural history studies in patients with HCMV have argued that adaptive immune responses including both antiviral antibodies and CD4+/CD8+ T lymphocyte responses play major roles in the modification of infection with HCMV by limiting virus dissemination, controlling virus within infected tissue, and eventually clearing replicating virus. Such responses could effectively limit virus replication following a primary or non-primary infection and thus convert the phenotypic manifestation of the infection from those characteristic of an acute HCMV infection with high levels of virus replication, dissemination, and end-organ disease to those of a chronic, tissue-restricted infection characterized by low level virus replication, limited dissemination, and a decrease in end-organ involvement. Approaches to induce such adaptive responses have (and continue to include) adjuvanted glycoprotein vaccines such as peptide based vaccines, glycoprotein B containing vaccines, DNA vaccines, vectored glycoprotein vaccines

**Table 6**
**Candidate vaccines for HCMV that have been tested in clinical trials**

| Candidate vaccine | Antigen composition | Clinical trial status | Reference |
|---|---|---|---|
| Adjuvanted protein | Glycoprotein B and others? | Phase II | [214, 217] |
| Adjuvanted peptide | pp65, IE-1 | Phase 1 | [179] |
| DNA vaccines | pp65, gB, IE-1 | Phase 1 | [216, 234] |
| Alphavirus replicon | pp65, IE-1, gB | Phase 1 | [218] |
| Chimeric replicating virus | NA | Phase 1 | [219] |

such as chimeric alphaviruses, and replication competent HCMV that have been attenuated by in vitro passage or through recombination between clinical and attenuated strains of HCMV [213, 214, 216–220] (Table 6). Each of these has been tested in early human trials, and each approach has provided some evidence of immunogenicity in terms of inducing antiviral antibodies and T lymphocyte responses (Table 6). In some cases such as the adjuvanted glycoprotein B vaccine, evidence of benefit has been provided in renal transplant patients as measured by modifications of antiviral medication management following transplantation [217]. However, it should be noted that the complexity of the clinical course of patients undergoing transplantation together with the highly effective antiviral treatment that is routinely provided to these patients has made it difficult to objectively define the added benefit of vaccine in this population. In addition, the enthusiasm for the development of vaccines for use in transplant recipients appears to be somewhat limited for several reasons, including the efficacy of current antiviral drugs and the expected efficacy of newer drugs that are entering clinical trials.

Studies in animal models, observations in patients infected with HCMV and at risk for invasive disease, and results from clinical trials using hyperimmune immune globulins to limit disease in several different clinical settings have all suggested that antiviral antibodies provide protection from disease associated with HCMV infections [221]. Furthermore, even though the mechanism of protection associated with the recently described trial of the adjuvanted gB vaccine in renal transplant recipients is unknown, the authors argued that antiviral antibodies, perhaps virus neutralizing antibodies, contributed to the efficacy of the vaccine [217]. Together these results have provided compelling evidence that an efficacious vaccine will likely induce antiviral antibodies and by extension of findings from studies in animal models, it is likely that these will include virus neutralizing antibodies. A complete discussion of supporting evidence for a critical role of antiviral antibodies in protection from end-organ disease associated with HCMV infections is beyond the scope of this chapter, but several findings warrant discussion. Passive acquisition of anti-HCMV antibodies, either acquired transplacentally by newborn infants or provided as part of prophylaxis for disease in transplant recipients, has been shown to modify the severity of HCMV infection but not to prevent infection [221–224]. More recently, treatment of pregnant women with recently acquired HCMV infections with HCMV immune globulins has been suggested to provide some benefit in limiting disease, but again there has not been convincing evidence that such treatment could prevent intrauterine transmission of HCMV [68]. Importantly, further analysis of maternal specimens obtained during the initial, uncontrolled trial suggested that some benefit that was ascribed to the hyperimmune globulin could be indirect and explained by its anti-inflammatory activity on the placentas from infected mothers [225–230]. Thus, there has been more controversy generated by this initial report of an uncontrolled trial that lacked definitive findings. In most studies that have linked protection from disease and/or infection following the transfer of antiviral antibodies, the nature of the protection provided by the transferred antibodies is undefined. Furthermore, because of the vast array of antiviral antibody specificities represented in the transferred antibodies, it is unclear if reactivity for specific viral protein such as envelope glycoproteins is in fact the protective component of the transferred antibodies. However, extensive studies in animal models and humans have identified targets of virus neutralizing antibodies in human serum. These include antibodies reactive with the envelope glycoproteins, gB, gH, gN, and the UL128-131/gH/gL pentameric complex. The latter envelope glycoprotein complex represents an entry mediator for infection of epithelial cells and monocytes and has been the subject of interest as a target for vaccine-induced immunity [231–233]. Wussow and colleagues have expressed the RhCMV homolog of the HCMV pentameric complex in a modified pox virus (MVA) and shown that immunization with this virus induces neutralizing antibodies that reach similar titers to those observed in naturally infected animals and that immunity induced by this candidate vaccine can modify the virologic parameters of a challenge infection with RhCMV [168]. This study provided the first definitive evidence of the in vivo relevance of antibodies directed at this complex of glycoproteins and suggested that additional studies on the role of antibodies reactive to the pentameric complex are warranted. Finally, current candidate vaccines either planned or currently in clinical trials contain gB and gH/gL [214, 218]. It is anticipated that preparations expressing the pentameric UL128-UL131 gH/gL complex will also be incorporated in next-generation glycoprotein based vaccine to produce a multivalent vaccine.

Although most investigators would argue that an ideal vaccine would induce a robust HCMV specific CD8+ T lymphocyte response and perhaps a HCMV specific CD4+ T lymphocyte response based on data generated from cell based immunotherapeutic clinical trials, thus far only the attenuated replicating virus and DNA based vaccines have been shown to induce CD8+ T lymphocyte responses in humans [216, 219]. It is notable that a bivalent DNA vaccine that included DNA encoding gB and pp65 (UL83) was shown to induce CD8+ T lymphocyte responses [216]. Moreover, a similar vaccine was shown to prime recipients of a replicating, attenuated HCMV vaccine resulting in a more consistent CD8+ and CD4+ T lymphocyte response [234]. As in the case of antibody based vaccine immunity, the quantitative parameters of protective immunity provided by CMV specific CD8+ T lymphocytes remains to be defined and merely achieving a response may be insufficient to provide protection from disease. Selection of candidate antigens for inclusion in multivalent vaccines to induce CD8+ T lymphocyte responses has been based on earlier studies that have cataloged the CD8+ T lymphocyte responses found in naturally immune individuals to a large number of viral gene products [79, 235–237]. Finally, it should be noted that adjuvanted vaccines such as the gB vaccine previously discussed induce antigen specific CD4+ T lymphocyte responses [220, 238, 239]. The role of these responses in any protective immunity generated by this class of vaccines other than a contribution to production of antibody remains uncertain.

### 4.3 Vaccine Strategies to Overcome Viral Immunoevasion Functions

As studies in murine models of CMV infection begin to demonstrate the importance of virus encoded immune evasion function, the potential value of vaccine strategies that could limit or eliminate these viral functions has been explored by several laboratories. By removing immunoevasive activities encoded by the virus, the virus could be attenuated for in vivo replication while maintaining or perhaps even increasing the immunogenicity of major targets of the adaptive immune response. Examples of this have included deletion of immune evasion functions in murine and Rhesus CMV followed by the demonstration of their attenuated phenotype in vivo [75, 119, 121, 123, 125, 128, 240, 241]. The importance of these immune evasion mechanisms was demonstrated in a novel study in which investigators demonstrated that deletion of viral functions in RhCMV that inhibited MHC class I presentation of viral antigens resulted in an attenuated virus that was incapable of reinfecting previously infected Rhesus macaques [75]. In contrast to mutation that crippled immune evasion functions, Jonjic and colleagues engineered a recombinant MCMV that expressed an NK activating ligand and demonstrated that the virus was attenuated in vivo presumably secondary to increased innate responses induced by the NK activating ligand [155] (Table 7).

**Table 7**
**Novel strategies for CMV vaccine development**

| Candidate vaccine | Strategy | Evidence of feasibility | Reference |
|---|---|---|---|
| RhCMV ΔUS2-7 (virus) | Attenuated secondary to absence of immune evasion function | Demonstrated that attenuated virus could not reinfect previously infected animals | [262] |
| MCMV Rae-1 (virus) | Attenuated secondary to presence of NK cell activating ligand | Demonstrated altered replication kinetics and salivary gland infection | [155] |
| RhCMV vIL-10 (protein) | Immunized with viral protein to block function of virus encoded vIL-10 | Altered virological parameters following challenge with wild-type virus | [243] |

Most recently, the potential importance of the viral IL-10 like molecule in primate CMVs to in vivo replication of RhCMV has been reported [136]. The primate CMV IL-10 like viral protein has been shown to provide important functions in vitro and thus far appears to interfere with innate responses very early after CMV infection [242]. In addition, this viral protein has been shown to dampen the interferon production by plasmacytoid dendritic cells [136]. The RhCMV IL-10 like viral protein does not appear to be antigenically cross reactive with the host IL-10, thus allowing animals to be immunized with this viral protein with the goal of producing antibodies that would neutralize the viral IL-10 activity in vivo [243, 244] (Table 7). Such as strategy has been tested in rhesus macaques and immunization with the RhCMV IL-10 like molecule resulted in antibodies reactive to the viral protein. Interestingly, this strategy led to decreased virus excretion in immunized animals challenged with wild-type virus as compared to control, unimmunized but challenged monkeys (PA Barry, personal communication). Thus, the plethora of immunoevasive functions that appear critical to the biology of CMVs could explain CMV persistence in various populations, and could eventually be exploited to enable induction of protective immunity by engineered vaccines that are truly attenuated in their capacity to infect and presumably spread within populations.

## 5 Target Populations for Evaluating HCMV Vaccines

It is important to identify appropriate target populations to be used for evaluation of candidate vaccines to prevent CMV-associated morbidity.

### 5.1 Women of Childbearing Age

The prevention of congenital HCMV infection and associated sequelae requires vaccination before pregnancy so that protective immune responses are developed prior to the first trimester.

However, there are significant challenges in vaccinating women of childbearing age who are planning a pregnancy because of the difficulty in identifying women at risk for HCMV acquisition prior to pregnancy and as such would be difficult to implement. An inadvertent administration of vaccine to pregnant women has the potential to adversely impact the pregnancy and the fetus. In addition, the risks and benefits of an HCMV vaccine in both seronegative and seropositive women have to be carefully evaluated.

### 5.2 Adolescents

Adolescent females could be another target population for HCMV vaccines. Immunizing this population also presents logistical challenges in obtaining high coverage. Screening for HCMV antibodies and immunizing only seronegative adolescent females is one potential approach. Even if the challenges of achieving high vaccine coverage could be overcome, immunizing adolescent females may only be effective in certain populations with lower HCMV seroprevalence and lower levels of exposures. Moreover, since the majority of congenitally infected infants are born to women with preexisting seroimmunity, any tactic that targets only seronegative individuals will have only limited impact on the prevalence and disease burden associated with congenital HCMV infection. Additionally, most children and adolescents in resource-poor settings are HCMV seropositive and therefore this strategy will not be effective in populations with the higher disease burden.

### 5.3 Early Childhood

Another approach is universal immunization of all children early in life, similar to the rubella vaccine. The advantages of universal immunization include the ability to achieve high coverage rates and the development of herd immunity that could decrease infection and possibly shedding in toddlers who are important sources HCMV transmission. However, a better understanding of the precise components of protective immunity against primary HCMV infections as well as reinfections in seropositive individuals is critical for this approach to be effective.

## 6 Prevention of Infection/Transmission/Disease

Prevention of primary HCMV infection in pregnant women in order to prevent fetal infection and disease using active and passive immunization has been an area of interest to many investigators. A study of recombinant gB vaccine to prevent maternal infections showed modest efficacy in preventing primary maternal HCMV infections [214]. It has also been suggested that administration of HCMV hyperimmune globulin to pregnant women with primary HCMV infection reduces intrauterine transmission and improves outcomes in infected children [68, 69]. However, strategies to prevent primary HCMV infection may not have significant impact on overall disease burden from congenital HCMV infection because a majority of congenitally infected children are born following non-primary infections [24, 41, 64, 245].

One of the important risk factors for the acquisition of HCMV infection is the exposure to young children shedding the virus and decreasing such exposure is considered to be central to reducing virus transmission (Table 1). Adler et al. showed that educating pregnant women in avoiding exposure to oral secretions of young children and practicing a frequent hand washing regimen significantly reduced the acquisition of HCMV infections during pregnancy [246]. A larger French study also demonstrated a significant decrease in primary HCMV infections during pregnancy when women were educated about various measures to reduce exposure to HCMV, especially from secretions of young children [247]. As the acquisition of a different strain of HCMV (reinfection) in seropositive pregnant women can lead to intrauterine transmission and damaging congenital infection, these behavioral interventions may help to reduce HCMV transmission in this group. However, such behavioral interventions are difficult to implement in resource-poor settings. HCMV transmission has also been associated with sexual activity, and therefore, limiting the number of sex partners and protective sex practices including the use of condoms may decrease the risk of HCMV acquisition [36, 37].

HCMV serological status of the donor and the recipient is an important factor for developing HCMV disease in solid organ transplant (SOT) recipients and in hematopoietic stem cell transplant (HSCT) patients. In SOT patients, antiviral prophylaxis with valganciclovir for 3–6 months after transplantation is a common practice for the prevention of HCMV infection and disease [248, 249]. In some transplant centers, preemptive therapy is also used in conjunction with intensive surveillance for the detection of HCMV infection [249, 250]. In allogeneic HSCT patients, the most common strategy to prevent HCMV disease is surveillance using real-time PCR and preemptive therapy [251, 252].

## 7 Roadblocks in the Development of Efficacious HCMV Vaccines

### 7.1 Introduction

As discussed above, groups of patients who could most benefit from HCMV vaccines include transplant recipients that are at risk for invasive HCMV infections and women of childbearing age. Successful vaccination of each these populations pose common as well as unique challenges for the design and clinical evaluation of candidate vaccines (Table 8). These challenges include the recently described genetic heterogeneity of HCMV strains isolated from immunocompetent individuals, a phenomenon appreciated in the past in studies of viral genetic diversity in immunocompromised patients and more recently in immunocompetent children and adults [42, 50, 51, 56, 58, 253–256]. Although in these recent studies,

**Table 8**
**Potential hurdles for successful deployment of HCMV candidate vaccines in two target populations**

| Target population | Potential hurdle | Comments |
|---|---|---|
| Women of childbearing age | Genetic variability of circulating viruses | [50, 57] |
| | Reinfection with new strains of virus | [15, 42, 56, 58] |
| | Prolonged window of maternal/fetal risk | Childbearing age |
| | Transplacental acquisition of protective immunity | Animal models [144, 205, 206]; limited data from IVIG trials [68] |
| | Uncertain role of virus load in fetal disease/immunopathology | [42, 281, 282] |
| | Potential size of clinical trial required to test efficacy of vaccine | Size of vaccine trial population dependent on end point of vaccine efficacy, i.e., prevention of sequelae, prevention of fetal transmission |
| Transplant recipients | Genetic variability of circulating viruses/ reinfection | *See* above |
| | Immunosuppression in recipients | Potent immunosuppression; intermittent corticosteroid use |
| | Proven efficacy of current antiviral therapy | |
| | Transmission of virus in transplanted allograft | [223, 283] |
| | Unknown role of vaccine in prevention of graft loss associated with HCMV | |

predominant genotypes were found in these individuals, the existence of multiple strains and the lack of an understanding of selective pressures that were (and could be) exerted on these populations and thus drive expansion of minor viral populations remain incompletely defined. Thus, these results would argue that conventional approaches to develop vaccines that target a single viral genotype could induce incomplete protection. Furthermore, the description of the natural history of HCMV infection in normal woman and in transplant recipients has provided evidence for the incomplete protection provided by preexisting immunity induced by naturally acquired infection. Admittedly in transplant patients and in the offspring of pregnant women, existing immunity offers some degree of protection from more severe disease and end-organ damage, but even in these patients, protection provided by existing immunity is incomplete in individual patients, suggesting that the parameters of protective immunity are far from understood.

Whether this failure of naturally acquired immunity reflects deficits unique to an individual host or reflects an as yet unexplored and more generalizable feature of the mixture of viruses or routes of exposure experienced by these patients remains to be explored. A major issue common to all HCMV vaccine programs is the lack of solid evidence that infection can be prevented by natural immunity or by existing vaccines, with the notable exception of the reported efficacy of the adjuvanted gB vaccine that was discussed in previous sections. The lack of a clear end point, such as prevention of infection, will complicate clinical trials and greatly increase the number of individuals that must be enrolled to demonstrate efficacy in the prevention of disease phenotypes associated with HCMV infections. In the case of a vaccine to prevent the sequelae associated with congenital HCMV, the number of women that must be enrolled could exceed almost any vaccine study thus far undertaken in the USA, and the duration of the follow-up of offspring from these women would extend into early childhood. Similarly, in transplant patients a vaccine trial must be conducted in the face of a very effective antiviral chemotherapy, both prophylactic and therapeutic, that has greatly reduced the incidence of severe HCMV infection in transplant recipients [257]. As a result, end points of vaccine efficacy must be measured in terms of modifying a relatively good virologic outcome during the early post-transplant period. Conversely, the efficacy of vaccines to limit the long-term contribution of HCMV to solid organ allograft loss could represent an important benefit of protective vaccines. However, such benefit would be apparent in a minority of patients and more importantly, would not be realized until between 5 and 10 years post transplantation, conditions that would likely make such a trial of lower priority for most pharmaceutical companies. A third issue that could lead to the lessening interest in the development of HCMV vaccines is that newer antiviral agents and potent antiviral antibodies could offer selectivity for the treatment (or prophylaxis) of patients at risk for HCMV infection and disease. Should such agents demonstrate favorable safety profiles and efficacy, improvements in early recognition of patients at risk for invasive HCMV disease or in the case of pregnant women, women with active HCMV replication, could lead to selective treatment and decrease a need for universal immunization. Lastly, it should also be recognized that most women in the world with the exception of maternal populations in North America and Northern Europe, are infected with HCMV and have immunity that is established following natural infection [4, 13, 25, 29]. Investigators in Brazil who previously reported a 1 % birth prevalence of congenital CMV and incidence of symptomatic infection and sequelae of 8 and 11 % respectively in infected infants recently published results from a study indicating that >96 % of this maternal population was infected with HCMV prior to childbearing age [13, 24, 47]. Thus, unless a candidate vaccine can be shown to induce protective immunity that exceeds that afforded by natural immunity, the target population for vaccination could potentially be limited to women of racial and socioeconomic backgrounds in the northern hemisphere.

In addition to the issues noted above, additional roadblocks exist for the development of an efficacious HCMV vaccine that could be utilized in allograft recipients. The first is the necessity to induce a protective immune response in patients that are often debilitated, sometimes receiving immunosuppressive agents, and with an underlying disease that may compromise normal immunity. In addition, the transplant population is aging and recipients of candidate vaccines could have age-related impairments in response to a vaccine and/or adjuvants. Furthermore, in the post-transplant period all solid organ transplant recipients will undergo some form of immunosuppression and many hematopoietic transplant recipients will be severely immunocompromised secondary to conditioning regimens for underlying disease. Thus, existing immunity induced by a vaccine will potentially be eliminated or severely depressed. Moreover, while vaccine immunity could be effective in limiting acquisition of virus following community exposures it could be significantly less effective when virus is acquired following transplantation of an infected organ, including hematopoietic grafts [257, 258]. Under conditions of organ transplantation, HCMV can essentially bypassed protective immune responses at host mucosal surfaces, responses that could provide some advantage to the host by slowing the kinetics of virus replication and spread and thus permitting timely development of protective immune responses. Lastly, the finding of HCMV in transplanted organs undergoing host versus graft responses, as well as in hematopoietic transplant recipients undergoing graft versus host reactions following hematopoietic transplantation suggest that the loss of normal immune regulation in these patients results in further immunosuppression and favors virus replication. Thus, induction of lasting protective immunity by vaccines in this population represents a formidable task.

In contrast to immunocompromised allograft recipients, women of childbearing age are immunologically competent and can be expected to respond to candidate HCMV vaccines. An efficacious vaccine for HCMV in this population must induce sufficient immunity to limit the transmission of HCMV to the developing offspring and/or modify the infection in the fetus. Although the virologic characteristics of maternal infection that result in fetal infection remain incompletely described, natural history studies of congenital HCMV infections indicate that vaccine immunity must be protective over several months as intrauterine transmission and disease in the offspring of pregnant women has been demonstrated following maternal infection in the early second trimester until the third trimester [10, 259–261]. Furthermore, it is well documented that women with preexisting immunity to HCMV can transmit virus to the fetus and that this infection can lead to disease in the newborn infant. Thus, and as noted above, requirements for vaccine-induced protective immunity in normal women could conceivably exceed the immunity induced following naturally acquired infection. A second potential hurdle in the development of a vaccine to limit damaging congenital HCMV infection is the necessity of vaccine immunity to be transferred across the placenta, a requirement that effectively limits potential mechanisms of protective immunity to antiviral antibodies acquired by the developing fetus. In order to prevent damaging intrauterine infection, vaccine-induced antiviral antibodies must restrict virus replication and dissemination in the fetus and particularly dissemination to the CNS. Although antiviral antibodies have been shown to limit virus entry and limit virus replication in models of CMV infection of the CNS, the nature of these antibodies and their mechanism of action in limiting CNS infection remain to be determined.

## References


1. Krech U, Konjajev Z, Jung M (1971) Congenital cytomegalovirus infection in siblings from consecutive pregnancies. Helv Paediatr Acta 26:355–362
2. Gold E, Nankervis GA (1976) Cytomegalovirus. In: Evans AS (ed) Viral infections of humans: epidemiology and control. Plenum, New York, pp 143–161
3. Pass RF et al (1981) Specific lymphocyte blastogenic responses in children with cytomegalovirus and herpes simplex virus infections acquired early in infancy. Infect Immun 34:166–170
4. Stagno S et al (1982) Maternal cytomegalovirus infection and perinatal transmission. Clin Obstet Gynecol 25:563–576
5. Marshall GS, Stout GG (2005) Cytomegalovirus seroprevalence among women of childbearing age during a 10-year period. Am J Perinatol 22:371–376
6. Kenneson A, Cannon MJ (2007) Review and meta-analysis of the epidemiology of congenital cytomegalovirus (CMV) infection. Rev Med Virol 17:253–276
7. Cannon MJ, Schmidt DS, Hyde TB (2010) Review of cytomegalovirus seroprevalence and demographic characteristics associated with infection. Rev Med Virol 20:202–213
8. Hyde TB, Schmidt DS, Cannon MJ (2010) Cytomegalovirus seroconversion rates and risk factors: implications for congenital CMV. Rev Med Virol 20:311–326
9. Hutto C et al (1985) Epidemiology of cytomegalovirus infections in young children: day care vs. home care. Pediatr Infect Dis 4:149–152
10. Stagno S et al (1986) Primary cytomegalovirus infection in pregnancy: incidence, transmission to fetus and clinical outcome. JAMA 256:1904–1908
11. Adler SP (1989) Cytomegalovirus and child day care. Evidence for an increased infection rate among day-care workers. N Engl J Med 321:1290–1296
12. Seo S, Cho Y, Park J (2009) Serologic screening of pregnant Korean women for primary human cytomegalovirus infection using IgG avidiyt test. Korean J Lab Med 29:557–562
13. Yamamoto AY et al (2012) Early high CMV seroprevalence in pregnant women from a population with a high rate of congenital infection. Epidemiol Infect 3:1–5
14. Saraswathy TS et al (2011) Seroprevalence of cytomegalovirus infection in pregnant women and associated role in obstetric complications: a preliminary study. Southeast Asian J Trop Med Public Health 42:320–322
15. Akinbami AA et al (2011) Seroprevalence of cytomegalovirus antibodies amongst normal pregnant women in Nigeria. Int J Womens Health 3:423–428
16. Hayes D, Danks M, Givas H, Jack I (1972) Cytomegalovirus in human milk. N Engl J Med 287:177
17. Lang DJ, Kummer JF (1975) Cytomegalovirus in semen: observations in selected populations. J Infect Dis 132:472–473
18. Reynolds DW et al (1973) Maternal cytomegalovirus excretion and perinatal infection. N Engl J Med 289:1–5
19. Alford CA, Stagno S, Pass RF (1980) Natural history of perinatal cytomegalovirus infection. In: Perinatal Infections. Excerpta Medica, Amsterdam, pp 125–147
20. Gold E, Nankervis GA (1982) Cytomegalovirus. In: Evans AS (ed) Viral infections of humans: epidemiology and control, 2nd edn. Plenum, New York, pp 167–186
21. Bate SL, Dollard SC, Cannon MJ (2010) Cytomegalovirus seroprevalence in the

United States: the national health and nutrition examination surveys, 1988–2004. Clin Infect Dis 50:1439–1447
22. Schopfer K, Lauber E, Krech U (1978) Congenital cytomegalovirus infection in newborn infants of mothers infected before pregnancy. Arch Dis Child 53:536–539
23. Vial P et al (1985) Serological study of cytomegalovirus, herpes simplex and rubella virus, hepatitis B and Toxoplasma gondii in 2 populations of pregnant women in Santiago, Chile. Bol Oficina Sanit Panam 99:528–538
24. Mussi-Pinhata MM et al (2009) Birth prevalence and natural history of congenital cytomegalovirus infection in a highly seroimmune population. Clin Infect Dis 49:522–528
25. Dar L et al (2008) Congenital cytomegalovirus infection in a highly seropositive semi-urban population in India. Pediatr Infect Dis J 27:841–843
26. Stagno S et al (1980) Breast milk and the risk of cytomegalovirus infection. N Engl J Med 302:1073–1076
27. Hamprecht K et al (2001) Epidemiology of transmission of cytomegalovirus from mother to preterm infants by breastfeeding. Lancet 357:513–518
28. Sohn YM et al (1992) Congenital cytomegalovirus infection in Korean population with very high prevalence of maternal immunity. J Korean Med Sci 7:47–51
29. van der Sande MAB et al (2007) Risk factors for and clinical outcome of congenital cytomegalovirus infection in a peri-urban West-African birth cohort. PLoS One 2:e492
30. Boppana SB et al (2001) Intrauterine transmission of cytomegalovirus to infants of women with preconceptional immunity. N Engl J Med 344:1366–1371
31. Yamamoto AY et al (2010) Human cytomegalovirus reinfection is associated with intrauterine transmission in a highly cytomegalovirus-immune maternal population. Am J Obstet Gynecol 202(3):297e1–297e8
32. Chandler SH, Handsfield HH, McDougall JK (1987) Isolation of multiple strains of cytomegalovirus from women attending a clinic for sexually transmitted diseases. J Infect Dis 155:655–660
33. Ishibashi K et al (2007) Association of the outcome of renal transplantation with antibody response to cytomegalovirus strain-specific glycoprotein H epitopes. Clin Infect Dis 45:60–67
34. Ross SA et al (2010) Cytomegalovirus reinfections in healthy seroimmune women. J Infect Dis 201:386–389
35. Fowler KB, Stagno S, Pass RF (1993) Maternal age and congenital cytomegalovirus infection: screening of two diverse newborn populations, 1980–1990. J Infect Dis 168:552–556
36. Preece PM et al (1986) Congenital cytomegalovirus infection: predisposing maternal factors. J Epidemiol Community Health 40: 205–209
37. Fowler KB, Stagno S, Pass RF (1991) Rates of congenital cytomegalovirus infection based on newborn screening in two populations over an eleven year interval. Pediatr Res 29:90A
38. Jordan MC et al (1973) Association of cervical cytomegaloviruses with venereal disease. N Engl J Med 288:932–934
39. Willmott FE (1975) Cytomegalovirus in female patients attending a VD clinic. Br J Vener Dis 51:278–280
40. Drew WL et al (1981) Prevalence of cytomegalovirus infection in homosexual men. J Infect Dis 143:188–192
41. Chandler SH et al (1985) The epidemiology of cytomegaloviral infection in women attending a sexually transmitted disease clinic. J Infect Dis 152:597–605
42. Sohn YM et al (1991) Cytomegalovirus infection in sexually active adolescents. J Infect Dis 163:460–463
43. Knox GE et al (1979) Comparative prevalence of subclinical cytomegalovirus and herpes simplex virus infections in the genital and urinary tracts of low income, urban females. J Infect Dis 140:419–422
44. Ahlfors K et al (1981) Secondary maternal cytomegalovirus infection causing symptomatic congenital infection. N Engl J Med 305:284
45. Stagno S et al (1982) Congenital cytomegalovirus infection: the relative importance of primary and recurrent maternal infection. N Engl J Med 306:945–949
46. Ahlfors K et al (1983) Congenital cytomegalovirus infection: on the relation between type and time of maternal infection and infant's symptoms. Scand J Infect Dis 15:129–138
47. Stagno S et al (1983) Congenital and perinatal cytomegaloviral infections. Semin Perinatol 7:31–42
48. Fowler KB et al (1992) The outcome of congenital cytomegalovirus infection in relation to maternal antibody status. N Engl J Med 326:663–667
49. Demmler GJ (1990) Infectious Diseases Society of America and Centers for Disease Control: summary of a workshop on surveillance for congenital cytomegalovirus disease. Rev Infect Dis 13:315–329

50. Boppana SB et al (1999) Symptomatic congenital cytomegalovirus infection in infants born to mothers with preexisting immunity to cytomegalovirus. Pediatrics 104:55–60
51. Ahlfors K, Ivarsson SA, Harris S (1999) Report on a long-term study of maternal and congenital cytomegalovirus infection in Sweden. Review of prospective studies available in the literature. Scand J Infect Dis 31:443–457
52. Yamamoto AY et al (2011) Congenital cytomegalovirus infection as a cause of sensorineural hearing loss in a highly seropositive population. Pediatr Infect Dis J 30:1043–1046
53. Townsend CL et al (2013) Long-term outcomes of congenital cytomegalovirus infection in Sweden and the United Kingdom. Clin Infect Dis 56:1232–1239
54. Huang ES et al (1980) Molecular epidemiology of cytomegalovirus infections in women and their infants. N Engl J Med 303: 958–962
55. Renzette N et al (2011) Extensive genome-wide variability of human cytomegalovirus in congenitally infected infants. PLoS Pathog 7:e1001344
56. Ross SA et al (2011) Mixed infection and strain diversity in congenital cytomegalovirus infection. J Infect Dis 204:1003–1007
57. Bale JF Jr et al (2001) Human cytomegalovirus a sequence and UL144 variability in strains from infected children. J Med Virol 65:90–96
58. Bale JF et al (1999) Cytomegalovirus transmission in child care homes. Arch Pediatr Adolesc Med 153:75–79
59. Chandler SH, McDougall JK (1986) Comparison of restriction site polymorphisms among clinical isolates and laboratory strains of human cytomegalovirus. J Gen Virol 67:2179–2192
60. Novak Z et al (2008) Cytomegalovirus strain diversity in seropositive women. J Clin Microbiol 46:882–886
61. Starr SE et al (1979) Impaired cellular immunity to cytomegalovirus in congenitally infected children and their mothers. J Infect Dis 140:500–505
62. Gehrz RC et al (1977) Specific cell-mediated immune defect in active cytomegalovirus infection of young children and their mothers. Lancet 2:844–847
63. Miles DJ et al (2008) CD4(+) T cell responses to cytomegalovirus in early life: a prospective birth cohort study. J Infect Dis 197:658–662
64. Tu W et al (2004) Persistent and selective deficiency of CD4+ T cell immunity to cytomegalovirus in immunocompetent young children. J Immunol 172:3260–3267
65. Marchant A et al (2003) Mature CD8(+) T lymphocyte response to viral infection during fetal life. J Clin Invest 111:1747–1755
66. Wang C et al (2011) Attribution of congenital cytomegalovirus infection to primary versus non-primary maternal infection. Clin Infect Dis 52:e11–13
67. Boppana SB, Britt WJ (1995) Antiviral antibody responses and intrauterine transmission after primary maternal cytomegalovirus infection. J Infect Dis 171:1115–1121
68. Britt WJ et al (1990) An immunodominant linear epitope on the major envelope glcoprotein complex (gB) of human cytomegalovirus, In: VIII international congress of virology, Berlin
69. Fouts AE et al (2012) Antibodies against the gH/gL/UL128/UL130/UL131 complex comprise the majority of the anti-cytomegalovirus (anti-CMV) neutralizing antibody response in CMV hyperimmune globulin. J Virol 86:7444–7447
70. Nigro G et al (2005) Passive immunization during pregnancy for congenital cytomegalovirus infection. N Engl J Med 353:1350–1362
71. Visentin S et al (2012) Early pregnancy cytomegalovirus infection in pregnancy: maternal hyperimmunoglobulin therapy improves outcomes among infants at 1 year of age. Clin Infect Dis 55:497–503
72. Maidji E et al (2006) Maternal antibodies enhance or prevent cytomegalovirus infection in the placenta by neonatal Fc receptor-mediated transcytosis. Am J Pathol 168:1210–1226
73. Pereira L et al (2003) Human cytomegalovirus transmission from the uterus to the placenta correlates with the presence of pathogenenic bacteria and maternal immunity. J Virol 77:13301–13314
74. Walker S et al (2007) Ex vivo monitoring of human cytomegalovirus-specific CD8+ T-cell responses using QuantiFeron-CMV. Transpl Infect Dis 9:165–170
75. Mattes FM et al (2008) Functional impairment of cytomegalovirus specific CD8 T cells predicts high-level replication after renal transplantation. Am J Transplant 8:990–999
76. Zanghellini F et al (1999) Asymptomatic primary cytomegalovirus infection: virologic and immunologic features. J Infect Dis 180: 702–707
77. Hansen SG et al (2010) Evasion of CD8+ T cells is critical for superinfection by cytomegalovirus. Science 328:102–106
78. Sester M et al (2002) Sustained high frequencies of specific CD4 T cells restricted to a single persistent virus. J Virol 76:3748–3755

79. Gamadia LE et al (2003) Primary immune responses to human CMV: a critical role for IFN-gamma-producing CD4+ T cells in protection against CMV disease. Blood 101:2686–2692
80. Moss P, Khan N (2004) CD8(+) T-cell immunity to cytomegalovirus. Hum Immunol 65:456–464
81. Sylwester AW et al (2005) Broadly targeted human cytomegalovirus-specific CD4+ and CD8+ T cells dominate the memory compartments of exposed subjects. J Exp Med 202:673–685
82. Gyulai Z et al (2000) Cytotoxic T lymphocyte (CTL) responses to human cytomegalovirus pp 65, IE1-Exaon4, gB, pp150, and pp28 in healthy individuals: reevaluation of prevalence of IE1-specific CTSs. J Infect Dis 181:1537–1546
83. Lilleri D et al (2007) Development of human cytomegalovirus-specific T cell immunity during primary infection of pregnant women and its correlation with virus transmission to the fetus. J Infect Dis 195:1062–1070
84. Baldanti F et al (2006) Human cytomegalovirus UL131A, UL130, and UL128 genes are highly conserved among field isolates. Arch Virol 151:1225–1233
85. Lissauer D et al (2011) Cytomegalovirus seropositivity dramatically alters the maternal CD8+ T cell repertoire and leads to the accumulation of highly differentiated memory cells during human pregnancy. Hum Reprod 26:3355–3365
86. Li CR et al (1994) Recovery of HLA-restricted cytomegalovirus (CMV)-specific T-cell responses after allogeneic bone marrow transplant: correlation with CMV disease and effect of ganciclovir prophylaxis. Blood 83:1971–1979
87. Reusser P et al (1991) Cytotoxic T-lymphocyte response to cytomegalovirus after human allogeneic bone marrow transplantation: pattern of recovery and correlation with cytomegalovirus infection and disease. Blood 78:1373–1380
88. Walter EA et al (1995) Reconstitution of cellular immunity against cytomegalovirus in recipients of allogeneic bone marrow by transfer of T-cell clones from the donor. N Engl J Med 333:1038–1044
89. Rosa L et al (2007) Longitudinal assessment of cytomegalovirus (CMV)-specific immune responses in liver transplant recipients at high risk for late CMV disease. J Infect Dis 195:633–644
90. Schooley RT et al (1983) Association of herpes virus infection with T-lymphocyte subset alterations, glomerulopathy, and opportunistic infections after renal transplantation. N Engl J Med 308:307–313
91. Sester M et al (2001) Levels of virus-specific CD4 T cells correlate with cytomegalovirus control and predict virus-induced disease after renal transplantation. Transplantation 71:1287–1294
92. Reusser P et al (1999) Cytomegalovirus (CMV)-specific T cell immunity after renal transplantation mediates protection from CMV disease by limiting the systemic virus load. J Infect Dis 180:247–253
93. Pourgheysari B et al (2009) Early reconstitution of effector memory CD4+ CMV-specific T cells protects against CMV reactivation following allogeneic SCT. Bone Marrow Transplant 43:853–861
94. Razonable RR (2008) Cytomegalovirus infection after liver transplantation: current concepts and challenges. World J Gastroenterol 14:4849–4860
95. Gallina A et al (1996) Human cytomegalovirus protein pp 65 lower matrix phosphoprotein harbours two transplantable nuclear localization signals. J Gen Virol 77:1151–1157
96. Riddell SR, Greenberg PD (1994) Therapeutic reconstruction of human viral immunity by adoptive transfer of cytotoxic T lymphocyte clones. Curr Top Microbiol Immunol 189:9–34
97. Riddell SR, Greenberg PD (2000) T-cell therapy of cytomegalovirus and human immunodeficiency virus infection. J Antimicrob Chemother 45:35–43
98. Komanduri KV et al (1998) Restoration of cytomegalovirus-specific CD4+ T-lymphocyte responses after ganciclovir and highly active antiretroviral therapy in individuals infected with HIV-1. Nat Med 4:953–956
99. Autran B et al (1997) Positive effects of combined antiretroviral therapy on CD4+ T cell homeostasis and function in advanced HIV disease. Science 277:112–116
100. Bronke C et al (2005) Dynamics of cytomegalovirus (CMV)-specific T cells in HIV-1-infected individuals progressing to AIDS with CMV end-organ disease. J Infect Dis 191:873–880
101. Boppana SB et al (1995) Virus specific antibody responses to human cytomegalovirus (HCMV) in human immunodeficiency virus type 1-infected individuals with HCMV retinitis. J Infect Dis 171:182–185
102. Dylewski J, Chou S, Merigan TC (1983) Absence of detectable IgM antibody during cytomegalovirus disease in patients with AIDS [letter]. N Engl J Med 309:493

103. Rasmussen L et al (1994) Deficiency in antibody response to human cytomegalovirus glycoprotein gH in human immunodeficiency virus-infected patients at risk for cytomegalovirus retinitis. J Infect Dis 170:673–677
104. Arribas JR et al (1996) Cytomegalovirus encephalitis. Ann Intern Med 125:577–587
105. Evans PC et al (1999) Cytomegalovirus infection of bile duct epithelial cells, hepatic artery and portal venous endothelium in relation to chronic rejection of liver grafts. J Hepatol 31:913–920
106. Lautenschlager I et al (2006) Cytomegalovirus infection of the liver transplant: virological, histological, immunological, and clinical observations. Transpl Infect Dis 8:21–30
107. Myerson D, Hackman RC, Meyers JD (1984) Diagnosis of cytomegaloviral pneumonia by in situ hybridization. J Infect Dis 150: 272–277
108. Myerson D et al (1984) Widespread presence of histologically occult cytomegalovirus. Hum Pathol 15:430–439
109. Schleiss MR (2006) Nonprimate models of congenital cytomegalovirus (CMV) infection: gaining insight into pathogenesis and prevention of disease in newborns. ILAR J 47: 65–72
110. Britt WJ, Boppana S (2004) Human cytomegalviurs virion proteins. Hum Immunol 65:395–402
111. Jonjic S et al (2008) Immune evasion of natural killer cells by viruses. Curr Opin Immunol 20:30–38
112. Jonjic S et al (1990) Efficacious control of cytomegalovirus infection after long-term depletion of CD8+ T lymphocytes. J Virol 64:5457–5464
113. Jonjic S et al (1994) Antibodies are not essential for the resolution of primary cytomegalovirus infection but limit dissemination of recurrent virus. J Exp Med 179:1713–1717
114. Koszinowski UH, Reddehase MJ, Jonjic S (1991) The role of CD4 and CD8 T cells in viral infections. Curr Opin Immunol 3: 471–475
115. Koszinowski UH et al (1987) Molecular analysis of herpesviral gene products recognized by protective cytolytic T lymphocytes. Immunol Lett 16:185–192
116. Krmpotic A et al (2003) Pathogenesis of murine cytomegalovirus infection. Microbes Infect 5:1263–1277
117. Polic B et al (1998) Hierarchical and redundant lymphocyte subset control precludes cytomegalovirus replication during latent infection. J Exp Med 188:1047–1054
118. Shanley JD (1991) Murine models of cytomegalovirus associated pneumonitis. Transplant Proc 23:S12–S16
119. Shanley JD, Biczak L, Formon SJ (1993) Acute murine cytomegalovirus infection induces lethal hepatitis. J Infect Dis 167: 264–269
120. Koszinowski UH, Del Val M, Reddehase MJ (1990) Cellular and molecular basis of the protective immune response to cytomegalovirus infection. Curr Top Microbiol Immunol 154:189–220
121. Alcami A, Koszinowski UH (2000) Viral mechanisms of immune evasion. Immunol Today 21:447–455
122. Krmpotic A et al (2005) NK cell activation through the NKG2D ligand MULT-1 is selectively prevented by the glycoprotein encoded by mouse cytomegalovirus gene m145. J Exp Med 201:211–220
123. Zimmermann A et al (2005) A cytomegaloviral protein reveals a dual role for STAT2 in IFN-{gamma} signaling and antiviral responses. J Exp Med 201:1543–1553
124. LoPiccolo DM et al (2003) Effective inhibition of K(b)- and D(b)-restricted antigen presentation in primary macrophages by murine cytomegalovirus. J Virol 77:301–308
125. Lenac T et al (2006) The herpesviral Fc receptor fcr-1 down-regulates the NKG2D ligands MULT-1 and H60. J Exp Med 203: 1843–1850
126. Krmpotic A et al (2001) The immunoevasive function encoded by the mouse cytomegalovirus gene m152 protects the virus against T cell control in vivo. J Exp Med 190:1285–1296
127. Krmpotic A et al (2002) MCMV glycoprotein gp40 confers virus resistance to CD8+ T cells and NK cells in vivo. Nat Immunol 3: 529–535
128. Kavanagh DG et al (2001) The multiple immune-evasion genes of murine cytomegalovirus are not redundant: m4 and m152 inhibit antigen presentation in a complementary and cooperative fashion. J Exp Med 194:967–978
129. Hasan M et al (2005) Selective down-regulation of the NKG2D ligand H60 by mouse cytomegalovirus m155 glycoprotein. J Virol 79:2920–2930
130. Hengel H et al (1999) Cytomegaloviral control of MHC class I function in the mouse. Immunol Rev 168:167–176
131. Smith HR et al (2002) Recognition of a virus-encoded ligand by a natural killer cell activation receptor. Proc Natl Acad Sci U S A 99:8826–8831

132. Upton JW, Kaiser WJ, Mocarski ES (2010) Virus inhibition of RIP3-dependent necrosis. Cell Host Microbe 7:302–313
133. Lisnic VJ, Krmpotic A, Jonjic S (2010) Modulation of natural killer cell activity by viruses. Curr Opin Microbiol 14:530–539
134. Humphreys IR et al (2007) Cytomegalovirus exploits IL-10-mediated immune regulation in the salivary glands. J Exp Med 204: 1217–1225
135. Reddehase MJ et al (1994) The conditions of primary infection define the load of latent viral genome in organs and the risk of recurrent cytomegalovirus disease. J Exp Med 179:185–193
136. Mathys S et al (2003) Dendritic cells under influence of mouse cytomegalovirus have a physiologic dual role: to initiate and to restrict T cell activation. J Infect Dis 187:988–999
137. Spencer JV et al (2002) Potent immunosuppressive activities of cytomegalovirus-encoded interleukin-10. J Virol 76:1285–1292
138. Chang WL, Barry PA (2009) Attenuation of innate immunity by cytomegalovirus IL-10 establishes a long-term deficit of adaptive antiviral immunity. Proc Natl Acad Sci U S A 107:22647–22652
139. Mandaric S et al (2012) IL-10 suppression of NK/DC crosstalk leads to poor priming of MCMV-specific CD4 T cells and prolonged MCMV persistence. PLoS Pathog 8:e1002846
140. Scalzo AA et al (2007) The interplay between host and viral factors in shaping the outcome of cytomegalovirus infection. Immunol Cell Biol 85:46–54
141. Jonjic S, Bubic I, Krmpotic A (2006) Innate immunity to cytomegaloviruses. In: Reddehase MJ (ed) Cytomegaloviruses: molecular biology and immunology. Caister, Norfolk, UK, pp 285–321
142. Klenovsek K et al (2007) Protection from CMV infection in immunodeficient hosts by adoptive transfer of memory B cells. Blood 110:3472–3479
143. Schlub TE et al (2011) Comparing the kinetics of NK cells, CD4, and CD8 T cells in murine cytomegalovirus infection. J Immunol 187:1385–1392
144. Walton SM et al (2008) The dynamics of mouse cytomegalovirus-specific CD4 T cell responses during acute and latent infection. J Immunol 181:1128–1134
145. Holtappels R et al (2008) CD8 T-cell-based immunotherapy of cytomegalovirus infection: "proof of concept" provided by the murine model. Med Microbiol Immunol 197:125–134
146. Cekinovic D et al (2008) Passive immunization reduces murine cytomegalovirus-induced brain pathology in newborn mice. J Virol 82:12172–12180
147. Pavic I et al (1993) Participation of endogenous tumour necrosis factor alpha in host resistance to cytomegalovirus infection. J Gen Virol 74(Pt 10):2215–2223
148. Scalzo AA, Yokoyama WM (2008) Cmv1 and natural killer cell responses to murine cytomegalovirus infection. Curr Top Microbiol Immunol 321:101–122
149. Vivier E et al (2011) Innate or adaptive immunity? The example of natural killer cells. Science 331:44–49
150. Rapp M et al (1993) In vivo protection studies with MCMV glycoproteins gB and gH expressed by vaccinia virus. In: Michelson S, Plotkin SA (eds) Multidisciplinary approach to understanding cytomegalovirus disease. Excerpta Medica, Amsterdam, pp 327–332
151. Reddehase MJ et al (1987) CD8-positive T lymphocytes specific for murine cytomegalovirus immediate-early antigens mediate protective immunity. J Virol 61:3102–3108
152. Del Val M et al (1991) Protection against lethal cytomegalovirus infection by a recombinant vaccine containing a single nonameric T-cell epitope. J Virol 65:3641–3646
153. Mohr CA et al (2010) A spread-deficient cytomegalovirus for assessment of first-target cells in vaccination. J Virol 84:7730–7742
154. Gonzalez Armas JC et al (1996) DNA immunization confers protection against murine cytomegalovirus infection. J Virol 70: 7921–7928
155. Morello CS, Cranmer LD, Spector DH (2000) Suppression of murine cytomegalovirus (MCMV) replication with a DNA vaccine encoding MCMV M84 (a homolog of human cytomegalovirus pp 65). J Virol 74: 3696–3708
156. Hakki M et al (2003) Immune reconstitution to cytomegalovirus after allogeneic hematopoietic stem cell transplantation: impact of host factors, drug therapy, and subclinical reactivation. Blood 102:3060–3067
157. Slavuljica I et al (2010) Recombinant mouse cytomegalovirus expressing a ligand for the NKG2D receptor is attenuated and has improved vaccine properties. J Clin Invest 120:4532–4545
158. Farrell HE, Shellam GR (1991) Protection against murine cytomegalovirus infection by passive transfer of neutralizing and non-neutralizing monoclonal antibodies. J Gen Virol 72:149–156

159. Farroway LN et al (2005) Transmission of two Australian strains of murine cytomegalovirus (MCMV) in enclosure populations of house mice (Mus domesticus). Epidemiol Infect 133:701–710
160. Gorman S et al (2006) Mixed infection with multiple strains of murine cytomegalovirus occurs following simultaneous or sequential infection of immunocompetent mice. J Gen Virol 87:1123–1132
161. Barry PA et al (2006) Nonhuman primate models of intrauterine cytomegalovirus infection. ILAR J 47:49–64
162. Assaf BT et al (2012) Patterns of acute rhesus cytomegalovirus (RhCMV) infection predict long-term RhCMV infection. J Virol 86: 6354–6357
163. Lockridge KM et al (1999) Pathogenesis of experimental rhesus cytomegalovirus infection. J Virol 73:9576–9583
164. Kaur A et al (1996) Cytotoxic T-lymphocyte responses to cytomegalovirus in normal and simian immunodeficiency virus-infected rhesus macaques. J Virol 70:7725–7733
165. Yue Y, Zhou SS, Barry PA (2003) Antibody responses to rhesus cytomegalovirus glycoprotein B in naturally infected rhesus macaques. J Gen Virol 84:3371–3379
166. Kaur A et al (2002) Decreased frequency of cytomegalovirus (CMV)-specific CD4+ T lymphocytes in simian immunodeficiency virus-infected rhesus macaques: inverse relationship with CMV viremia. J Virol 76:3646–3658
167. Kaur A et al (2003) Direct relationship between suppression of virus-specific immunity and emergence of cytomegalovirus disease in simian AIDS. J Virol 77:5749–5758
168. Yue Y et al (2008) Evaluation of recombinant modified vaccinia Ankara virus-based rhesus cytomegalovirus vaccines in rhesus macaques. Med Microbiol Immunol 197:117–123
169. Yue Y et al (2007) Immunogenicity and protective efficacy of DNA vaccines expressing rhesus cytomegalovirus glycoprotein B, phosphoprotein 65-2, and viral interleukin-10 in rhesus macaques. J Virol 81:1095–1109
170. Wussow F et al (2013) A vaccine based on rhesus cytomegalovirus UL128 complex induces broadly neutralizing antibodies in rhesus macaques. J Virol 87:1322–1332
171. Hansen SG et al (2009) Effector memory T cell responses are associated with protection of rhesus monkeys from mucosal simian immunodeficiency virus challenge. Nat Med 15:293–299
172. Holtappels R et al (2001) Experimental preemptive immunotherapy of murine cytomegalovirus disease with CD8 T-cell lines specific for pp M83 and pM84, the two homologs of human cytomegalovirus tegument protein ppUL83 (pp65). J Virol 75:6584–6600
173. Podlech J et al (2000) Murine model of interstitial cytomegalovirus pneumona in syngenic bone marrow transplantation: persistence of protective pulmonary CD8-T-cell infiltrates after clearance of acute infection. J Virol 74:7496–7507
174. Holtappels R et al (2006) CD8 T-cell-based immunotherapy of cytomegalovirus disease in the mouse model of the immunocompromised bone marrow transplantation recipient. In: Reddehase MJ (ed) Cytomegaloviruses: molecular biology and immunology. Caister, Norfolk, UK, pp 383–419
175. Scalzo AA et al (1992) The effect of the Cmv-1 resistance gene, which is linked to the natural killer cell gene complex, is mediated by natural killer cells. J Immunol 149:581–589
176. Scalzo AA et al (1990) Cmv-1, a genetic locus that controls murine cytomegalovirus replication in the spleen. J Exp Med 171:1469–1483
177. Brown MG et al (1999) Localization on a physical map of the NKC-linked Cmv1 locus between Ly49b and the Prp gene cluster on mouse chromosome 6. J Immunol 163:1991–1999
178. Webb JR, Lee SH, Vidal SM (2002) Genetic control of innate immune responses against cytomegalovirus: MCMV meets its match. Genes Immun 3:250–262
179. Lee SH et al (2001) Susceptibility to mouse cytomegalovirus is associated with deletion of an activating natural killer cell receptor of the C-type lectin superfamily. Nat Genet 28:42–45
180. Depatie C et al (1999) Assessment of Cmv1 candidates by genetic mapping and in vivo antibody depletion of NK cell subsets. Int Immunol 11:1541–1551
181. Cheng TP et al (2008) Ly49h is necessary for genetic resistance to murine cytomegalovirus. Immunogenetics 60:565–573
182. Cooper MA, Yokoyama WM (2010) Memory-like responses of natural killer cells. Immunol Rev 235:297–305
183. Elliott JM, Yokoyama WM (2011) Unifying concepts of MHC-dependent natural killer cell education. Trends Immunol 32:364–372
184. Fodil-Cornu N et al (2008) Ly49h-deficient C57BL/6 mice: a new mouse cytomegalovirus-susceptible model remains resistant to unrelated pathogens controlled by

the NK gene complex. J Immunol 181: 6394–6405

185. Hadaya K et al (2008) Natural killer cell receptor repertoire and their ligands, and the risk of CMV infection after kidney transplantation. Am J Transplant 8:2674–2683
186. Bacon L et al (2004) Two human ULBP/ RAET1 molecules with transmembrane regions are ligands for NKG2D. J Immunol 173:1078–1084
187. Bauer S et al (1999) Activation of NK cells and T cells by NKG2D, a receptor for stress-inducible MICA. Science 285:727–729
188. Chalupny NJ et al (2003) ULBP4 is a novel ligand for human NKG2D. Biochem Biophys Res Commun 305:129–135
189. Eagle RA et al (2006) Regulation of NKG2D ligand gene expression. Hum Immunol 67: 159–169
190. Eagle RA et al (2009) ULBP6/RAET1L is an additional human NKG2D ligand. Eur J Immunol 39:3207–3216
191. Rolle A et al (2003) Effects of human cytomegalovirus infection on ligands for the activating NKG2D receptor of NK cells: up-regulation of UL16-binding protein (ULBP)1 and ULBP2 is counteracted by the viral UL16 protein. J Immunol 171:902–908
192. Jonjic S et al (1989) Site-restricted persistent cytomegalovirus infection after selective long-term depletion of CD4+ T lymphocytes. J Exp Med 169:1199–1212
193. Lucin P et al (1992) Gamma interferon-dependent clearance of cytomegalovirus infection in salivary glands. J Virol 66:1977–1984
194. Holtappels R et al (1998) Control of murine cytomegalovirus in the lungs: relative but not absolute immunodominance of the immediate-early 1 nonapeptide during the antiviral cytolytic T-lymphocyte response in pulmonary infiltrates. J Virol 72:7201–7212
195. Reddehase MJ et al (1985) Interstitial murine cytomegalovirus pneumonia after irradiation: characterization of cells that limit viral replication during established infection of the lungs. J Virol 55:264–273
196. Kurz S et al (1997) Latency versus persistence or intermittent recurrences: evidence for a latent state of murine cytomegalovirus in the lungs. J Virol 71:2980–2987
197. Baroncelli S et al (1997) Cytomegalovirus and simian immunodeficiency virus coinfection: longitudinal study of antibody responses and disease progression. J Acquir Immune Defic Syndr Hum Retrovirol 15:5–15
198. Sequar G et al (2002) Experimental coinfection of rhesus macaques with rhesus cytomegalovirus and simian immunodeficiency virus: pathogenesis. J Virol 76:7661–7671
199. Reuter JD et al (2004) Systemic immune deficiency necessary for cytomegalovirus invasion of the mature brain. J Virol 78:1473–1487
200. Woolf NK, Jaquish DV, Koehrn FJ (2007) Transplacental murine cytomegalovirus infection in the brain of SCID mice. Virol J 4:26
201. Bia FJ et al (1983) Cytomegaloviral infections in the guinea pig: experimental models for human disease. Rev Infect Dis 5:177–195
202. Bia FJ, Miller SA, Davidson KH (1984) The guinea pig cytomegalovirus model of congenital human cytomegalovirus infection. Birth Defects 20:233–241
203. Griffith BP et al (1986) Inbred guinea pig model of intrauterine infection with cytomegalovirus. Am J Pathol 122:112–119
204. Borune N et al (2001) Preconception immunization with a cytomegalovirus (CMV) glycoprotein vaccine improves pregnancy outcome in a guniea pig model of congenital CMV infection. J Infect Dis 183:59–64
205. Schleiss MR (2007) Comparison of vaccine strategies against congenital CMV infection in the guinea pig model. J Clin Virol 41:224–230
206. Harrison CJ et al (1995) Reduced congenital cytomegalovirus (CMV) infection after maternal immunization with a guinea pig CMV glycoprotein before gestational primary CMV infection in the guinea pig model. J Infect Dis 172:1212–1220
207. Bratcher DF et al (1995) Effect of passive antibody on congenital cytomegalovirus infection in guinea pigs. J Infect Dis 172:944–950
208. Chatterjee A et al (2001) Modification of maternal and congenital cytomegalovirus infection by anti-glycoprotein b antibody transfer in guinea pigs. J Infect Dis 183:1547–1553
209. Woolf NK (1991) Guinea pig model of congenital CMV-induced hearing loss: a review. Transplant Proc 23:32–34
210. Park AH et al (2010) Development of cytomegalovirus-mediated sensorineural hearing loss in a guinea pig model. Arch Otolaryngol Head Neck Surg 136:48–53
211. Britt WJ, Cekinovic D, Jonjic S (2013) Murine model of neonatal cytomegalovirus infection. In: Reddehasse M (ed) Cytomegaloviruses. Cassister, London
212. Koontz T et al (2008) Altered development of the brain after focal herpesvirus infection of the central nervous system. J Exp Med 205:423–435
213. Kosmac K et al (2013) Glucocortiocoid treatment of MCMV infected newborn mice

attenuates CNS inflammation and limits deficits in cerebellar development. PLoS Pathog 9:e1003200

214. Stagno S et al (1977) Congenital cytomegalovirus infection: occurrence in an immune population. N Engl J Med 296:1254–1258
215. Adler SP et al (1995) Immunity induced by primary human cytomegalovirus infection protects against secondary infection among women of childbearing age. J Infect Dis 171:26–32
216. Pass RF et al (2009) Vaccine prevention of maternal cytomegalovirus infection. N Engl J Med 360:1191–1199
217. Hansen SG et al (2011) Profound early control of highly pathogenic SIV by an effector memory T-cell vaccine. Nature 473:523–527
218. Wloch MK et al (2008) Safety and immunogenicity of a bivalent cytomegalovirus DNA vaccine in healthy adult subjects. J Infect Dis 197:1634–1642
219. Griffiths PD et al (2011) Cytomegalovirus glycoprotein-B vaccine with MF59 adjuvant in transplant recipients: a phase 2 randomised placebo-controlled trial. Lancet 377: 1256–1263
220. Bernstein DI et al (2009) Randomized, double-blind, phase I trial of alphavirus replicon vaccine for cytomegalovirus in CMV seronegative adult volunteers. Vaccine 28:484–493
221. Heineman TC et al (2006) A phase 1 study of 4 live, recombinant human cytomegalovirus Towne/Toledo chimeric vaccines. J Infect Dis 193:1350–1360
222. La Rosa C et al (2012) Clinical evaluation of safety and immunogenicity of PADRE-cytomegalovirus (CMV) and tetanus-CMV fusion peptide vaccines with or without PF03512676 adjuvant. J Infect Dis 205: 1294–1304
223. Snydman DR (1993) Review of the efficacy of cytomegalovirus immune globulin in the prophylaxis of CMV disease in renal transplant recipients. Transplant Proc 25:25–26
224. Yeager AS et al (1981) Prevention of transfusion-acquired cytomegalovirus infections in newborn infants. J Pediatr 98: 281–287
225. Rubin R (2002) Clinical approach to infection in the compromised host. In: Rubin R, Young LS (eds) Infection in the organ transplant recipient. Kluwer Academic, New York, pp 573–679
226. Snydman DR et al (1987) Use of cytomegalovirus immune globulin to prevent cytomegalovirus disease in renal transplant recipients. N Engl J Med 317:1049–1054
227. Nigro G, Adler SP (2011) Cytomegalovirus infections during pregnancy. Curr Opin Obstet Gynecol 23:123–128
228. Hedlund M et al (2009) Human placenta expresses and secretes NKG2D ligands via exosomes that down-modulate the cognate receptor expression: evidence for immunosuppressive function. J Immunol 183:340–351
229. Iwasenko JM et al (2011) Human cytomegalovirus infection is detected frequently in stillbirths and is associated with fetal thrombotic vasculopathy. J Infect Dis 203:1526–1533
230. La Torre R et al (2006) Placental enlargement in women with primary maternal cytomegalovirus infection is associated with fetal and neonatal disease. Clin Infect Dis 43:994–1000
231. Pereira L, Maidji E (2008) Cytomegalovirus infection in the human placenta: maternal immunity and developmentally regulated receptors on trophoblasts converge. Curr Top Microbiol Immunol 325:383–395
232. Scott GM et al (2012) Cytomegalovirus infection during pregnancy with maternofetal transmission induces a proinflammatory cytokine bias in placenta and amniotic fluid. J Infect Dis 205:1305–1310
233. Cui X et al (2008) Cytomegalovirus vaccines fail to induce epithelial entry neutralzing antibodies comparable to natural infection. Vaccine 26:5760–5766
234. Gerna G et al (2008) Human cytomegalvirus serum neutralizing antibodies block virus infection of endothelial/epithelial cells but not fibroblasts, early during primary infection. J Gen Virol 89:853–865
235. Lilleri D et al (2012) Antibodies against neutralization epitopes of human cytomegalovirus gH/gL/pUL128-130-131 complex and virus spreading may correlate with virus control in vivo. J Clin Immunol 32:1324–1331
236. Jacobson MA et al (2009) A CMV DNA vaccine primes for memory immune responses to live-attenuated CMV (Towne strain). Vaccine 27:1540–1548
237. Boppana SB, Britt WJ (1996) Recognition of human cytomegalovirus gene products by HCMV-specific cytotoxic T cells. Virology 222:293–296
238. Borysiewicz LK et al (1988) Human cytomegalovirus-specific cytotoxic T cells: their precursor frequency and stage specificity. Eur J Immunol 18:269–275
239. Wills MR et al (1996) The human cytotoxic T lymphocyte (CTL) response to cytomegalovirus is dominated by structural protein pp 65: frequency, specificity and T-cell receptor

usage of pp65-specific CTL. J Virol 70: 7569–7579

240. Pass RF et al (1995) A phase I trial of BIOCINE CMV gB vaccine in seronegative adults. In: Fifth international cytomegalovirus conference, Stockholm, Sweden
241. Sabbaj S et al (2011) Glycoprotein B vaccine is capable of boosting both antibody and CD4 T-cell responses to cytomegalovirus in chronically infected women. J Infect Dis 203:1534–1541
242. Bubic I et al (2004) Gain of virulence caused by loss of a gene in murine cytomegalovirus. J Virol 78:7536–7544
243. Crnkovic-Mertens I et al (1998) Virus attenuation after deletion of the cytomegalovirus Fc receptor gene is not due to antibody control. J Virol 72:1377–1382
244. Chang WL et al (2009) Human cytomegalovirus suppresses type I interferon secretion by plasmacytoid dendritic cells through its interleukin 10 homolog. Virology 390:330–337
245. Logsdon NJ et al (2011) Design and analysis of rhesus cytomegalovirus IL-10 mutants as a model for novel vaccines against human cytomegalovirus. PLoS One 6:e28127
246. Jones BC et al (2002) Crystal structure of human cytomegalovirus IL-10 bound to soluble human IL-10R1. Proc Natl Acad Sci U S A 99:9404–9409
247. Stagno S et al (1982) Prevalence and importance of congenital cytomegalovirus infection in three different populations. J Pediatr 101:897–900
248. Adler SP et al (1996) Prevention of child-to-mother transmission of cytomegalovirus by changing behaviors: a randomized controlled trial. Pediatr Infect Dis J 15:240–246
249. Picone O et al (2009) A 2-year study on cytomegalovirus infection during pregnancy in a French hospital. BJOC 116:818–823
250. Paya CV et al (2004) Efficacy and safety of valganciclovir vs oral ganciclovir for prevention of cytomegalovirus disease in solid organ transplant recipients. Am J Transplant 4: 611–620
251. Kotton CN et al (2010) International consensus guidelines on the management of cytomegalovirus in solid organ transplantation. Transplantation 89:779–795
252. Asberg A et al (2009) Long-term outcomes of CMV disease treatment with valganciclovir versus IV ganciclovir in solid organ transplant recipients. Am J Transplant 9:1205–1213
253. Ljungman P et al (2008) Manamagement of CMV, HHV-6, HHV-7 and Kaposi-sarcoma associated herpesvirus (HHV-8) infections in patients with hematological malignancies and after SCT. Bone Marrow Transplant 42: 227–240
254. Tomblyn M et al (2009) Guidelines for preventing infectious complications among hematopoietic cell transplant recipients: a global perspective. Biol Blood Marrow Transplant 15:1143–1238
255. Drew WL et al (1984) Multiple infections by cytomegalovirus in patients with acquired immune deficiency syndrome: documentation by Southern blot hybridization. J Infect Dis 150:952–953
256. Burkhardt C et al (2009) Glycoprotein N subtypes of human cytomegalovirus induce a strain-specific antibody response during natural infection. J Gen Virol 90:1951–1961
257. Dal Monte P et al (2001) The product of human cytomegalovirus UL73 is a new polymorphic structural glycoprotein (gpUL73). J Hum Virol 4:26–34
258. Gnann JW et al (1988) Inflammatory cells in transplanted kidneys are infected by human cytomegalovirus. Am J Pathol 132:239–248
259. Rasmussen L et al (2002) The genes encoding the gCIII complex of human cytomegalovirus exist in highly diverse combinations in clinical isolates. J Virol 76: 10841–10848
260. Pass RF et al (2006) Congenital cytomegalovirus infection following first trimester maternal infection: symptoms at birth and outcome. J Clin Virol 35:216–220
261. Hodson EM et al (2005) Antiviral medications for preventing cytomegalovirus disease in solid organ transplant recipients. Cochrane Database Syst Rev CD003774
262. Enders G et al (2011) Intrauterine transmission and clinical outcome of 248 pregnancies with primary cytomegalovirus infection in relation to gestational age. J Clin Virol 52:244–246
263. Andersen H et al (1979) A prosepctive study on the incidence and significance of congenital cytomegalovirus infection. Acta Paediatr Scand 68:329–336
264. Larke RBP et al (1980) Congenital cytomegalovirus infection in an urban Canadian community. J Infect Dis 142:647–653
265. Peckham CS et al (1983) Cytomegalovirus infection in pregnancy: preliminary findings from a prospective study. Lancet 1:1352–1355
266. Child SJ et al (2004) Evasion of cellular antiviral responses by human cytomegalovirus TRS1 and IRS1. J Virol 78:197–205
267. Brune W, Nevels M, Shenk T (2003) Murine cytomegalovirus m41 open reading frame

encodes a Golgi-localized antiapoptotic progein. J Virol 77:11633–11643

268. Jurak I et al (2008) Murine cytomegalovirus m38.5 protein inhibits Bax-mediated cell death. J Virol 82:4812–4822
269. Cam M et al (2010) Cytomegaloviruses inhibit Bak- and Bax-mediated apoptosis with two separate viral proteins. Cell Death Differ 17:655–665
270. Skaletskaya A et al (2001) A cytomegalovirus-encoded inhibitor of apoptosis that suppresses caspase-8 activation. Proc Natl Acad Sci U S A 98:7829–7834
271. Brune W et al (2001) Secreted virus-encoded proteins reflect murinr cytomegalovirus productivity in organs. J Infect Dis 184: 1320–1324
272. Lenac T et al (2008) Murine cytomegalovirus regulation of NKG2D ligands. Med Microbiol Immunol 197:159–166
273. Lodoen MB et al (2004) The cytomegalovirus m155 gene product subverts natural killer cell antiviral protection by disruption of H60-NKG2D interactions. J Exp Med 200: 1075–1081
274. McCormick AL et al (2003) Differential function and expression of the viral inhibitor of caspase 8-induced apoptosis (vICA) and the viral mitochondria-localized inhibitor of apoptosis (vMIA) cell death suppressors conserved in primate and rodent cytomegaloviruses. Virology 316:221–233
275. Chalupny NJ et al (2006) Down-regulation of the NKG2D ligand MICA by the human cytomegalovirus glycoprotein UL142. Biochem Biophys Res Commun 346:175–181
276. Fruh K, Yang Y (1999) Antigen presentation by MHC class I and its regulation by interferon gamma. Curr Opin Immunol 11: 76–81
277. Bubeck A et al (2002) The glycoprotein gp48 of murine cytomegalovirusL proteasome-dependent cytosolic dislocation and degradation. J Biol Chem 277:2216–2224
278. Vidal SM, Lanier LL (2006) NK cell recognition of moouse cytomegalovirus-infected cells. Curr Top Microbiol Immunol 298: 183–206
279. Babic M et al (2010) Cytomegalovirus immunoevasion reveals the physiological role of "missing self" recognition in natural killer cell dependent virus control in vivo. J Exp Med 207:2663–2673
280. Tomazin R et al (1999) Cytomegalovirus US2 destroys two components of the MHC class II pathway, preventing recognition by CD+ T cells. Nat Med 5:1039–1043
281. Jackson SE, Mason GM, Wills MR (2011) Human cytomegalovirus immunity and immune evasion. Virus Res 157:151–160
282. Arora N et al (2010) Cytomegalovirus viruria and DNAemia in healthy seropositive women. J Infect Dis 202:1800–1803
283. Rivera LB et al (2002) Predictors of hearing loss in children with symptomatic congenital cytomegalovirus infection. Pediatrics 110: 762–767

# Chapter 18

# Approaches for the Generation of New Anti-cytomegalovirus Agents: Identification of Protein–Protein Interaction Inhibitors and Compounds Against the HCMV IE2 Protein

**Beatrice Mercorelli, Giorgio Gribaudo, Giorgio Palù, and Arianna Loregian**

## Abstract

Human cytomegalovirus (HCMV) infection is responsible for severe, often even fatal, diseases in immunocompromised subjects and also represents the major cause of viral-associated congenital malformations in newborn children. The few drugs licensed for anti-HCMV therapy suffer from many drawbacks and none of them have been approved for the treatment of congenital infections. Furthermore, the emergence of drug-resistant viral strains represents a major concern for disease management. Thus, there is a strong need for new anti-HCMV drugs. Here we describe three different assays for the discovery of novel anti-HCMV compounds: two are in vitro assays, i.e., a fluorescence polarization (FP)-based assay and an enzyme-linked immunosorbent assay (ELISA), which are designed to search for compounds that act by disrupting the interactions between the HCMV DNA polymerase subunits, but in general can be employed to find inhibitors of any protein–protein interaction of interest; the third is a cell-based assay designed to identify inhibitors of the viral immediate-early 2 (IE2) protein activities.

**Key words** Antivirals, Cell-based assay, DNA polymerase inhibitors, Enzyme-linked immunosorbent assay, Fluorescence polarization, Human cytomegalovirus, IE2 protein, Protein–protein interaction

## 1 Introduction

Human cytomegalovirus (HCMV) infection is associated with severe morbidity and mortality in immunocompromised individuals, such as transplant recipients and AIDS patients, and is also the most frequent cause of viral-associated congenital defects in newborn children. The few drugs currently licensed for anti-HCMV therapy, most of which target the catalytic activity of the viral DNA polymerase, suffer from many drawbacks, including low potency, poor bioavailability, and long-term toxicity; furthermore, none of these drugs have been approved for the treatment of congenital

Andrew D. Yurochko and William E. Miller (eds.), *Human Cytomegaloviruses: Methods and Protocols*,
Methods in Molecular Biology, vol. 1119, DOI 10.1007/978-1-62703-788-4_18, 

infections. Additionally, due to the fact that most of the current anti-HCMV drugs share the same mechanism of action, the emergence of drug-resistant viral strains is becoming an increasing problem for disease management. Thus, there is a strong need for novel anti-HCMV drugs that are more potent, less toxic, and possibly act by a mechanism of action different from that of the currently available drugs.

The viral DNA polymerase represents a major target for antivirals [1], and new compounds that act by disrupting the interactions between the polymerase subunits rather than inhibiting the enzyme catalytic activity have been recently discovered, thus demonstrating the feasibility of alternative antiviral strategies [2]. In addition, the earliest events of the virus replication cycle, in particular the functions of the immediate-early 2 (IE2) protein, represent attractive targets for the development of novel antiviral compounds [3]. In fact, the multifunctional IE2 protein is crucial for regulation of viral early (E) gene expression and there is compelling evidence that it plays a direct role in the pathogenesis of HCMV infection by inducing a broad dysregulation of host gene expression. This leads to changes in host cell physiology and contributes to HCMV-induced cell cycle alterations, immunomodulation, and proinflammatory responses. Thus, molecules that inhibit IE2-dependent activities may be effective in blocking the virus-induced pathological phenomena at an early stage of infection. Moreover, inhibitors of IE2 could be of particular importance for the treatment of patients who do not respond to currently available inhibitors of viral DNA replication.

In this chapter, we will describe novel approaches to develop new compounds against HCMV. First, we will focus on two kinds of in vitro assays to identify compounds able to inhibit protein–protein interactions essential for HCMV replication. As an example, we will describe two different methods for identifying inhibitors of the interactions occurring between the two subunits of HCMV DNA polymerase, i.e., UL54 and UL44, which are essential for viral genome replication. However, these approaches can be virtually applied to every molecular interaction for which the interacting domains have been identified.

The first assay is based on measurement of fluorescence polarization (FP), which is a simple and rapid method to study molecular interactions and their inhibition. The basic principle of an FP-based assay is that a peptide or another molecule of small size (e.g., an oligodeoxynucleotide) labelled with a fluorophore rotates rapidly when free in solution. When the conjugated fluorophore is excited with plane polarized light, the emitted fluorescence is depolarized (i.e., in a plane different from the excitation light). When the rotation is slowed down due to an interaction with a molecule of greater size (e.g., a protein or another kind of ligand), the fluorophore tumbles slowly with respect to fluorescence lifetime

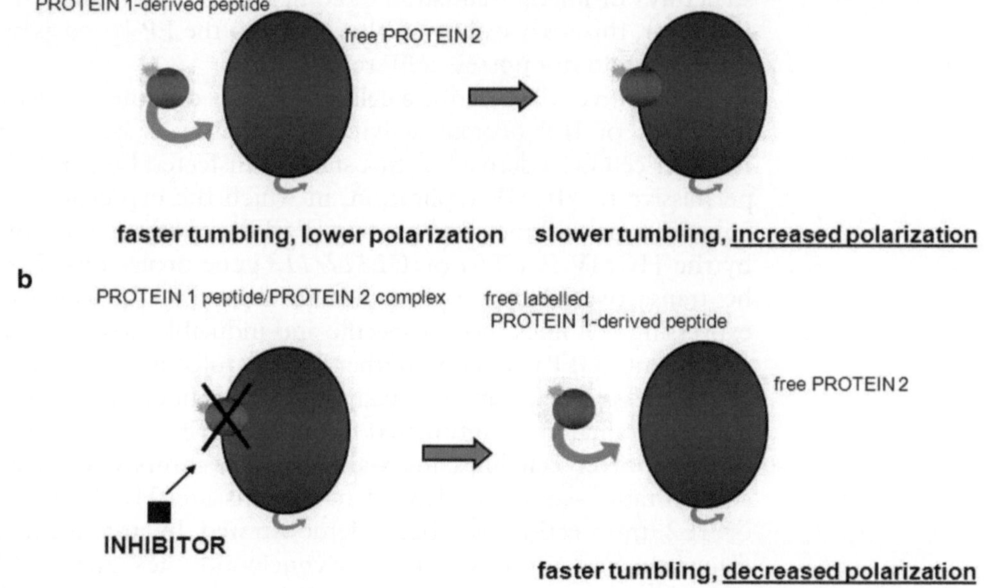


**Fig. 1** Principles of the FP-based protein–protein interaction assay. (**a**) A synthetic peptide that corresponds to the binding region of one of the two protein partners (PROTEIN 1) and is labelled with a fluorophore is mixed with the other protein partner (PROTEIN 2). The unbound labelled peptide tumbles relatively rapidly; thus, if it is excited with polarized light, by the time emission occurs, the polarization of the emitted light is low (*left panel*). However, upon binding to the interacting partner, the peptide will tumble much more slowly, so the emitted light remains relatively polarized, resulting in an increase in FP (*right panel*). (**b**) Compounds that interfere with the protein–protein interaction will produce a quantifiable reduction in FP

and hence the emitted light remains polarized. The presence of an inhibitor of the molecular interaction between the small, labelled molecule and the large ligand causes a depolarization that can be quantitatively detected by using an FP reader. The advantages of FP assays are that the detection limits are in the sub-nanomolar range, it is performed in solution, it does not require immobilization or washing steps, it is very rapid, and it permits quantitative measurements. Altogether, these features render an FP-based assay suitable for the screening of a large number of compounds, and thus, it is very useful for high-throughput screenings (HTS). In addition, FP-based assays can be used for the identification of inhibitors not only of protein–protein interactions but also of protein–nucleic acid interactions (e.g., binding of transcription factors to specific promoters or other viral DNA or RNA sequences) [4, 5]. Figure 1 illustrates the principles of an FP-based assay to discover inhibitors of protein–protein interactions.

The second approach is based on the enzyme-linked immunosorbent assay (ELISA) and is recommended for the screening of a restricted number of compounds, for example, for testing compounds identified from in silico screenings of small-molecule structures or for the validation of compounds discovered by HTS. However, this assay can be an alternative to the FP-based assay for those who do not possess a FP reader.

Finally, we will describe a cell-based assay designed to discover inhibitors of IE2 protein activities. This assay is based on new reporter cell lines derived from a stably transfected human cell line permissive to HCMV replication, in which the expression of the enhanced green fluorescent protein (EGFP) reporter gene is driven by the HCMV E *UL54* or *UL112/113* gene promoters that can be transactivated by the IE2 protein alone [6, 7]. The EGFP-expressing cell lines show a specific and inducible dose- and time-dependent EGFP response to either HCMV infection or constitutive IE2 expression that can be visually assessed by fluorescence microscopy or evaluated by automated fluorometry [8]. The reliability of these reporter cell lines for assessing virus-inhibitory effects in concentration- and time-dependent fashions after HCMV infection or IE2 transfection has been demonstrated by treatment with fomivirsen, an antisense oligodeoxynucleotide designed to block IE2 expression [8, 9], and with WC5, a novel 6-aminoquinolone endowed with potent inhibitory activity for HCMV replication [10]. Thus, these EGFP-based cell lines can be exploited as a tool to screen antiviral compounds that specifically target IE2 expression and/or function.

## 2 Materials

### 2.1 Fluorescence Polarization (FP) Assay

1. Recombinant glutathione S-transferase (GST)-UL44ΔC290 fusion protein expressed from *Escherichia coli* BL21 (DE3) pLysS (Invitrogen) harboring the appropriate plasmid and purified as previously described [11].
2. Synthetic peptides corresponding to the 22 C-terminal residues (LPRRLHLEPAFLPYSVKAHECC) of HCMV UL54 protein, which have been described in ref. 12. Both unlabelled peptide and peptide labelled at the N-terminus with the pentafluorofluorescein-derivative Oregon Green 514 (molecular probes) will be needed for these assays.
3. FP buffer: 50 mM Tris–HCl (pH 7.5), 2 mM dithiothreitol (DTT), 0.5 mM EDTA, 150 mM NaCl, 4 % glycerol, and 100 μg/ml of bovine serum albumin (BSA).
4. Collection of small-molecule compounds to be tested in the FP assay, preferably resuspended in dimethyl sulfoxide (DMSO) at concentrations between 10 and 50 mM (*see* **Note 1**).
5. 384- or 96-well black plates.
6. Automated dispenser for 384- or 96-well plates (*see* **Note 2**).
7. FP reader (e.g., Analyst Plate Reader, LJL Biosystems).

### 2.2 Enzyme-Linked Immunosorbent Assay (ELISA)

1. ELISA microtiter plates. Microtiter plates with high protein-binding capacity are recommended (e.g., Immulon 1, Dynatech).
2. Recombinant HCMV UL54 and UL44 proteins expressed from baculovirus-infected *Sf9* insect cells and purified as described in ref. 12. The UL44 protein has an EEF epitope tag fused at the C-terminus.
3. 2 % BSA solution in phosphate-buffered saline (PBS: 137 mM NaCl, 2.7 mM KCl, 4.3 mM $Na_2HPO_4$, 1.47 mM $KH_2PO_4$, pH 7.4).
4. Wash buffer: 0.3 % Tween 20 in PBS.
5. Synthetic peptide corresponding to the 22 C-terminal residues (LPRRLHLEPAFLPYSVKAHECC) of HCMV UL54 protein, which is a known inhibitor of UL54–UL44 interaction [12], dissolved in 100 mM Tris–HCl pH 8.0 with 0.1 % Tween 20.
6. Collection of small-molecule compounds to be tested in ELISA, preferably resuspended in DMSO at concentrations between 10 and 50 mM (*see* **Note 1**).
7. Monoclonal antibody (MAb) YL1/2 (Serotec Ltd.), which recognizes an EEF epitope tag [13] inserted at the C-terminus of UL44 protein.
8. Horseradish peroxidase (HRP)-conjugated anti-rat antibody.
9. Chromogenic substrate 2,2′-azino-bis(3-ethylbenzthiazinzoline-6-sulfonic acid) (ABTS) in citrate phosphate buffer (pH 4.0) containing 0.01 % hydrogen peroxide.
10. 3 N NaOH.
11. ELISA plate reader with absorbance filter at 405 nm (e.g., Sunrise Reader, Tecan).

### 2.3 Cell-Based Assay

#### 2.3.1 Cultivation of the Reporter Cell Lines 2F7 and 1B4

1. U373-MG UL54-EGFP (clone 2F7) and U373-MG UL112/113-EGFP (clone 1B4) cell lines are derived from the astrocytoma/glioblastoma cell line U373-MG (*see* **Note 3**).
2. G418 stock solution (50 mg/ml).
3. Growth medium for 2F7 and 1B4 cells: Dulbecco's Modified Eagle Medium (DMEM) supplemented with 10 % fetal bovine serum (FBS), 2 mM glutamine, 100 U/ml penicillin, 100 μg/ml streptomycin sulfate, and 750 μg/ml G418.
4. PBS.
5. 100 μg/ml trypsin and 2 mM EDTA in PBS.

#### 2.3.2 Infection of the Reporter Cell Lines 2F7 and 1B4

1. HCMV stocks: HCMV laboratory strain AD169 (ATCC, #VR-538) is propagated and titered in low-passage human embryonic lung fibroblasts (HELF) by standard protocols. HCMV clinical isolates are propagated in low-passage human

umbilical vein endothelial cells (HUVEC) and titered by the indirect immunoperoxidase staining procedure on HELFs using a MAb reactive to the HCMV IE1 and IE2 proteins (*see* **Note 4**).

2. Maintenance medium for HCMV-infected 2F7 and 1B4 cells: DMEM supplemented with 5 % FBS, 2 mM glutamine, 100 U/ml penicillin, and 100 μg/ml streptomycin sulfate.

*2.3.3 Quantification of EGFP Expression in HCMV-Infected 2F7 and 1B4 Cells by Fluorescence Microscopy*

1. Glass coverslips in 24-well plates.
2. PBS.
3. Fixing solution: 4 % paraformaldehyde in PBS.
4. Mounting medium (e.g., Vectashield).

*2.3.4 Quantification of EGFP Expression in HCMV-Infected 2F7 and 1B4 Cells by Automated Fluorometry*

1. 24- or 96-well culture plates.
2. PBS.
3. NP-40 lysis buffer: 50 mM Tris–HCl pH 7.8, 150 mM NaCl, 5 mM EDTA, 15 mM $MgCl_2$, 0.15 % NP-40.
4. 96-well plate suitable for fluorescence measurement (e.g., Nunc #136101).

*2.3.5 Validation of 2F7 and 1B4 Reporter Cell Lines for Selection of Inhibitors of IE2 Protein Activity*

1. Antiviral stock solutions: 1 mM fomivirsen (phosphorothioate oligodeoxynucleotide 5′-GCGTTTGCTCTTCTTCTTGCG-3′, dissolved in10 mM Tris–HCl pH 8.0, 1 mM EDTA); 25 mM WC5 synthesized and dissolved in DMSO as described in ref. 10. Fomivirsen (ISIS 2922) is 21-base phosphorothioate oligodeoxynucleotide complementary to the IE2 mRNA [9] that has been approved for intraocular application in patients with HCMV retinitis. WC5 is a 6-aminoquinolone derivative that has been shown to exert an inhibitory activity on HCMV replication by interfering with the IE2-dependent transactivation of the early genes *UL54* and *UL112/113* [10].

## 3 Methods

***3.1 Fluorescence Polarization (FP) Assay to Search for Inhibitors of HCMV DNA Polymerase Subunit Interactions***

1. Mix 2.5 μM GST-UL44ΔC290 and 3 nM of the Oregon Green-labelled UL54 peptide in FP buffer and incubate on ice (*see* **Note 5**).
2. Dispense the appropriate volume of the reaction mixture containing the labelled peptide and GST-UL44ΔC290 protein into 384-well black plates (40 μl of reaction mixture) or into 96-well plates (160 μl of reaction mixture) (*see* **Note 6**).
3. Add potential small-molecule inhibitors to the 96- or 384-well plates at a final concentration of 12.5 μg/ml, or the same volume of compound vehicle (e.g., DMSO; *see* **Note 7**), to

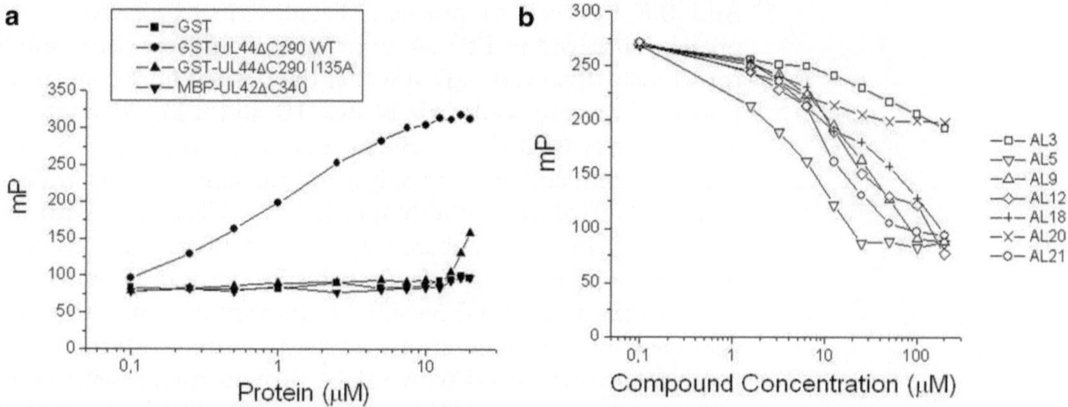


**Fig. 2** Typical results of an FP-based assay to detect the interaction between the two subunits, UL54 and UL44, of HCMV DNA polymerase and to search for inhibitors of such interaction. (**a**) Increasing concentrations of wild-type GST-UL44ΔC290 (*filled circle*), of GST-UL44ΔC290 I135A, a mutant UL44 which does not bind UL54 (*filled triangle*), of GST (*filled square*), or of an unrelated protein (MBP-UL42ΔC340) (*filled inverted triangle*) were added to 3 nM of a fluorescently labelled peptide corresponding to the C-terminal 22 residues of HCMV UL54 (peptide 1), and FP (as millipolarization units, mP) was measured (reproduced from ref. 2 with permission from Elsevier). (**b**) Increasing concentrations of the indicated small-molecule compounds (named AL3, AL5, AL9, AL12, AL18, AL20, and AL21) were added to reaction mixtures containing 2.5 μM GST-UL44ΔC290 and 3 nM labelled peptide 1, and FP (as millipolarization units, mP) was measured (reproduced from ref. 2 with permission from Elsevier)

exclude potential inhibitory effects due to compound vehicle. It is preferable to test each potential inhibitor in duplicate. As controls, include wells with the UL54-derived peptide and GST-UL44ΔC290 protein mixture with no compound and also include wells with the UL54-derived peptide only (*see* **Note 8**). In addition, as a positive control for inhibition of the interaction between UL44 and UL54, add unlabelled UL54-derived peptide at a final concentration of 20 μM in duplicate wells.

4. Incubate at room temperature (RT) for 15 min.
5. Measure FP values from each well (expressed as millipolarization units) using an appropriate FP plate reader.
6. Typical results of an FP-based assay designed to search for inhibitors of the interaction between the two subunits, UL54 and UL44, of HCMV DNA polymerase are shown in Fig. 2.

### 3.2 Enzyme-Linked Immunosorbent Assay (ELISA) to Search for Inhibitors of HCMV DNA Polymerase Subunit Interactions

1. Use 0.2 μg/well of purified, baculovirus-expressed UL54 protein dissolved in PBS for the coating of ELISA microtiter plates and incubate at 37 °C overnight.
2. Aspirate solution containing unbound protein and wash each well five times with 0.2 ml of wash buffer (*see* **Note 9**).
3. Block with 0.1 ml/well of 2 % BSA in PBS at RT for 1 h.
4. Aspirate and wash as described in **step 2**.

5. Add 0.5 μg/well of purified, baculovirus-expressed UL44 protein dissolved in PBS alone or mixed with test compounds (previously dissolved in DMSO) at the desired final concentration in duplicate wells (*see* **Notes 10** and **11**). As controls, include wells with the UL54 protein alone (i.e., do not add UL44 protein and inhibitors, but add the same volume of PBS; this is a control for antibody specificity), wells coated with the UL54 protein and incubated with UL44 protein mixed with the UL54-derived peptide (the UL54-derived peptide is a known inhibitor of UL54–UL44 interaction; thus, this is a positive control for inhibition), and wells coated with the UL54 protein and incubated with UL44 protein mixed with compound vehicle (e.g., DMSO; *see* **Note 10**). The latter control is to exclude inhibitory effects due to compound vehicle.
6. Incubate the plates at 37 °C for 1 h to allow protein–protein interaction.
7. Aspirate and wash as described in **step 2**.
8. Add 50 μl/well of YL1/2 MAb diluted 1:50 in PBS containing 2 % FBS and incubate at 37 °C for 1 h in the dark.
9. Aspirate and wash as described in **step 2**.
10. Add 50 μl/well of HRP-conjugated anti-rat antibody diluted 1:500 in PBS with 2 % FBS and incubate at RT for 1 h.
11. Aspirate and wash as described in **step 2**.
12. Add 100 μl/well of ABTS solution (prepared as described in Subheading 2) and incubate at RT for 15 min in dark (*see* **Note 11**). Block the reaction by adding 50 μl/well of 3 N NaOH (*see* **Note 12**).
13. Measure absorbance at 405 nm using an appropriate ELISA reader.
14. Typical results of an ELISA-based assay designed to search for inhibitors of the interaction between the two subunits, UL54 and UL44, of HCMV DNA polymerase are shown in Fig. 3.

### 3.3 Cell-Based Assay

#### 3.3.1 Cultivation of the Reporter Cell Lines 2F7 and 1B4

1. For maintenance, 2F7 and 1B4 cells are routinely cultured in DMEM-10 % FBS containing 750 μg/ml G418. Cells up to 60th passage can be used for HCMV quantification and antiviral assays.
2. Cells are subcultured by trypsinization with a solution of 100 μg/ml trypsin in PBS-2 mM EDTA solution to obtain new maintenance cultures using a split ratio of 1:3. It is recommended to subculture the cells prior to reaching confluence.

#### 3.3.2 Infection of the Reporter Cell Lines 2F7 and 1B4

1. Seed 2F7 and 1B4 cells in 24- or 96-well plates at a density of $70 \times 10^4$ or $1 \times 10^4$ cells/well, respectively.
2. After 24 h, remove the growth medium by aspiration, and immediately add a small volume of viral inoculum (0.25 ml for

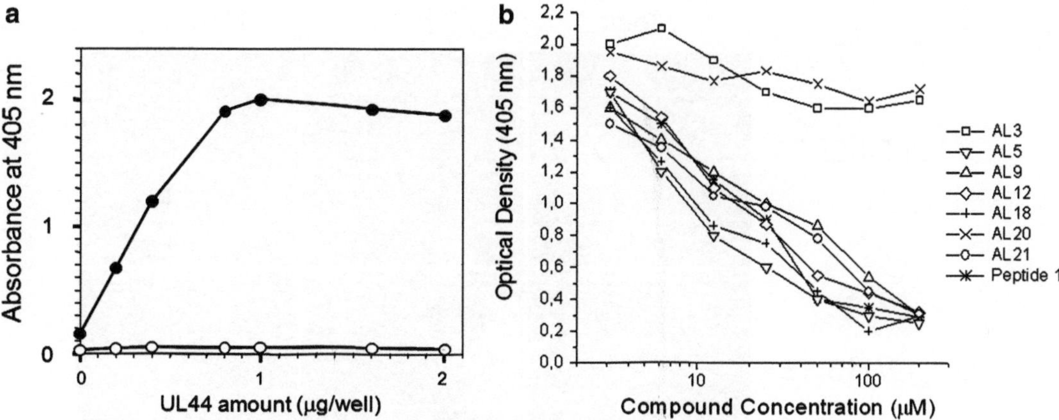


**Fig. 3** Typical results of an ELISA-based assay to detect the interaction between the two subunits, UL54 and UL44, of HCMV DNA polymerase and to search for inhibitors of such interaction. (**a**) Binding of UL44 to UL54 as detected by ELISA. The UL44 protein was added to microtiter wells precoated with 0.2 μg of purified UL54 (*filled circle*) or uncoated wells (*open circle*). Bound UL44 was detected with MAb YL1/2, which was in turn detected with an HRP-conjugated anti-rat antibody (reproduced from ref. 12 with permission from American Society for Microbiology). (**b**) The inhibition of the interaction between UL54 and UL44 proteins in the presence of varying concentrations of the indicated small-molecule compounds (named AL3, AL5, AL9, AL12, AL18, AL20, and AL21) and of a synthetic peptide corresponding to the C-terminal 22 residues of HCMV UL54 (peptide 1) as a positive control for UL54–UL44 inhibition was detected by ELISA as described in (**a**) (reproduced from ref. 2 with permission from Elsevier)

24-well plate or 0.05 ml for 96-well plate) that is prepared by diluting viral stock solutions in serum-free medium for HCMV-infected cells to obtain the appropriate multiplicity of infection (MOI). Plates should be gently rocked to achieve even distribution of the viral inoculum.

3. Incubate infected cells for 2 h at 37 °C in a humidified incubator with 5 % $CO_2$ to allow virus adsorption.
4. Remove viral inoculum by aspiration and add an appropriate volume of DMEM with 5 % FBS to each well and incubate in a $CO_2$ incubator at 37 °C.

#### 3.3.3 Quantification of EGFP Expression in HCMV-Infected 2F7 and 1B4 Cells by Fluorescence Microscopy

1. Prepare 2F7 and 1B4 cells on glass coverslips in 24-well plates and infect cells with HCMV AD169 or HCMV clinical isolates as described in Subheading 3.3.2.
2. Prepare in advance the following solutions: precooled (on ice) PBS and fixing solution.
3. At the appropriate times postinfection (p.i.), remove medium from mock- and HCMV-infected cultures and gently wash cells twice with ice-cold PBS (*see* **Note 13**).
4. Aspirate the residual PBS from cell cultures and place the plates on ice and add 0.25 ml of fixing solution to each well.

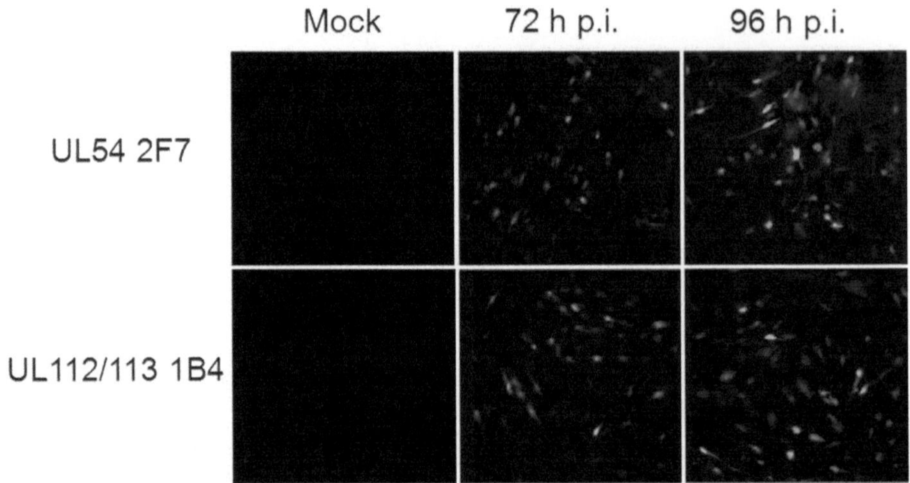


**Fig. 4** Typical results of a fluorescent microscopy assay to detect IE2-driven GFP reporter gene activity during HCMV infection of U373-2F7 and U373-1B4 reporter cells lines. Cells were infected with HCMV AD169 (MOI 5) and the number of EGFP-expressing cells was examined by fluorescence microscopy at multiple times postinfection (shown are 72 and 96 h p.i.). Representative fluorescence microscopy images at 72 and 96 h p.i. are shown (modified from ref. 8)

5. Incubate the plates at RT for 20 min and gently wash cells twice with ice-cold PBS, and then mount coverslips with mounting medium.
6. Examine EGFP-expressing cells by fluorescence microscopy in combination with conventional phase contrast (e.g., Olympus Fluoview-IX70 inverted confocal laser scanning microscope). A 10× objective lens and a fixed data collection time of 0.2 s are used for the conventional assay.
7. Evaluate the numbers of EGFP-expressing cells in captured images using an image-processing software package (e.g., Image J).
8. Figure 4 shows a typical result of a fluorescent microscopy evaluation of EGFP expression in 2F7 and 1B4 cells infected with HCMV AD169 for 72 and 96 h (*see* **Note 13**).

*3.3.4 Quantification of EGFP Expression in HCMV-Infected 2F7 and 1B4 Cells by Automated Fluorometry*

1. Prepare 2F7 and 1B4 cells seeded in 24- or 96-well plates and infect cells with HCMV AD169 or HCMV clinical isolates as described in Subheading 3.3.2.
2. On the day of the experiment, prepare precooled (on ice) PBS and NP-40 lysis buffer.
3. At the appropriate times p.i., remove medium from mock- and HCMV-infected cultures and gently wash cells twice with ice-cold PBS.
4. Aspirate the residual PBS from cell cultures and place the plates on ice. Add 0.05 ml or 0.25 ml of NP-40 lysis buffer to each well of 96-well or 24-well plates, respectively.

5. Incubate at 4 °C for 30 min. Carefully transfer the lysates to 0.5 ml labelled microcentrifugation tubes, and clarify by centrifugation at 13,500 × *g* for 10 min at 4 °C.
6. Transfer 0.05 ml of the supernatants to a 96-well plate suitable for fluorescence measurement.
7. Measure the EGFP fluorescence content in a multiwell fluorescence plate reader (e.g., Perkin Elmer Victor[3] 1420 Multilabel Counter) with excitation and emission filters set at 485 and 530 nm, respectively.
8. Calculate the net EGFP intensity of infected cells by subtracting the fluorescence intensity of the mock-infected cells from that of HCMV-infected cells. Express the HCMV infectivity as fluorescence units (*see* **Note 14**).

*3.3.5 Validation of 2F7 and 1B4 Reporter Cell Lines for Selection of Inhibitors of IE2 Protein Activity*

1. Prepare 2F7 and 1B4 cells seeded in 24- or 96-well plates as described in Subheading 3.3.2.
2. For cultures to be treated with fomivirsen, add different concentrations of the antisense oligodeoxynucleotide (0.01–5 μM) 1 h prior to infection.
3. Infect cultures with HCMV AD169 or HCMV clinical isolate as described in Subheading 3.3.2.
4. Remove viral inoculum by aspiration; add DMEM 5 % FBS containing an appropriate concentration of fomivirsen, or WC5 (50 μM), or 0.2 % DMSO as a control; and incubate in a $CO_2$ incubator at 37 °C for 48 h p.i. It would be appropriate to include a positive control for the inhibition of IE2-dependent EGFP expression in each assay. To this end, a selected concentration of fomivirsen would give a consistent positive control.
5. Quantify the HCMV-induced EGFP expression by automated fluorometry assay as described in Subheading 3.3.4.
6. Estimate the IE2-dependent EGFP expression at each drug concentration.
7. Determine the 50 % inhibitory concentration ($IC_{50}$) values using dose–response curves (*see* **Note 15**).
8. An example of the concentration-dependent inhibitory effect of the IE2-targeting antisense oligodeoxynucleotide fomivirsen on EGFP expression measured in HCMV AD169-infected 2F7 and 1B4 cells at 48 h p.i. is shown in Fig. 5 (left panel). In comparison, the same concentrations of fomivirsen were evaluated by a conventional plaque reduction assay in HELFs (Fig. 5, right panel). Figure 6 shows the employment of the EGFP-based cell assay to test the ability of the anti-HCMV 6-aminoquinolone WC5 to inhibit the IE2-dependent transactivating activity [10].

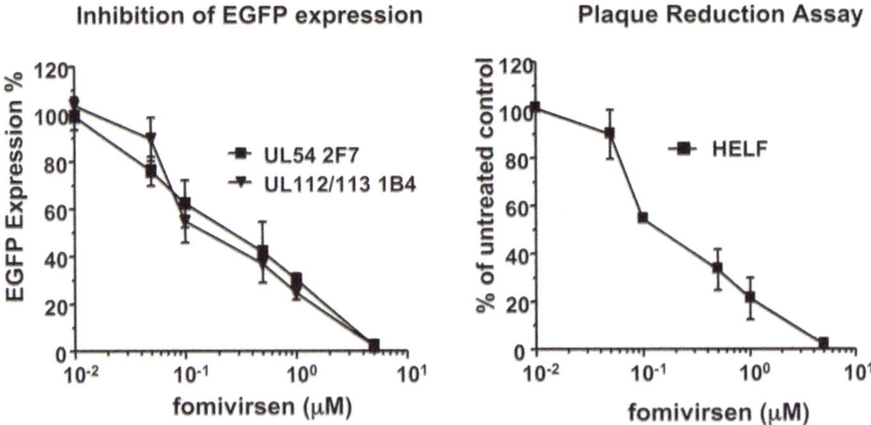


**Fig. 5** Validation of the EGFP-based reporter cell lines U373-2F7 and U373-1B4 for the assessment of the inhibitory activity of fomivirsen. U373-2F7 and U373-1B4 cells were infected with HCMV AD169 (MOI 5) or mock-infected. Where indicated, cells were treated with different concentrations of fomivirsen (0.01–5 μM) (ISIS 2922) prior to and during infection. At 48 h p.i., cells were lysed and assayed for EGFP expression by quantitative automated fluorometry (*left panel*). In comparison, a conventional plaque reduction assay was performed in HELF cell (*right panel*). The calculated $IC_{50}$ of fomivirsen in fluorescence-based experiments was 0.26 μM in 2F7 cells and 0.17 μM in 1B4 cells. The calculated $IC_{50}$ of fomivirsen in traditional plaque assays was 0.16 μM in HELFs. Taken together, these studies indicate that the sensitivity of the EGFP-based fluorescence assay is similar to that of the plaque reduction assays in measuring the antiviral activity of fomivirsen [8]

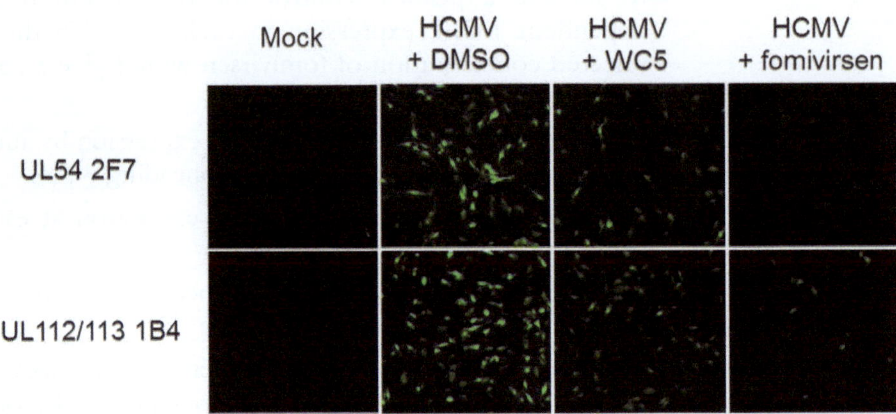


**Fig. 6** Investigation of the ability of the 6-aminoquinolone WC5 to inhibit the IE2-dependent transactivation of HCMV viral gene promoters by using the EGFP-expressing U373-2F7 and U373-1B4 reporter cells. 2F7 and 1B4 cells were mock-infected or infected with HCMV AD169 (MOI 0.1) and then treated with WC5 (50 μM), fomivirsen (5 μM), or DMSO as a control. At 48 h p.i., the cells were examined by confocal fluorescence microscopy to evaluate the number of EGFP-expressing cells (reproduced from ref. 10 with permission from American Society for Microbiology)

## 4 Notes

1. Small-molecule libraries are either commercially available or can be found at academic facilities. For an example, *see* ref. 2. Compounds can be purchased in lyophilized form and dissolved in DMSO at the desired concentration.
2. Automated dispensers are preferable for maximal standardization of the operations. When the screening is manually performed, a 96-well format is recommended to minimize differences between samples, in particular between the first and last samples dispensed.
3. The astrocytoma/glioblastoma cell line U373-MG (ATCC HTB-17) was chosen, as these cells can be routinely infected with HCMV to >90 % and are permissive for viral replication [14]. The cells were transfected with plasmids pUL54-EGFP and pUL112/113-EGFP, which contain the UL54 (positions −425/+15) or the UL112/113 (positions −353/+32) viral early gene promoters cloned upstream from the EGFP reporter gene. These viral E gene promoters were chosen on the basis of the negligible basal activity in uninfected cells, strong inducibility upon HCMV infection, and transactivation in response to constitutive IE2 expression. The EGFP plasmid harbors the neomycin-resistance gene that confers drug resistance to transfected cells, so stable transfectants were isolated by G418 selection. One pUL54-EGFP-containing clone designated 2F7 and one pUL112/113-EGFP-containing clone designated 1B4 were selected, on the basis of low-background expression levels and a constant increase in EGFP emission upon either HCMV infection or IE2 expression.
4. To minimize the risk of contamination, biosafety level 2 (BSL-2) practices, containment equipment, and facilities are required for all procedures involving HCMV cultivation and manipulation.
5. Always calculate volumes to be aliquoted in excess of what is actually needed for the experiment, particularly when automated operations are carried out.
6. Samples can be either automatically dispensed using a 384-pin array or manually dispensed. When manually dispensed, a 96-well format is recommended in order to avoid variability among first and last samples.
7. Ensure that the final DMSO concentration in each well is not higher than 1 % to avoid nonspecific effects of DMSO.
8. If using an automated dispenser, 0.1 μl of each compound at 5 mg/ml is transferred to individual wells using a 384-pin array.
9. To avoid differences among the wells, the use of a multichannel pipette is recommended.

10. Mix equal volume of UL44 and peptides, e.g., 60 μl of UL44 and 60 μl of peptide solution.
11. Prior to incubation with HRP substrate, if there are bubbles in the sample, eliminate them by punching with a needle.
12. Alternatively, other commercially available substrates of horseradish peroxidase can be used, such as 3,3,5,5-tetramethylbenzidine (TMB), 5-aminosalicylic acid (5AS), and O-phenylenediamine (OPD).
13. HCMV infection induces EGFP expression in 2F7 and 1B4 cells in a time- and dose-dependent manner. Under fluorescent microscopy observation, fluorescent cells can be detected as early as 48 h p.i., with a gradual increase at 72 and 96 h p.i.
14. Analysis of quantitative EGFP expression by automated fluorometry provides a simple and rapid protocol in which there are only two steps: cell lysis and fluorescence measurement by a plate reader. EGFP quantification by automated fluorometry is sensitive, specific, and less expensive than other analytical methods, as it does not require additional extraction procedures, enzymes, or immunological reagents. The adaptability of 2F7 and 1B4 cell lines to a 96-well format makes EGFP quantification by automated fluorometry suitable for HTS of large collections of small molecules.
15. Concentrations producing 50 % reductions in plaque formation and HCMV-induced EGFP expression ($IC_{50}$) may be calculated by nonlinear regression using a computer program (e.g., PRISM, version 4.0, GraphPad Software, Inc).

## Acknowledgments

The research in our laboratories was supported by MURST EX60%, Progetto di Ricerca di Ateneo 2007 (grant no. CPDA074945), and PRIN 2008 (grant no. 20085FF4J4) to A.L., by PRIN and the Piedmont Region (Ricerca Sanitaria Finalizzata) to G.G., and by Regione Veneto and Progetto Strategico di Ateneo 2008 to G.P.

### References

1. Mercorelli B, Sinigalia E, Loregian A, Palù G (2008) Human cytomegalovirus DNA replication: antiviral targets and drugs. Rev Med Virol 18:177–210
2. Loregian A, Coen DM (2006) Selective anti-cytomegalovirus compounds discovered by screening for inhibitors of subunit interactions of the viral polymerase. Chem Biol 13:191–200
3. Mercorelli B, Lembo D, Palù G, Loregian A (2011) Early inhibitors of human cytomegalovirus: state-of-art and therapeutic perspectives. Pharmacol Ther 131:309–329
4. Lundblad JR, Laurance M, Goodman RH (1996) Fluorescence polarization analysis of protein-DNA and protein-protein interactions. Mol Endocrinol 10:607–612

5. Moerke NJ (2009) Fluorescence polarization (FP) assays for monitoring peptide-protein or nucleic acid-protein binding. Curr Protoc Chem Biol 1:1–15
6. Wu J, O'Neill J, Barbosa MS (1998) Transcription factor Sp1 mediates cell-specific trans-activation of the human cytomegalovirus DNA polymerase gene promoter by immediate-early protein IE86 in glioblastoma U373MG cells. J Virol 72:236–244
7. Asmar J, Wiebusch L, Truss M, Hagemeier C (2004) The putative zinc finger of the human cytomegalovirus IE2 86-kilodalton protein is dispensable for DNA binding and autorepression, thereby demarcating a concise core domain in the C-terminus of the protein. J Virol 78:11853–11864
8. Luganini A, Caposio P, Mondini M, Landolfo S, Gribaudo G (2008) New cell-based indicator assays for the detection of human cytomegalovirus infection and screening of inhibitors of viral immediate-early 2 protein activity. J Appl Microbiol 105:1791–1801
9. Azad RF, Driver VB, Tanaka K, Crooke RM, Anderson KP (1993) Antiviral activity of a phosphorothioate oligonucleotide complementary to RNA of the human cytomegalovirus major immediate-early region. Antimicrob Agents Chemother 37: 1945–1954
10. Loregian A, Mercorelli B, Muratore G, Sinigalia E, Massari S, Gribaudo G, Gatto B, Tabarrini O, Palumbo M, Cecchetti V, Palù G (2010) The 6-aminoquinolone WC5 inhibits human cytomegalovirus replication at an early stage by interfering with the transactivating activity of viral immediate-early 2 protein. Antimicrob Agents Chemother 54:1930–1940
11. Loregian A, Appleton BA, Hogle JM, Coen DM (2004) Residues of human cytomegalovirus DNA polymerase catalytic subunit, UL54, that are necessary and sufficient for interaction with the accessory protein UL44. J Virol 78:158–167
12. Loregian A, Rigatti R, Murphy M, Schievano E, Palù G, Marsden HS (2003) Inhibition of human cytomegalovirus DNA polymerase by C-terminal peptides from the UL54 subunit. J Virol 77:8336–8344
13. Stammers DK, Tisdale M, Court S, Parmar V, Bradley C, Ross CK (1991) Rapid purification and characterisation of HIV-1 reverse transcriptase and RNaseH engineered to incorporate a C-terminal tripeptide alpha-tubulin epitope. FEBS Lett 283:298–302
14. Jault FM, Spector SA, Spector DH (1994) The effects of cytomegalovirus on human immunodeficiency virus replication in brain-derived cells correlates with permissiveness of the cells for each virus. J Virol 68:959–973

# Index



Andrew D. Yurochko and William E. Miller (eds.), *Human Cytomegaloviruses: Methods and Protocols*, Methods in Molecular Biology, vol. 1119, DOI 10.1007/978-1-62703-788-4, © Springer Science+Business Media New York 2014

## G



## H



## I



## L



## M

MIX
Papier aus verantwortungsvollen Quellen
Paper from responsible sources
FSC® C105338

If you have any concerns about our products,
you can contact us on
**ProductSafety@springernature.com**

In case Publisher is established outside the EU,
the EU authorized representative is:
**Springer Nature Customer Service Center GmbH**
**Europaplatz 3, 69115 Heidelberg, Germany**

Printed by Libri Plureos GmbH
in Hamburg, Germany